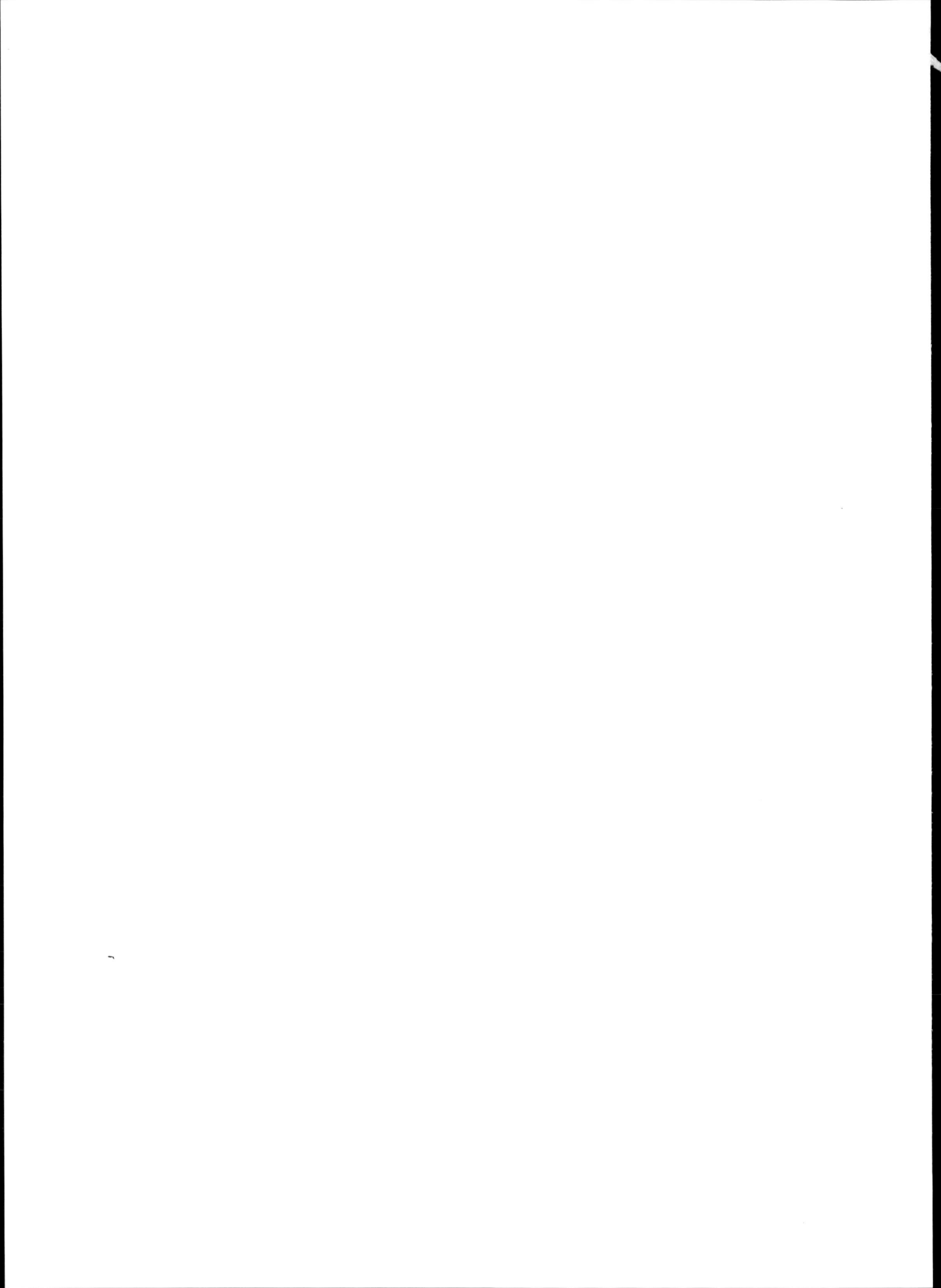

Communication Technologies: Theory and Practice

Communication Technologies: Theory and Practice

Edited by Akira Hanako

CLANRYE
INTERNATIONAL
www.clanryeinternational.com

Clanrye International,
750 Third Avenue, 9th Floor,
New York, NY 10017, USA

ISBN: 978-1-63240-631-6

Cataloging-in-Publication Data

Communication technologies : theory and practice / edited by Akira Hanako.
   p. cm.
Includes bibliographical references and index.
ISBN 978-1-63240-631-6
1. Telecommunication. 2. Communication. 3. Information technology. I. Hanako, Akira.
TK5101 .C66 2017
621.382--dc23

For information on all Clanrye International publications
visit our website at www.clanryeinternational.com

$\mathcal{CL}$ANRYE
$\mathcal{IN}$TERNATIONAL

Printed in the United States of America.

Contents

Preface

This book on communication technologies discusses topics related to the implementation of software and hardware for communication and information sharing practices. Communication technologies facilitate the quick transmission of data to multiple users. Technical innovation in this sector determines the future progress in this field. The various advancements in communication technologies are glanced at and their applications as well as ramifications are looked at in detail. This book is meant for students who are looking for an elaborate reference text on communication technologies. It is a vital tool for all researching or studying this field as it gives incredible insights into emerging trends and concepts.

This book has been the outcome of endless efforts put in by authors and researchers on various issues and topics within the field. The book is a comprehensive collection of significant researches that are addressed in a variety of chapters. It will surely enhance the knowledge of the field among readers across the globe.

It gives us an immense pleasure to thank our researchers and authors for their efforts to submit their piece of writing before the deadlines. Finally in the end, I would like to thank my family and colleagues who have been a great source of inspiration and support

Editor

A Transformative Collegiate Discourse

Evan Ortlieb[1]

Abstract

The languages, attitudes, experiences, and interests of groups of people are often interconnected through discourses. Within collegiate courses, systems of connected individuals progress through learning opportunities, often led by their professors. During one graduate course at Louisiana State University, two students acted as research investigators studying the unique characteristics of a collegiate discourse. The findings indicate that authors of the study were members of the discourse, topics of discussion were guided by only those students in attendance, and there was reciprocity in teaching and learning between active members of the discourse.

Keywords

collegiate, discourse, doctoral, educational community

The languages, attitudes, experiences, and interests of groups of people are often interconnected through discourses. These social structures are, as Gee (1989) explained, a "sort of 'identity kit' which comes complete with the appropriate costume and instructions on how to act, talk, and often write, so as to take on a particular role that others will recognize" (p. 7). As its members shape a discourse, one can determine many unique aspects of one's own discourse by taking a step back and studying its uniqueness. Palmquist (n.d.) pointed out that "the new perspective provided by discourse analysis allows personal growth and a high level of creative fulfillment" (p. 1). Thus, the context of the discourse becomes more evident and conceivably meaningful.

Understanding these networks on a theoretical level though is limiting. Assigned with the task of finding and studying any discourse, two graduate students set out to investigate their own class as a system of connected individuals or discourse. They acted as research investigators studying the characteristics of a collegiate discourse without anyone being the wiser. What follows is a depiction of this specific discourse in an attempt to specify what makes this discourse distinctive.

Since the initial class meeting of EDCI 7910, the participants have been fascinating to both inquirers. Members of EDCI 7910 are PhD candidates with three exceptions, the professor, and two students seeking a master's degree. These participants come from a variety of backgrounds, including educational administration, nursing, special education, literacy education, and several other fields. Even though members have different career-oriented interests, they share the desire to be part of the EDCI 7910 discourse community. This allows the community to have the opportunity to learn from all participants, which is significantly exceptional.

Through discourse analysis, one should act as an outsider to identify characteristics of one's own discourse community. Often, people are part of many discourses at a time but function unaffectedly because of their quiet membership. Looking at one's own discourse can be beneficial for several reasons. First, identifying acceptable and unacceptable behavior within the discourse can lead to continued belonging to the faction. Once it is known that certain behaviors are looked down upon, one would likely act in accordance. Second, one may determine, after analyzing a discourse community, whether to associate or disassociate with it. Many times, certain attitudes and rules of conduct are required of the participant. In addition, being identified with one discourse may be assumed to be unacceptable by another discourse to which one belongs. All of these rationales are at the basis for the analysis of the discourse community, EDCI 7910.

Method

Being part of the discourse, it was somewhat simple to record specific aspects of the community. Initially, recorders took notes on observed data of various information, including member eating habits, posture, speech patterns, and much more. As repetitive behaviors were noticed, infor

[1]Texas A&M University–Corpus Christi, TX, USA

Corresponding Author:
Evan Ortlieb, Department of Curriculum and Instruction, Texas A&M University—Corpus Christi, 6300 Ocean Drive, Corpus Christi, TX 78412
Email: evan.ortlieb@tamucc.edu

Question One

What do you enjoy about being part of the EDCI 7910 discourse community?

- Dr. Kellen's willingness to help us.

- Humor of the class is refreshing

- We are encouraged to say something smart and I like that it is important to the professor that we come prepared and ready to discuss the material.

Question Two

What do you dislike about being part of the EDCI 7910 discourse community?

- Chatting, there seems to be a constant conversation. Sometimes, it seems as though people are off task.

- A single cell phone ring is heard most classes.

- People are late. I feel like the project was hard because people kept trickling in. It reminded me of being an undergrad.

Question Three

What are your perceptions of the investments of other members of the discourse community?

- I am not sure. It appears that some people expect a grade without working for it.

- Some members would never say that they think that the material is unimportant, but they say that in their behavior of reading newspapers, eating, and not paying attention to the person who is speaking. This is especially noticeable when some have trouble expressing themselves.

- Everyone is nice. It seems like they really want to help you do well.

Question Four

Other Comments

- There is a strong sense of work ethic. For years, I was in retail and I learned that we could tell about a person and their ability by their ethic. I want the class to see that the discourse members are willing to make others be ethical and report when they feel as though results or work is not presented fairly.

- I think some people should look at character education. Some of the behaviors in the class are rude. You do not cough on others, use poor language, or disrespect the teacher. Maybe this is a bit old-fas hioned, but this teacher is an expert and we need to gain as much as we can from her.

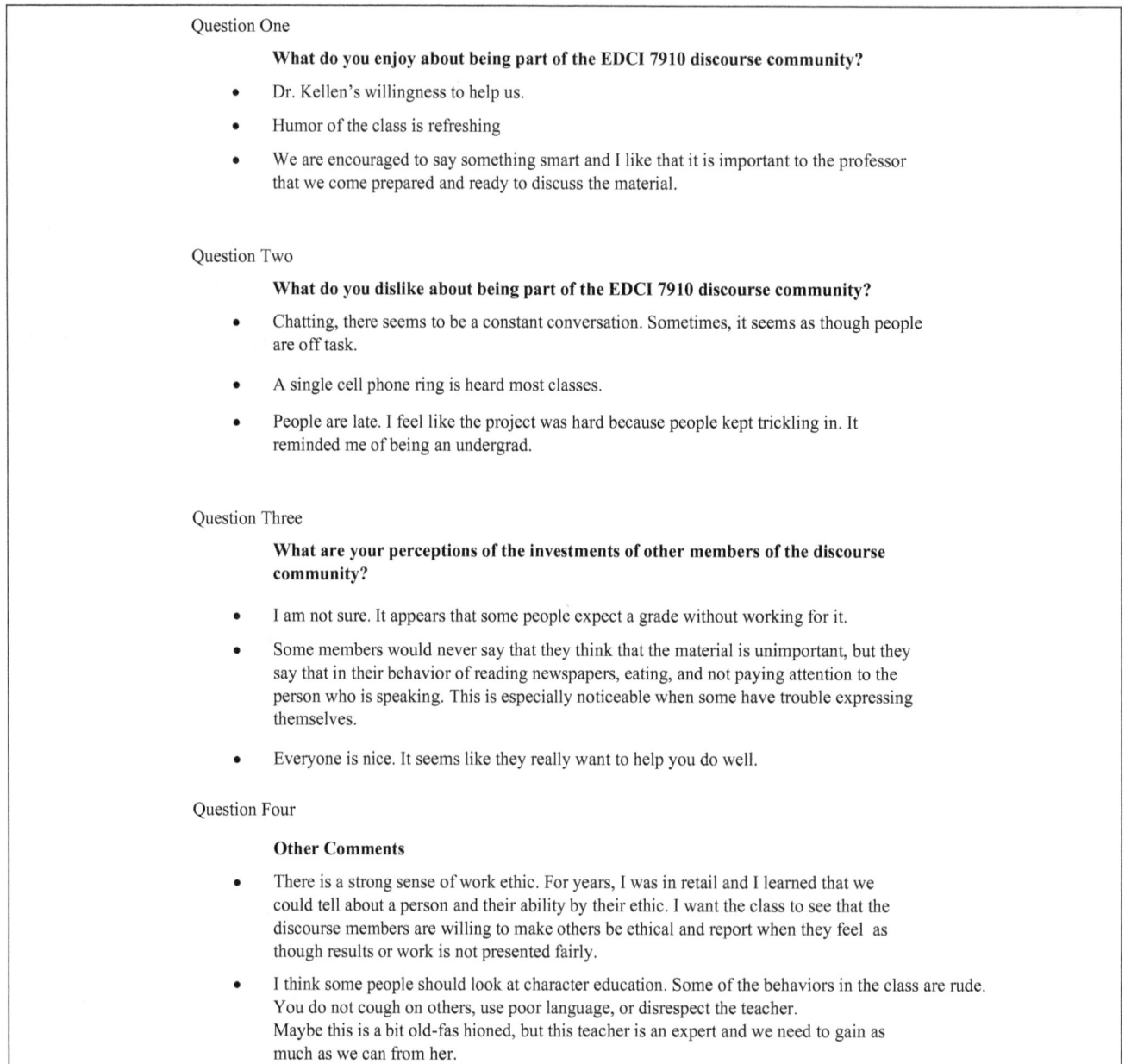

Figure 1. Interview responses

mal question and answer surveys served to solicit student feedback and eliminate observational bias. For specific student comments, see Figure 1.

Once several characteristics of the discourse were determined from analyses of student surveys and other collected data, a chart permitted recorders to tally each time a particular action occurred. For 5 weeks, recorders collected data by writing tally marks of members' attendance, participation, cellular phone usage, and dialogue styles. These characteristics were the foci of analysis because "the professional language . . . of a disciplinary group plays a key role in establishing its cultural identity" (Becher, 1989, p. 24).

Data were gathered using Spradley's (1980) ethnographic research cycle, beginning with broad ethnographic questions while collecting data through observation in the classroom, speaking with students before and after class, and observing their interactions with each other. Several key findings emerged from these data analyses. Participants were asked the following three questions and invited to make any additional comments:

1. What do you enjoy about being part of the EDCI 7910 discourse community?
2. What do you dislike about being part of the EDCI 7910 discourse community?

3. What are your perceptions of the investments of other members of the discourse community?

The responses to these three questions varied. When asked the first question about their likes, some felt they most enjoyed Dr. Kellen's willingness to help. Others found the humor of the class refreshing. One student even responded that "we are encouraged to say something smart, and I like that it is important to the professor that we come prepared and ready to discuss the material."

When asked what they disliked, some respondents stated they did not appreciate the chatting because it seems to be a constant conversation, and people are, at times, digressing or off task. Others answers included a problem with cell phones ringing during classes and the tardiness of some of the classmates. In fact, one student shared, "I feel like the project was hard because people kept trickling in. It reminded me of being an undergrad."

The third question, regarding the students' perceptions of their classmates' investments in the course, provoked a strong variance in responses. Some students were unsure of their opinion on the matter but felt that some students expect a certain grade without working for it. Other students felt that although their classmates may never say that they think the material is unimportant, they clearly show through their behavior that they do not see the importance. For example, they read the newspapers and eat during class and do not pay attention to the person who is speaking. Those students who feel this way about their classmates believe their sense of apathy toward the course is especially noticeable when the apathetic students have trouble expressing themselves in class. Another group of students responded that they feel everyone is nice, and they feel people really want to help them do well.

Additional comments made by the participants included,

There is a strong sense of work ethic. For years, I was in retail and I learned that we could tell about a person and their ability by their ethic. I want the class to see that the discourse members are willing to make others be ethical and report when they feel as though results or work is not presented fairly.

I think some people should look at character education. Some of the behaviors in the class are rude. You do not cough on others, use poor language, or disrespect the teacher. Maybe this is a bit old-fashioned, but this teacher is an expert and we need to gain as much as we can from her.

Using Glaser and Strauss' (1967) constant comparative method, data were collected and analyzed simultaneously, which included inductive behavior-category coding with a comparison of all incidents observed. To reduce premature assumptions derived from these transcriptions, we analyzed the data from these sources (Erickson, 1986). After the transcription of the raw notes from observations, associations were sought between multiple observers.

Results

Findings indicate that the leader, Nancy Kellen, scaffolds members of the class who have not already mastered the discourse. All members are encouraged to maximize learning opportunities by engaging actively in the community. This is unusual in that "tribes of academe . . . define their own identities and defend their own patches of intellectual ground" (Becher, 1989, p. 24). Through this "enculturation into social practices," students can acquire a new discourse: academia (Gee, 1990, p. 7). According to a data analysis of recorded dialogue patterns, Kellen often states, "OK. I think we will start by" to lead conversations in regard to specific topics. At other times, she poses the question, "Well, what did you all think about?" to stimulate topical discourse. Through sharing her insight with other nonelitists of academia, students can learn from her experiences. Furthermore, Kellen gently maintains authority by halting off-topic discussions to facilitate effective use of discourse time.

Nonetheless, some members of the class discourse often do not attend the seminar. Perhaps, they have other duties that take precedence over the class meeting (see Figure 2). Nevertheless, these members miss ideal opportunities to learn from the discourse's insights and experiences. The leader, Kellen, is one of the members' most valuable resources; yet, many of the members do not reach this realization. They apparently skip class when they wish and participate at their convenience. In addition, those members who attend regularly also suffer from not having the entire class present. Those in attendance cannot always discuss, question, and engage in discourse with others from their particular fields of study (i.e., literacy, curriculum theory, and mathematics). Thus, some students' hidden motivations cause the entire class discourse not to reach its potential.

Along with many unseen members, authors from various sources (e.g., Becher, 1989; Gee1989; Porter 1992), nonclass members are welcomed into the discourse community to share their insights from other academic discourses. A guest panel on publication, consisting of three academics, changed the makeup of the discourse as they shared ideas on publication. EDCI 7910 members took on a listening role within the discourse, as the guests spoke uninterruptedly. Following each panelist's speech, class members asked questions. Based on informal conversations with discourse participants, some members view the guests as role models; thus, they will likely use some of the panelists' suggestions in an attempt to become published. Guests like Madeline Grumet and the publishing panel transform the discourse, bringing new insights and thus, alter the knowledge and sense of academic direction of the discourse community—EDCI 7910.

Through the analysis of EDCI 7910, it is evident that the majority of the discourse does not prioritize attending class for 3 hr per week for the sake of its members. If one is to be considered part of the discourse community, one must fulfill certain prescribed obligations, as outlined in the course syllabus, including attending, participating, and completing

Percentages of participants from a class of 15 students				
Date	Absences	Tardies	Early exits	Cellular phone usage
2/03/2005	7%	7%		
2/10/2005	20%			7%
2/17/2005		20%	7%	
2/24/2005	20%		7%	20%
3/03/2005	13%	33%		20%

* Percentages were rounded to the nearest whole number.

Category Definitions:

Absences note that a participant did not attend.
Tardies mark that a participant arrived late to class.
Early exits indicate that a participant left class prior to its culmination.
Cellular phone usage represents incidents of incoming and outgoing calls.

Figure 2. Studied behaviors

coursework, both in class and outside of class. Although these duties may seem daunting to some, EDCI 7910 is a graduate-level education class in which considerable effort is necessary for one to become indoctrinated in traditions of theory. Every participant has outside duties, obligations, and concerns; nevertheless, most members still fulfill their roles. Essentially, members of EDCI 7910 can only expect to learn as much as they put forth effort.

The following recorded results are percentages of participants from a class of 15 students. All the percentages reported in the results were rounded to the nearest whole number. Absences note that a student did not attend the class at all. Tardies mark that a participant arrived late to class on that day. Early exists indicate that participants left the class prior to its designated culmination. Cellular phone usage represents incidents of incoming or outgoing calls that occurred during the class period.

On February 3, 7% of the students were absent or tardy. On February 10, 20% of the students were absent, and 7% used their cellular phones at some point during the class period. On February 17, 20% of the students were tardy to class, and 7% left the class period early. On February 24, 20% of the students were absent, 7% left the class period early, and 20% used their cellular phones at some point during the class period. On March 3, 13% of the students were absent, 33% of the students were tardy to class, and 20% of the students used their cellular phones at some point during the class period.

Declaration of Conflicting Interests

The author(s) declared no potential conflicts of interest with respect to the research, authorship, and/or publication of this article.

Funding

The author(s) received no financial support for the research, authorship, and/or publication of this article.

References

Becher, T. (1989). *Academic tribes and territories*. Milton Keynes, UK: Open University Press.

Erickson, F. (1986). Qualitative methods in research on teaching. In M. C. Wittrock (Ed.), *Handbook of research on teaching* (3rd ed., pp. 119-160.). New York, NY: Macmillan.

Gee, J. P. (1989). Literacy, discourse, and linguistics: Introduction. *Journal of Education, 17*, 5-25.

Gee, J. P. (1990). *Social linguistics and literacies: Ideology in discourses, critical perspectives on literacy and education*. London: Falmer Press.

Glaser, B., & Strauss, A. (1967). *The discovery of grounded theory: Strategies for qualitative research*. Chicago, IL: Aldine.

Palmquist, R. (n.d.). *Discourse analysis*. Retrieved from http://www.gslis.utexas.edu/~palmquis/courses/discourse.htm

Porter, J. E. (1992). *Audience and rhetoric: An archaeological composition of the discourse community*. Englewood Cliffs, NJ: Prentice Hall.

Spradley, J. (1980). *Participant observation*. New York, NY: Holt, Rinehart & Winston.

Bio

Evan Ortlieb, PhD, is an assistant professor in the Department of Curriculum and Instruction. His interests include reading clinics, struggling readers, and teacher education.

2

Culture Matters: Norwegian Cultural Identity Within a Scandinavian Context

Gillian Warner-Søderholm[1]

Abstract

Whether managers are concerned with financial issues, marketing, or human resource management (HRM), cultural values and practices do matter. The purpose of this article is to understand Norwegian managers' cultural values within the cross-cultural landscape of her neighbors in the "Scandinavian cluster." Clearly, subtle but disturbing differences may surface even when representatives from similar cultures work together. As a follow on from the GLOBE project, data based on the GLOBE instrument were collected on culture and communication values in Norway from 710 Norwegian middle managers for this present study. Although the Scandinavian cultures appear ostensibly similar, the results illustrate that research can reveal subtle but important cultural differences in nations that are similar yet dissimilar. All three Scandinavian societies appear intrinsically egalitarian; they appear to value low Power Distance, directness, and consensus in decision making and to promote Gender Egalitarianism. Nevertheless, there are significant differences in the degrees of commitment to these values by each individual Scandinavian partner. These differences need to be understood and appreciated to avoid misunderstandings.

Keywords

cross-cultural management, Scandinavian cultural practices, project GLOBE

Introduction

Clearly, culture matters in both business practices and academic research. Indeed a Google search for "culture" provides more than half a billion hits, and major social science electronic databases provide links to 100,000 to 700,000 articles when "culture" is used as the search key word (Taras, Rowney, & Steel, 2009, p. 357). Furthermore, a recent search by this author of published studies based on Scandinavian culture alone during the last decade shows more than 100 papers with culture as an independent variable (please see appendix). Indeed, academic studies published in the last two decades have indicated some distinctive aspects of Scandinavian management. Smith, Andersen, Ekelund, Gravesen, and Ropo's (2003) seminal paper, "In search of Nordic management styles," offers a valuable contribution to this field. Nevertheless, little has been published in the last decade to update these findings or to highlight the unique element of each individual Nordic society. Hence, the debate of how culture matters is ongoing. Consequently, this article will compare the findings from the GLOBE study of societal cultural practices from the Danish and Swedish data, to the Norwegian data from the author's doctoral research, to understand any such cultural differences in Scandinavia.

The methodology and quantitative instrument applied in this present study of Scandinavian societal culture is based on Project GLOBE's (Global Leadership and Organizational Behaviour Effectiveness) study (House, Javidan, Dorfman, & Gupta, 2004). This ongoing research is a multiphase,

multimethod project examining the interrelationships between societal culture, organizational culture, and leadership. A total of 170 social scientists and management scholars from 62 cultures representing all major regions of the world are engaged in this long-term programmatic series of intercultural studies. Approximately 17,300 middle managers from 950 organizations took part in the original study by House et al. (2004), and 710 middle managers took part in this researcher's doctorate study of Norwegian societal cultural practices. The meta-goal of GLOBE is to develop empirically based theories to describe, understand, and predict the impact of specific cultural variables on leadership and organizational process and to determine their effectiveness. More detailed information can be found on GLOBE's website at http://mgmt3.ucalgary.ca/web/globe.nsf/index.

Intercultural research supports the belief that to be able to understand cultural and communication values, business people first need to be open and aware of their own values and then be able to compare their own cultural norms with those of their business partners. In terms of societies in the Nordic region, if business people appreciate how their behavior is perceived by others and how they can expect the other party will act, cultural clashes that could otherwise

[1]Norwegian Business School BI, Oslo, Norway

Corresponding Author:
Gillian Warner-Søderholm, Norwegian Business School BI, Nydalveien 37, 0484, Oslo, Norway
Email: Gillian.warner.soderholm@bi.no

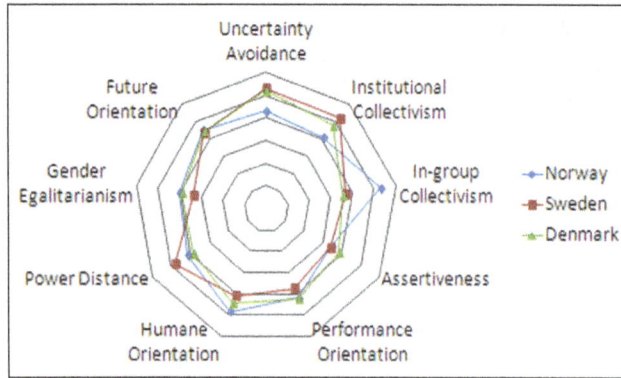

Figure 1. Societal cultural practices scores from Norway, Sweden, and Denmark
Source: Adapted from House et al., 2004; Warner-Søderholm, 2010b.

threaten the business relationship can be avoided (Warner-Søderholm, 2010b).

Participants in the Norwegian study unanimously saw societal culture and behavior in business in Scandinavia as regionally different (Warner-Søderholm, 2010a). Figure 1 below summarizes the findings of this "Scandinavian cluster study" from this researcher's doctoral research. The following section then discusses the GLOBE dimensions of Uncertainty Avoidance, Collectivism, Assertiveness, Performance Orientation, Humane Orientation, Power Distance, Gender Egalitarianism, and Future Orientation, in a Norwegian context within a Scandinavian framework.

Project GLOBE's Performance Orientation Within a Scandinavian Context

Performance Orientation is an important cultural dimension that has not been sufficiently examined in past theoretical or empirical cultural research (House et al., 2004). The few researchers such as Parsons and Shils (1952) and Trompenaars (1993) who have discussed elements of Performance Orientation have explored this in relation to other society values such as how status is accorded. Project GLOBE presents Performance Orientation as the degree to which an organization or society encourages and rewards group members for performance improvement and excellence (House et al., 2004). Kahn (1979) and Hofstede and Bond (1988) argue that performance values are not limited to regions with

Protestant beliefs. They posit that Confucian principles of perseverance, working hard, and learning new skills mirror performance, thus proposing that Confucius teachings and Weber's emphasis on hard work reach similar conclusions: that cultural values of performance exist more strongly among some cultural and regional groups than others (Hofstede & Bond, 1988).

High-scoring cultures tend to focus on achievement, the future, taking initiative, and job-related accomplishments. Low-scoring countries, however, tend to focus on tradition, family, affiliation, and social ties. Hence, social relationships are valued more than achieving. Additional predictors of Performance Orientation were found to be related to a country's level of economic prosperity, higher levels of human development, and a stronger social support for competitiveness (House et al., 2004). Data gathered from the IMD's Global Competitiveness Ranking (1998), the United Nation's Human Development Report (1998), and The World Values Survey (Inglehart, 1997) supported these hypothesized correlations. Table 9 offers a summary of economic indicators related to the intercultural focus of this study.

As can be seen in Table 1, Performance Orientation evaluates the extent to which a community encourages and rewards setting challenging goals, striving for innovation, and performance improvement versus valuing family relations, traditions, sensitivity, and experience. Norway, along with all her Scandinavian partners are categorized within band "B" of the GLOBE scores (moderate to high Performance Orientation), with scores as follows: highest in Denmark (4.22), followed by Norway (4.18) and Sweden (3.72).

Project GLOBE's Future Orientation Within a Scandinavian Context

Future Orientation is the degree to which individuals in organizations or societies engage in future-oriented behaviors such as planning, investing in the future, and delaying individual or collective gratification (House et al., 2004). With regard to the economic and social predictors of this dimension, project GLOBE applied relevant data from the IMD Global Competitiveness Ranking (1994), the World Economic Forum's Competitiveness Ranking (1998), the United Nation's Human Development Report (1998), and

Table 1. Key Elements of Performance Orientation

Societies with stronger Performance Orientation	Societies with weaker Performance Orientation
Value education and learning	Value social and family relations
Emphasize results	Value loyalty and traditions
Set high performance targets	Value sensitivity
Value initiative taking	Value seniority and experience
Prefer explicit, direct communication	Value subtle, indirect language

Source: Based on House, Javidan, Dorfman, and Gupta (2004, p. 245).

Table 2. Key Elements of Future Orientation

Societies with stronger Future Orientation	Societies with weaker Future Orientation
Emphasize visionary leadership that is capable of seeing patterns in the face of chaos and uncertainty	Emphasize leadership that focuses on repetition of reproducible and routine sequences
Value deferment of gratification placing a higher priority on long-term success	Value instant gratification and place higher priorities on immediate rewards
Have organizations with a longer strategic orientation	Have organizations with a shorter strategic orientation
Value collective safety nets in society	Do not see a specific need for society to provide a collective safety net.

Source: Based on House et al. (2004, p. 302).

the World Values Survey (Inglehart, 1997). This data show the following factors as predictors of high levels of Future Orientation: economic prosperity, high levels of society health, active political ideology, and positive attitudes toward gender issues. Norway, along with Denmark, is placed in band "A" and ranks highest of all the Scandinavian countries in this study, while Sweden is placed in band B (moderately high). The mean scores for Future Orientation in the Scandinavian cluster are for Norway: 4.48, followed by Denmark: 4.44 and Sweden: 4.39

These results correlate with Hofstede and Bond's study of long-term orientation (Hofstede & Bond, 1988). Government planning, tax contributions, public investment, and a commitment to a strong welfare state are elements of all the Scandinavian nations' long-term approach to planning for a stable future. Table 2 offers a summary of the key elements recognized in the Future Orientation dimension.

Project GLOBE's Gender Egalitarianism Within a Scandinavian Context

One of the most fundamental ways in which societies differ is the extent to which each prescribes and proscribes different roles for women and men (Hofstede, 2001). Some societies are more gender egalitarian and seek to minimize gender role differences (House et al., 2004). Researchers of this dimension support these claims (Barry, Bacon, & Child, 1957/1967; Coltrane, 1996; Williams & Best, 1982). Gender Egalitarianism is defined as the degree to which an organization or society minimizes gender role differences while promoting gender equality. Predictors of gender egalitarian values include national wealth per capita (Hofstede, 2001). The figures in Table 9 summarize GDP, with Norway's figures indicating a stable national wealth and a variation from her main trading nations. Schwartz (1994) posits a positive correlation between levels of mastery or self-assertiveness in society and lower levels of Gender Egalitarianism. Coltrane (1996) claims that an important predictor of strong Gender Egalitarianism is the degree of men's parental investment. He claims that a greater investment in child rearing by men in society exposes children to nontraditional parent roles, which then encourages their efforts in balancing gender

roles in society later. Furthermore, this increased parental investment by men gives women the time to pursue nontraditional careers. Nordic countries have introduced state incentives and regulations that help fathers to achieve a work–life balance (Kvande, 2009).

Project GLOBE claims to have discovered meaningful variances in societal values such as Gender Egalitarianism as a function of a society's climate. Project GLOBE found a negative relationship between ambient temperatures and a society's placement on the Gender Egalitarianism scale. The lower the average daytime temperature, the more a gender egalitarian a society will be (House et al., 2004). Hofstede's masculine/feminine construct research correlates closely to these findings. Hofstede claims that in less hospitable climates with stable economies, both men and women must take part in daily tasks to manage or even ensure survival (Hofstede, 2001). Scandinavian countries experience harsh winters with average winter temperatures falling as low as -20 degrees. Hence Hofstede's research posits that Scandinavian countries, with harsh climates, value high levels of Gender Egalitarianism.

The findings of the GLOBE study of societies correlate closely to Hofstede's masculinity index, with Norway's score of 4.03 placing her in the top band A, with Denmark's score of 3.93 and Sweden's score of 3.38 placing them also in band A.

A strong correlation is also posited to be that of GNP per capita and societal values such as Gender Egalitarianism. Norway's post–World War II regional development policy has encouraged industry and work opportunities in all regions of Norway. Table 3 summarizes the key elements of Gender Egalitarianism.

Project GLOBE's Assertiveness Within a Scandinavian Context

The concept of Assertiveness originates in part from Hofstede's cultural dimension of masculinity versus femininity (House et al., 2004). In masculine societies, men are supposed to be assertive and tough and women are expected to be modest and tender (House et al., 2004). Kluckhohn and Strodtbeck (1961) discussed dominance as an element of

Table 3. Key Elements of Gender Egalitarianism

Societies with higher Gender Egalitarianism	Societies with lower Gender Egalitarianism
Have a higher percentage of women in the workplace and more women in positions of authority	Have a lower percentage of women in the workplace and fewer women in positions of authority
Accord women a higher status	Accord women a lower status
Promote equal opportunities for all people in society	Tolerate inequality

Source: Based on House et al. (2004, p. 359).

Table 4. Key Elements of Assertiveness

Societies with higher Assertiveness	Societies with lower Assertiveness
Value assertive, dominant, tough behavior for everyone in societies	Value modesty and tenderness
Value competition and believe that anyone can succeed if he or she tries hard enough	Value cooperation, people, and warm relations
Value direct and unambiguous communication	Emphasize the importance of "face-saving," ambiguity and subtlety in communication
Emphasize results over relationships and reward performance	Are concerned that "merit pay" can be destructive to harmony

Source: Based on House et al. (2004, p. 405).

Assertiveness in relation to the nature of the relationship of individuals, groups, and societies with the outside world. In line with Kluckhohn and Strodtbeck (1961), Trompenaars and Hampden-Turner (1998) researched societies' orientation to nature. They posit that certain cultural groups believe that they can and should control or dominate nature (Trompenaars & Hampden-Turner, 1998). Hence, assertive societies will thus view relations in terms of dominance.

All elements of previous research into cultural Assertiveness share a common understanding that Assertiveness is the degree to which individuals in organizations or societies are assertive, confrontational, and aggressive in social relationships (House et al., 2004). Foreigners often regard Norwegians and indeed all Scandinavian societies as somewhat reserved and "cold-hearted" due to the fact that many Scandinavians are nondominant and do not reveal their emotions openly. This way of showing feelings is very culture specific. A Scandinavian person's insular approach does not mean that individuals do not feel emotions: It is an indication of the sense of order and of keeping control in an interdependent society such as Norway. In Scandinavia, it is rare to have a heated argument or strong disagreement at work or in private life. In traffic, it is unusual to blow your horn or to push into a queue. A sense of order and fairness is prevalent from the time one starts in nursery schools and learns to take turns to the custom of patiently waiting in line in the company canteen. Country mean scores for Norway and her Scandinavian partners are as follows: Denmark showing slightly higher scores of 3.80, placing them in band B, and Sweden and Norway showing slightly lower scores of 3.38 and 3.37, placing them in band C, indicating moderate to low levels of Assertiveness in comparison. Table 4 summarizes key elements of Assertiveness.

Project GLOBE's Collectivism Within a Scandinavian Context

Across all disciplines researched, the importance of the individual versus the importance of the group has been studied. Hofstede (2001) posits a correlation between individualism and wealth, with industrialized countries such as England, the United States, and Australia as the most individualistic cultures (Hofstede, 2001). He also suggests that an increase in national wealth in a developing country causes an increase in individualism (Hofstede, 2001). Levine and Norenzayan (1999) posit that a more accurate predictor of individualism is pace of life. They argue that individualistic cultures have a faster pace of life than collectivist cultures as they focus on achievement. Triandis (1995) discusses intrinsic family systems as an antecedent to Collectivist values.

Project GLOBE divided this dimension into two constructs at the second stage of their pilot study (House et al., 2004): Institutional Collectivism and In-Group Collectivism. Institutional Collectivism (Collectivism 1) is the degree to which organizational and societal institutional practices encourage and reward collective distribution of resources and collective action (House et al., 2004). In-Group Collectivism (Collectivism 2), however, is defined as the degree to which individuals express pride, loyalty, and cohesiveness in their organization or families (House et al., 2004). Table 5 offers a summary of the key traits of these two dimensions of Collectivism.

To conclude, mean country scores for Institutional Collectivism practices show rather high levels for all Scandinavian nations: Sweden: 5.22 (band A), Denmark: 4.80 (band A), and Norway: 4.07 (band B). An interesting reflection from Lindell and Sigfrids (2007) is that Institutional

Table 5. Key Elements of Collectivism-Individualism

Societies with higher Collectivism	Societies higher individualism
Individuals are integrated into strong cohesive groups	Individuals look after themselves or their immediate families
The self is viewed as interdependent with groups	The self is viewed as autonomous and independent of groups
Duties and obligations are important determents of social behavior	Individual goals take precedence over group goals and commitments
Individuals make greater distinctions between in-groups and out-groups	Few distinctions are made between in-groups and out-groups

Source: Based on House et al. (2004, p. 463).

Collectivism seems, at least to some extent, to go in waves. When times are good in society, individuality will be stronger, and in bad times Collectivism is stronger.

A specific anomaly in Norway's In-Group Collective culture lies in the Norwegian tradition for expecting the state to take care of old people and the sick rather than expecting the family to take this collective responsibility. A culture that scores high on In-Group Collectivism is traditionally a culture that values a home where many generations live together and where the family collectively assumes responsibility for the elderly or infirm within the home and where young people at work and university traditionally stay at home until they start their own family. In Norway, the high taxation system supports a comprehensive welfare state that in turn provides for state care of the elderly or sick—Thus, the collective responsibility is not to provide a home for all generations but to contribute to the welfare state via paying one's taxes. It is not therefore the norm to take care of elderly parents personally. The state pension scheme alleviates the financial responsibility of taking care of the elderly. To sum up, mean In-Group Collectivism scores for Norway and her Scandinavian partners are as follows: Norway: 5.34 (band B), Sweden: 3.66, Denmark: 3.53 (both in band C).

Project GLOBE's Power Distance Within a Scandinavian Framework

Power Distance is defined as the degree to which members of an organization or society expect and agree that power should be stratified and concentrated at higher levels of an organization or government (House et al., 2004). McClelland (1961) and McClelland and Burnham (1976) determined that effective managers were characterized primarily by their need for power. Bass (1985) posits that charismatic attributes and behaviors play an important role in Power Distance. Haire, Ghiselli, and Porter (1966) explored the differences in preferences for power among different cultures. Hofstede (2001) posits that the three most significant predictors of relative position on the Power Distance continuum are (a) latitude—distance from the equator, (b) population size, and (c) wealth—GNP per capita. Geographical latitude in a region predicts the variance in Power Distance, with a decrease in Power Distance as distance from the equator increases.

As population increases, its members accept a political power that is more distant and less accessible than smaller

societies. In terms of the third predictor of Power Distance, namely, wealth per capita, this relationship is explained by Hofstede as the fact that societal wealth is directly correlated to the growth of the middle class, which can act as a bridge between the powerless and the powerful. As presented above, Norway's post–World War II regional development policy has encouraged industry and work opportunities in all regions in Norway.

Norway's mean score of 4.13 in the Power Distance dimension depicts Norway as a low Power Distance society, in keeping with her Scandinavian profile. Such low Power Distance values in Scandinavia are manifested in certain aspects of Norwegian business practices such as little use of formal titles, dress codes, and attitudes to practical tasks in the workplace. Most organizations do not adhere to strict dress codes to show status. Even in institutions such as parliamentary offices and legal institutions, a senior member may be dressed as informally as a junior staff member. Titles or last names are rarely used when addressing others, even if they are of senior rank. Another element of Power Distance— the roles and hierarchy within a society—is mirrored in the egalitarian practices at work. In company canteens, for example, all staff, whether directors or junior staff, pick up their food themselves in the staff canteen in an orderly manner and dispose of uneaten food and used plates. Canteens are almost never segregated on a basis of position.

As a society that expects and agrees that power should be equally shared, the Norwegian progressive and comprehensive tax system, the high union membership, and the generous welfare state exemplify systems that are in place to protect and promote egalitarian values. Holmberg and Åkerblom (2007) discuss elements of low Power Distance seen outside the workplace in Sweden. These cultural elements can be seen to be equally pertinent to the discussion of Power Distance within a Norwegian context. Burial grounds for instance, are generally similar for everyone, regardless of family wealth or social status. Nor is the ability to get on a bus or any other public transportation helped by personal status as such: Everybody is obliged to queue (Holmberg & Åkerblom, 2007). Thus, egalitarian values are promoted on a national and company level.

To sum up, the mean scores in Power Distance for the countries in this study are as follows: Denmark: 3.89 (band "D," low), Norway: 4.13 (band D, low), and Sweden:v 4.85 (band B, moderate). Indeed, these results support the findings from Lindell and Arvonen (1996) that a dominant

Table 6. Key Elements of Power Distance

Societies with higher Power Distance	Societies with lower Power Distance
Society differentiated into social classes based on several criteria	Society has a large middle class
Clear power is seen as offering social order, relational harmony, and role stability	Power is seen as a source of corruption, coercion, and dominance
Civil liberties are weaker, public corruption can be higher	Civil liberties are strong and public corruption low
Democracy does not ensure equal opportunities	Civil liberties are stronger and public corruption lower

Source: Based on House et al. (2004).

Table 7. Key Elements of Humane Orientation

Societies with higher Humane Orientation	Societies with lower Humane Orientation
People are urged to provide social support for each other	People are expected to solve personal problems on their own
Offspring are expected and are prepared to give material support to their parents in their old age	Offspring are not expected to give material support to their parents in their old age
Values of altruism, benevolence, kindness, and generosity have high priority	Values of pleasure, comfort, and self-enjoyment have high priority
Others are important, i.e., family, friends, community, strangers	Self-interest is important

Note: Based on House et al. (2004, p. 570).

feature of the Scandinavian management style is delegation of responsibility. Table 6 offers a summary of Power Distance.

Project GLOBE's Humane Orientation Within a Scandinavian Framework

The concept of "urban helpfulness" is an important concept in relation to this cultural dimension. Project GLOBE claims that there is a correlation between decreasing unhelpfulness and urbanization with increasing population density (House et al., 2004). According to Triandis (1995), values of altruism, benevolence, kindness, love, and generosity are salient motivating factors guiding people's behavior in societies characterized by a strong Humane Orientation. Rokeach's Value Theory (1968) and Schwartz' Value Theory (1994) support this proposition. Schwartz posits that the central norms of a society can be categorized based on their position on the bipolar dimensions of Self-Transcendence (including values of tolerance and protection of others) versus Self-Enhancement (self-gratification; House et al., 2004). A supporting theory to this is Espen-Anderson's studies of welfare states and their relationship to Humane Orientation (Espen-Anderson & Korpi, 1987). The social democratic model of a welfare state in Scandinavia is claimed to be more humanely oriented compared with the liberal model exemplified in England and the United States. The social democratic model in Scandinavia recognizes the personal rights of all individuals as an element of universal welfare rights, whereas the liberal model found in England and the United States offers welfare state on a proportionally means-tested basis (House et al., 2004). Table 7 below offers an overview of the main traits of this dimension.

The scores in Humane Orientation for the countries in this study are highest for Norway: 4.81 (band A, high), followed by Denmark: 4.44 (band B, moderate to high) and Sweden: 4.10 (band C, moderate to low). The social concern that is characteristic of Norway is captured in part by this dimension, where Norway scores rather highly. At a regional level, historically, as a predominately agricultural and fishing nation, local communities living in harsh climates have traditionally helped each other in times of need. Even today, the philosophy of "civic duty" and taking part in a *dugnad* (voluntary local help projects) remain a part of daily life in most regions in Norway, but to a lesser extent in large cities.

Project GLOBE's Uncertainty Avoidance Within a Scandinavian Context

Uncertainty Avoidance is the extent to which members of an organization or society strive to avoid uncertainty by relying on established social norms, rituals, and bureaucratic practices. People in high Uncertainty Avoidance cultures actively seek to decrease the probability of unpredictable future events that could adversely affect the operation of an organization or society, and remedy the success of such adverse effects (House et al., 2004). Hofstede (2001) posits that people in societies create coping mechanisms to handle the anxiety produced by excessive uncertainty. He notes that technology has helped us to defend ourselves against uncertainties caused by nature and the law, and against uncertainty in the behavior of others and religion, and to accept the uncertainties we cannot defend ourselves against (Hofstede, 2001).

The GLOBE study of Uncertainty Avoidance practices revealed high correlations between high Uncertainty

Table 8. Key Elements of Uncertainty Avoidance

Societies with higher Uncertainty Avoidance	Societies with lower Uncertainty Avoidance
Show a stronger desire to establish rules and have less tolerance for the breaking of rules	Have a higher tolerance for risk taking
Document agreements in legal contracts	Rely on the word of others they trust rather than the contractual arrangement
Take more moderate risks	Show more tolerance for breaking rules
Show a stronger desire to establish rules, agendas, and routines	Are less concerned with orderliness and the maintenance of records

Source: Based on House et al. (2004, p. 618).

Avoidance scores, and high economic prosperity and extremely supportive governments in terms of economic initiatives. The Norwegian government has introduced supportive initiatives countrywide to support industry and welfare in all districts, and Norway scored high on Future Orientation. Table 8 below offers a summary of the key traits associated with this dimension:

Examples of uncertainty reduction and protection measures in Norway and also in her Scandinavian neighbors include the high value placed on the comprehensive welfare system with generous social security payments for sick leave, long-term disability, unemployment, and maternity and paternity pay. A second Norwegian institution that mirrors the value placed on Uncertainty Avoidance is the ombudsman system—a system of checks and balances that protects individuals against misgovernment in the legal or public administration systems. Another element of Norwegian culture that is reflected in the sense of order in society is people's approach to time. The social norm is to always be "on time" for both business meetings and social gatherings. Agendas are frequently distributed in business meetings and social club meetings, and even for birthdays, weddings, and christenings, to ensure a sense of order. In this way, good time keeping and the ethos of sticking to agreed times is important in the maintenance of good social relations, both in working and private life in Norway. The mean country scores for Uncertainty Avoidance showed a high level for all Scandinavian countries with Denmark and Sweden falling into band A with the following scores: Denmark: 5.22, Sweden: 5.32. Norway falls into band B (moderate to high) with 4.31.

Key Economic Data

Table 9 below summarizes key data for the countries in this study in terms of Gross Domestic Product, population density, Corruption Perception Index, and Global Competitiveness. The Global Competitiveness Report (2010), published by the World Economic Forum " assesses the ability of countries to provide high levels of prosperity to their citizens, as it measures the set of institutions, policies and factors that set the sustainable current and medium-term levels of prosperity"(World Economic Forum, 2010).

Summary of Findings on Scandinavian Cultural Practices

To summarize, although the Scandinavian cultures studied in this intercultural research appear ostensibly similar, the results illustrate Ashkansy's (1997) point, that such research can reveal subtle but important cultural differences in nations that are similar yet dissimilar. All three societies appear intrinsically egalitarian, they appear to value low Power Distance, directness, and consensus in decision making and to promote Gender Egalitarianism. Nevertheless, there are significant differences in the degrees of commitment to these values by each individual Scandinavian partner. These differences need to be understood and appreciated to avoid misunderstandings. The hallmark of Norwegian cultural practices within a Scandinavian context was seen to be higher Gender Egalitarianism. The most pronounced Norwegian cultural values within a Nordic framework were also lower Power Distance and higher Humane Orientation values.

Conclusion

The purpose of this article was twofold: first, to map contemporary Norwegian cultural practices within a Scandinavian context in relation to the author's recent doctoral research and project GLOBE findings (House et al., 2004), and second, to briefly introduce the argument that culture matters in Scandinavian academic research, by offering a summary of published research in the last decade where culture has been a variable. A future direction for research could be a more in-depth meta-analysis of such studies.

There is no shortage of advice available to managers about how to succeed with a new project in a culturally different region. Unfortunately much of this can be superficial as it only focuses on the practical elements of etiquette, such as "speak in a low voice in Asia and avoid eye contact with females in Japan." On the other end of the scale, we might be given only general advice such as "always be ethical." What is important to bear in mind here, however, is that what might be seen as ethical business behavior in one country may not be the same in another (Warner-Søderholm, 2010a).

Table 9. Key Economic, Population, and Transparency Indicators in Sweden, Norway, and Denmark

Country	GDP: USD, millions	Population density/km² (Wikipedia, 2010)	Global Competitiveness Report (World Economic Forum, 2010; all the nations below are ranked in the top 15 globally)	Corruption perception ranking (Transparency International, 1996) Ranking number
Sweden	478,961	21	5.51	3
Norway	451,836	12	5.17	9
Denmark	341,255	128	5.46	1

As discussed earlier in this paper, culture matters. Subtle but disturbing differences may surface when representatives from neighboring regions in Scandinavia work together. As depicted in the literature review of Scandinavian cultural research during the last decade (appendix), culture impacts a wide range of business issues from investment activities, environmental commitment, social responsibility, management practices, and dealing with crises to employee relations. Management practices in Norway differ in subtle ways from those of her Scandinavian counterparts because Norwegians tend to place greater value on directness, stronger Gender Egalitarianism values, and a greater tolerance for uncertainty. Furthermore, Norway's higher scores in Humane Orientation and In-Group Collectivist scores indicate a stronger focus on a more paternal, inclusive style of management in Norway. While these differences are not major, their combined effects may have consequences in certain situations. If a manager from Sweden operates in a more hierarchical, assertive manner in Norway, this could have a negative impact on the outcome of the negotiation. Furthermore, a statement could be made by a Danish consultant such as, "I'm afraid that we believe that the company should consider cutbacks." The manager from Norway might have phrased this message as, "You, you have make cutbacks!" Knowledge about such specific cultural characteristics can help team members from different regions value and anticipate differences. Potential problems such as interpersonal conflict that leads to stress and unnecessary personal strain and potential loss of revenues can then be minimized. Clearly, culture does matter.

Appendix

Academic Articles in the Last Decade With Scandinavian Culture as a Variable

Authors	Article	Nordic cultural context
Smith et al. (2011)	Individualism—Collectivism and business context as predictors of behaviours in cross-national work settings: Incidence and outcomes. *International Journal of Intercultural Relations.*	Multicountry study of cross-national work settings
Chiang and Birtch (2010)	Appraising performance across borders: An empirical examination of the purposes and practices of performance appraisal in a multicountry context. *Journal of Management Studies.*	Cultural issues related to performance appraisals
Saleh, Hassan, Jaffar, and Shukor (2010)	Intellectual Capital Disclosure quality: Lessons from selected Scandinavian countries. *Journal of Knowledge Management.*	Intellectual capital and corporate cultures in Scandinavia
Gjølberg (2010)	Varieties of corporate social responsibility (CSR): CSR meets the Nordic Model. *Regulation and Governance.*	CSR and corporate cultures
Peltokorpi (2010)	Intercultural communication in foreign subsidiaries: The influence of expatriates' language and cultural competencies. *Scandinavian Journal of Management.*	Intercultural communication in foreign subsidiaries: the influence of expatriates' language and cultural competencies
Stavrou and Kilaniotos (2010)	Flexible work and turnover: An empirical investigation across cultures. *British Journal of Management.*	Overtime and shift systems: Scandinavian culture
Kvande (2009)	Work-life balance for fathers in globalized knowledge work. Some insights from a Norwegian context. *Work and Organization.*	Scandinavian culture and gender issues
Clemmensen, Hertzum, Hornbeck, Shi, and Yammiyavar (2009)	Interacting with computers: Cultural cognition in usability evaluation. *Proceedings Interacting With Computers.*	International systems development: Human-computer interaction and culture

(continued)

Appendix Continued

Authors	Article	Nordic cultural context
Sanchez-Franco, Martinez-Lopez, and Martin-Velicia (2009)	Exploring the impact of individualism and Uncertainty Avoidance in web-based electronic learning: An empirical analysis. *Computers and Education.*	Cross-cultural differences in learning
Sandhu, Helo, and Kekale (2009)	The influence of culture on the management of project business: A case study. *International Journal of Innovation and Learning.*	National cultures and management style
Muller, Spang, and Ozcan (2009)	Cultural differences in decision making in project teams. *International Journal of managing Projects in Business.*	Decision making: cultural differences in German and Swedish teams
Sippola (2009)	The two faces of Nordic management? Nordic firms and their employee relations in the Baltic states. *International Journal of Resource management.*	Corporate culture, management by objectives
Ekelund (2009)	Cultural perspectives on team consultation in Scandinavia: Experiences and reflections. *Scandinavian Journal of Organizational Psychology.*	Cultural values and feedback styles in teams
Gheorghiu, Vignoles, and Smith (2009)	Beyond the United States and Japan: Testing Yamagishi's emancipation theory of trust across 31 nations. *Social Psychology Quarterly.*	Individualism, Collectivism, and trust
Bengtsson (2008)	Socially responsible investing in Scandinavia—a comparative analysis. *Sustainable Development.*	Business ethics and corporate cultures in Scandinavia
Hookana (2008)	Organizational culture and the adoption of new public-management practices. *Management.*	Culture and management practices
Rubin, Schultz, and Hatch (2008)	Coming to America: Can Nordic brand values engage American stakeholders? *Journal of Brand Management.*	Branding and business ethics
Chandrasekaran and Tellis (2008)	Global takeoff of new products: Culture, wealth, or vanishing differences? *Marketing Science.*	Innovation and communication
Uslaner (2008)	Where you stand depends on where your grandparents sat. *Public Opinion Quarterly.*	Cultural values
Brannback and Carsrud (2008)	Do they see what we see? A critical Nordic tale about perceptions of entrepreneurial opportunities, goals, and growth. *Journal of Enterprising Culture.*	Entrepreneurial values
Ramstrom (2008)	Interorganizational meets interpersonal: An exploratory study of social capital processes in relationships between Northern European and ethnic Chinese firms. *Industrial Marketing Management.*	HRM values between Northern Europe and China
Langvasbråten (2008)	A Scandinavian model? Gender equality discourses on multiculturalism. *Social Politics.*	Gender equality and inequality in Scandinavia
Halldorsson, Larson, and Poist (2008)	Supply chain management: A comparison of Scandinavian and American perspectives. *International Journal of Physical Distribution & Logistics management.*	Cross-cultural studies of supply chain management
Gotcheva (2008)	A cross-cultural study of leadership styles among executives in Bulgaria and Finland. *Journal of Human Resources Development and Management*	Cultural differences in leadership styles
Hjorth (2008)	Nordic entrepreneurship research. *Entrepreneurship Theory and Practice.*	The Nordic context
Håvold (2007)	National cultures and safety orientation: A study of seafarers working for Norwegian shipping companies. *Work and stress.*	Safety orientation culture on Norwegian vessels
Marnburg (2007)	Management principles in hospitality and tourism. *Journal of Human Resources in Hospitality and Tourism.*	Conflict vs. harmony values, functionalism vs. idealism, and organic vs. mechanic values
Waldstrøm and Madsen (2007)	Social relations among managers: Old boys and young women's networks. *Women in management Review.*	Gender values and networking
Demir and Søderman (2007)	Skills and complexity in management of IJVs: Exploring Swedish managers' experiences in China. *International Business Review*	Management values in China versus Sweden
Lawler (2007)	Janus-faced solidarity—Danish internationalism reconsidered. *Cooperation and Conflict.*	Internationalism, politics and culture

(continued)

Appendix Continued

Authors	Article	Nordic cultural context
Nafstad, Carlquist, and Blaker (2007)	Community and care in a world of changing ideologies. *Community, Work and Family*.	Consumers, ideological conflict
Kjeldgaard and Ostberg (2007)	Coffee grounds and the global cup: Global consumer culture in Scandinavia. *Consumption, Markets and Culture*.	Scandinavia consumption culture
Ekman and Emami (2007)	Cultural diversity in health care. *Scandinavian Journal of Caring*.	Culture's impact of care provision
Lutgen-Sandvik, Tracy, and Alberts (2007)	Burned by bullying in the American workplaces: Prevalence, perception, degree, and impact. *Journal of Management Studies*.	Masculine/feminine cultural values and bullying: the United States vs. Scandinavia
Pollanen (2006)	Northern European Leadership in transition—a survey of the insurance industry. *Journal of General management*.	Cultural issues in leadership in the insurance industry
Peltokorpi (2006)	Japanese organizational behaviour in Nordic subsidiaries: A Nordic expatriate perspective. *Employee relations*.	Insights into Japanese culture through Nordic eyes
Lynes and Dredge (2006)	Going green: Motivations for environmental commitment in the airline industry. A case study of Scandinavian airlines. *Journal of Sustainable Tourism*	Environmental commitment in the airline industry in Scandinavia
Peltokorpi (2006)	Japanese organizational behaviour in Nordic subsidiaries. *Employee Relations*.	Collectivism and culture
Zander (2005)	Communication and country clusters: A study of language and leadership preferences. *International studies of Management and Organization*	Leadership styles across groups of countries
Maenpåå, Antii, Hannu, and Pallab (2006)	More hedonic versus less hedonic consumption behaviour in advanced Internet bank services. *Journal of Financial Services Marketing*.	Consumer behaviour in different cultures
Wurtz (2005)	A cross-cultural analysis of websites from high-context and low-context cultures. *Journal of Computer-Mediated Communication*.	Direct vs. indirect communication in different cultures
Ka (2005)	Cultural traditions and the Scandinavian social policy model *Social Policy and Administration*.	Scandinavian traditions and social policy
Haner (2005)	Spaces for creativity and innovation in two established organizations. *Creativity and Innovation Management*	Work environment and corporate culture
Åselius (2005)	Swedish strategic culture after 1945. *Cooperation and conflict*.	Culture and national security
Neumann and Heikka (2005)	Grand strategy, strategic culture, practice: The social roots of the Nordic defence. *Cooperation and conflict*.	Culture and national security
Howlett and Glenn (2005)	Epilogue: Nordic strategic culture. *Scandinavian Journal of Hospitality and Tourism*.	International relations
Asheim and Coenen (2005)	Knowledge bases and regional innovation systems: Comparing Nordic Clusters. *Research Policy*	Technological innovations and forestry products
Gyekye and Salminen (2005)	Responsibility assignment at the workplace: A Finnish and Ghanaian perspective. *Scandinavian Journal of Psychology*.	Power Distance issues, assignment of responsibility, and accidents in the workplace
Roper (2005)	Marketing standardisation: Tour operators in the Nordic region. *European Journal of Marketing*.	Culture and marketing decisions
Arellano-Gault and Del (2004)	Maturation of public administration in a multicultural environment: Lessons from the Anglo-Saxon, Latin, and Scandinavian political traditions. *International Journal of Public Administration*.	Scandinavian political culture in public administration
Holmqvist (2004)	Experiential learning processes of exploitation and exploration within and between organizations. *Organization Science*.	Interorganizational relations, corporate culture
Ulijin and Lincke (2004)	The effect of CMC and FTF on negotiation outcomes between R&D and manufacturing partners in the supply chain: An Anglo/Nordic/Latin comparison. *International Negotiation*.	Negotiation cultures

(continued)

Appendix Continued

Authors	Article	Nordic cultural context
Aperia, Simcic, and Schultz (2004)	A reputation analysis of the most visible companies in the Scandinavian countries. *Corporate Reputation Review*	Social responsibility of Scandinavian businesses
Metcalf et al. (2004)	Cultural tendencies in negotiation: A comparison of Finland, India, Mexico, Turkey and the US. *Journal of World Business*	Cultural differences in negotiation styles
Smith, Andersen, Ekelund, Gravesen, and Ropo (2003)	In search of Nordic management styles. *Scandinavian Journal of Management.*	Cultural values of Nordic managers
De Geer, Borglund, and Frostenson (2003)	Anglo-Saxification of Swedish business. *Business Ethics: A European Review.*	International business
Gregory (2003)	Scandinavian approaches to participatory design. *International Journal ENGNG*	Distinctive Scandinavian approaches to participatory design
Tellis, Stremersch, and Yin (2003)	The international takeoff of new products: The role of economics, culture, and country innovativeness. *Marketing Science.*	International business enterprises
Webster Rudmin, Ferradanoli, and Skolbekken (2003)	Questions of culture, age, and gender in the epidemiology of suicide. *Scandinavian Journal of Psychology.*	Cultural values as predictors of suicide
Grennes (2003)	Scandinavian managers on Scandinavian management. *International Journal of Value-Based Management.*	Cultural values of Scandinavian managers
Kasper (2002)	Culture and leadership in market-oriented service organizations. *European Journal of Marketing.*	Achievement oriented managers and culture
Suutari and Riusala (2001)	Leadership styles in Central Eastern Europe: Experiences of Finnish expatriates in the Czech Republic, Hungary, and Poland. *Scandinavian Journal of Management.*	Cultural differences in management in Central Eastern Europe
Kurzer (2001)	Cultural diversity in post-Maastricht Europe. *Journal of European Policy*	Culture, social policy, and liquor laws
Mykletun and Haukeland (2001)	Scope and background: The Nordic context. *Journal of Hospitality and Tourism.*	Hospitality industry and culture

Note: HRM = human resource management; IJVs = International Joint Ventures; CMC = computer mediated communication; FTF = face-to-face communication.

Declaration of Conflicting Interests

The author(s) declared no potential conflicts of interest with respect to the research, authorship, and/or publication of this article.

Funding

The author(s) received no financial support for the research and/or authorship of this article.

References

Ashkanasy, N. M. (1997). A cross-national comparison of Australian and Canadian supervisors: Attributional and evaluative responses to subordinate performance. *Australian Psychologist*, *32*, 29-36.

Barry, H. I., Bacon, M. K., & Child, I. L. (1967). Definitions, ratings and bibliographic sources for child training practices of 110 cultures. In C. S. Ford (Ed.), *Cross cultural approaches* (pp. 293-331). New Haven, CT: HRAF Press. Original work published (1957)

Bass, B. M. (1985). *Leadership and performance beyond expectations*. New York, NY: Free Press.

Coltrane, S. (1996). *Family man, fatherhood, housework and gender equity*. New York, NY: Oxford University Press.

Espen-Andersen, B., & Korpi, M. (1987). *The Scandinavian model: Welfare state and welfare research*. Oslo, Norway: Scandinavian University Press.

Global Competitiveness Report. (2010). Retrieved from http://www3.weforum.org/docs/WEF_GlobalCompetitivenessReport_2010-11.pdf

Haire, G., Ghiselli, E. E., & Porter, L. W. (1966). *Managerial thinking: An international study*. New York. NY: Wiley.

Hofstede, G. (2001). *Culture's consequences: Comparing values, behaviours, institutions and organisations across nations*. Thousand Oaks, CA: Sage.

Hofstede, G., & Bond, M. H. (1988). The confucius connection: From cultural roots to economic growth. *Organizational Dynamics*, *16*, 5-21.

Holmberg, I., & Åkerblom, S. (2007). "Primus inter pares" leadership and culture in Sweden. In J. S. Chhokar, F. C. Brodbeck, & R. J. House (Eds.), *Culture and leadership across the world. The GLOBE book of in-depth studies of 25 societies* (33-76). New York, NY: Lawrence Erlbaum.

House, R. J., Javidan, M., Dorfman, P., & Gupta, V. (2004). *Culture, leadership and organisations: The globe study of 62 societies*. Thousand Oaks, CA: Sage.

IMD Global Competitiveness Ranking. (1994). Retrieved from http://www.imd.org/research/centers/wcc/index.cfm

IMD Global Competitiveness Ranking. (1998). Retrieved from http://www.imd.org/news/IMD-announces-its-2012-World-Competitiveness-Rankings.cfm

Inglehart, R. (1997). *Modernization and postmodernization: Cultural, economic, and political Change in 43 societies*. Princeton, NJ: Princeton University Press.

Kahn, H. (1979). *World economic development: 1979 and beyond.* Boulder, CO: Groom Helm.

Kluckhohn, F., & Strodtbeck, F. (1961). *Variations in value orientations*. Evanston, IL: Row, Peterson.

Kvande, E. (2009). Work-life balance of fathers in globalized knowledge work. *Some insights from the Norwegian context.* Retrieved from http://wwwsearch.ebsohost.com/login.asp?direct=

Levine, R. J., & Norenzayan, A. (1999). The pace of life in 31 countries. *Journal of Cross Cultural Psychology, 30,* 178-192.

Lindell, M., & Arvonen, J. (1996). The Nordic management style of investigation. In S. Jonsson (Ed.), *Perspectives of Scandinavian management* (pp. 11-36). Gothenburg, Sweden: Gothenburg School of Economics and Commercial Law.

Lindell, M., & Sigfrids, C. (2007). Culture and leadership in Finland. In J. S. Chhokar, F. C. Brodbeck, & R. J. House (Eds.), *Culture and leadership across the world. The GLOBE book of in-depth studies of 25 societies* (75-106). New York, NY: Lawrence Erlbaum.

McClelland, D. C. (1961). *The achieving society.* Princeton, NJ: D. Van Nostrand.

McClelland, D. C., & Burnham, D. H. (1976). Power is the great motivator. *Harvard Business Review, 54,* 100-110.

Parsons, T., & Shils, E. (Eds.). (1952). *Toward a general theory of action.* Evanston, IL: Row, Peterson.

Rokeach, M. (1968). *Beliefs, attitudes and values: A theory of organisation and change.* San Francisco, CA: Jossey-Bass.

Schwartz, S. H. (1994). Cultural dimensions of values: Towards an understanding of national differences. In U. Kim, H. C. Triandis, C. Kagitcibasi, S. C. Choi, & G. Yoon (Eds.), *Individualism and collectivism: Theoretical and methodological issues* (pp. 85-119). Thousand Oaks, CA: Sage.

Smith, P., Andersen, J. A., Ekelund, B., Gravesen, G., & Ropo, A. (2003). In search of Nordic management styles. *Scandinavian Journal of Management, 19,* 491-507.

Taras, V., Rowney, J., & Steel, P. (2009). Half a century of measuring culture: Review of approaches, challenges and limitations based on the analysis of 21 instruments for quantifying culture. *Journal of International Management, 15,* 357-373.

Transparency International. (1996). Retrieved from http://archive.transparency.org/publications/gcr

Triandis, H. C. (1995). *The analysis of subjective culture.* New York, NY: Wiley.

Trompenaars, F. (1993). *Riding the waves of culture: Understanding diversity in global business.* Chicago, IL: Irwin.

Trompenaars, F., & Hampden-Turner, C. (1998). *Riding the waves of culture: Understanding diversity in global business* (2nd ed.). New York, NY: McGraw-Hill.

United Nations Development Program. (1998). *Human development report.* New York, NY: Oxford University Press. Mannen Van (1983).

Warner-Søderholm, G. (2010a). Kommunikasjon og Kultur: Regionale Forskjeller i Forretningskultur i Norge [Communication and culture: Regional differences in business culture within a Norwegian context]. *MAGMA, 310,* 20-28.

Warner-Søderholm, G. (2010b). *Understanding perceptions of cultural and intracultural societal practices and values of Norwegian managers* (Doctoral dissertation). PowerGen Library, Henley Business School, University of Reading, UK.

Wikipedia on Nordic Countries. (2010). Retrieved from http://www.en.wikipedia.org/wike/Nordic_countries

Williams, J. E., & Best, D. L. (1982). *Measuring sex stereotypes: A thirty nation study.* Beverly Hills, CA: SAGE.

World Economic Forum. (1998). Retrieved from http://www.weforum.org/issues/global-competitiveness

World Economic Forum. (2010). Retrieved from http://www.weforum.org/events/world-economic-forum-annual-meeting-2010

Bio

Gillian Warner-Søderholm is a full-time faculty member of the BI Norwegian Business School and is also the head of the Department of Communication, Culture, and Languages. In addition, she holds the position of associate professor, teaching on a number of programs in intercultural communication, international business, Scandinavian culture, study strategies, negotiations, and presentations.

3

I Have to Give an "I Can" Attitude: Gender Patterns in Beeping Practices

Caroline Victoria Wamala[1]

Abstract

Intentional missed calling, referred to as beeping through the mobile phone, is a popular communication practice among Africans. Targeting young mobile phone users in Uganda, this article builds on previous research on beeping, but focuses on gender as a point of analysis. Data informing this article are based on 76 qualitative interviews with university students and recent graduates who are currently employed, and the results indicate that beeping practices are embedded in sociocultural, normative, gender patterns. The data also show that beeping is a multilayered exercise that each individual at some social-relational level engages in: It is the relationship to the beep recipient that negotiates this practice. Mapping local, diverse expressions of masculinities and femininity at the intersection of beeping activities, the study offers some recommendations on how Information Communication Technologies (ICT) in general can be useful signals of understanding sociological order.

Keywords

mobile phone, ICT, beeping, gender, Uganda

Background

In September 2007, I came across an online Reuter's article titled, "Phone credit low? Africans go for beeping." The article went on to reveal that on a daily basis, approximately 130 million missed calls swarm the circuits of mobile phone service providers in Africa.[1] What is apparent is that a large percentage of these missed calls are intentional (Castells, Fernandez-Ardevol, Linchuan Qiu, & Sey, 2006; Donner, 2007; Mckemey et al., 2003). This is a practice that does not always imply a "call me back" action but is an exercise that has been developed and perfected by users into an effective means of communication, with the added bonus of being absolutely free of costs. These intentional missed calls are referred to across Africa as buzzing, flashing, miskin, pitiful, menacing, boom-call, fishing, bipage, beeping, missed calling, and, although the term of reference to intentional missed calling may differ from country to country, the action is the same. In Uganda, the country under focus, the practice is referred to as beeping, and this reference will be maintained for the rest of this article.

Beeping someone is an action that involves calling the other user's mobile phone and hanging up before they have an opportunity to pick up the call. The messages emitted through this action are usually prenegotiated between users prior to the action (Donner, 2007), such as "I have arrived home safely," or "goodnight," or "I'm thinking of you," "I love you," and even "please call me back." The phone features such as the call log and phone book records enable the receiver to discern who has "called" them. Beeping on the

part of the service providers has created two (among others) glaring challenges that will be explored here. First, communication through the mobile phone is supposed to earn the service providers money whether it is a call or a text message. The general idea has been and continues to be that all communication through this technology is charged a fee. But how and what constitutes as communication is a constant power play between the end user and the service provider because what producers offer does not always tie in with the consumers' uses of a product[2] (cf. Cockburn & Ormrod, 1993; Johnson, 1988; Lie & Sørensen, 1996; Oudshoorn & Pinch, 2003). The other challenge with beeping is that these missed calls swarm the circuits, clogging up the communication channels. Service providers across the continent have stepped up to respond to this challenge by creating "call me back" services that are circuited through different channels, and some of these have been limited to a specific number a day.[3] Not every country though has managed to curb the practice, and an effort to discern why this practice persists has instigated research interests that have produced varying reasons behind the practice.

Even though Africa boasts the world's highest mobile telephony penetration figures in the world, some of the world's poorest are encountered in the same region (Coyle, 2005) and

[1]Karlstad University, Sweden

Corresponding Author:
Caroline Victoria Wamala, Center for Gender Studies, Karlstad University, Karlstad 65188, Sweden
Email: caroline.wamala@kau.se

the predominant reasoning behind beeping is that the economic situation for most Africans has pushed them to devise a communication practice through the technology that fits in with their finances (Castells et al., 2006; Donner, 2007; Goodman, 2005; Horst & Miller, 2006). This article builds on these debates, specifically highlighting gender performances through the practice of beeping. Instead of thinking of beeping as purely informed by economics, this article forges an understanding of how the identities constructed in relation to gender are aided by finances or lack of, and made visible, through beeping practices. Thus, the economic argument put forth by some of the previous work in beeping may not be the principal negotiator of beeping practices, but rather gender.

In this article, beeping is contextualized as a relational practice, a communicative exercise that is understood by individuals who are engaged in some sort of relationship. The social relationships alluded to here involve class negotiations, cultural negotiations, generational negotiations, intellectual negotiations, economic negotiations, as well as gender-based negotiations; a cluster of categories that reveal beeping as a practice imbued with intersectional social orders. These are enacted on multiple levels simultaneously with an overall impact on social exercises, existences, and expressions. I return to a theoretical outline for this article shortly but will at this juncture reemphasize that gender as a social category is the focus for this article. Generational, kinship, or class forms of beeping can be subjects for further research, not to mention the notion of users being active agents in what a technology becomes.

To understand who constitutes the 130 million daily missed calls, this study analyses diverse gender-based negotiations that confront beeping practices. Mapping local, varied expressions of masculinities and femininity at the intersection of beeping activities, the study offers some recommendations on how Information Communication Technologies (ICT) in general can be useful signals of understanding sociological order.

In terms of organization, this article is structured as follows. Immediately following this introduction are the objectives that informed this study. The reader is then introduced to the methods used before an outline of previous research related to beeping is provided. The cultural landscape within which beeping occurs receives mention within the theoretical outline that is lodged within social constructivist approaches to gender and technology. Six empirical sections follow the theoretical discussion, the first section details a quantitative side of the data that serve as a backdrop to the subsequent qualitative analyses before rounding up the discussion with some conclusions on who constitutes the daily 130 million beeps.

Objective

My goal going into the field was to investigate to what level the economic argument put forth by the literature on beeping

influenced the practice. Having engaged in beeping exercises myself, I knew there was more to the process than one's financial ability or lack thereof. I therefore sought to establish, under what circumstances, how often, and to whom the 130 million daily beeps were directed. Looking at my notes from the field, and the transcriptions from the recorded interviews, I noticed specific themes relating to expressions of gender, emerging in the data, many of which were aided by but not altogether overridden by the respondent's socioeconomic status. The objective of this article based on these identified themes, is to ascertain through the specific practices of beeping, how gender relations have been affected by and have affected mobile phone use in Uganda. As more than 70% of Uganda's population is below the age of 30,[4] the study focus aimed at this segment of society can also establish how modern technologies such as mobile phones have influenced or are being influenced by the gender structure in place.

In the field of ICT for development (ICT4D), there is the understanding that these technologies of which the mobile phone is a prominent member have ushered in an era of empowered development (Plant, 1998). Cyberfeminists view the ICT era as "providing the technological basis for a new form of society that is potentially liberating for women" (Wajcman, 2010, p. 148). As the most prominent and fastest-growing technology in ICT4D, studying the communication practices with the technology may help answer the question whether ICTs are in fact liberating women in the developing regions. McGuigan (2005) asserts that "[i]f you want to understand any kind of society you should look at how its members communicate with one another" (p. 47; see also Bray, 2007, p. 37). Symptomatic to these analyses are cyberfeminists (cf. Kirkup et al., 2000; Plant, 1998) who allude to a reembodiment of gender in the cyber realm as mentioned earlier. Wajcman's (2004) *Techno feminism* argues differently as she takes seriously the ways in which technologies are often framed to distance or subordinate women. Bending toward this reasoning, this article approaches the beeping practice with critical interrogations toward how gender relations are renegotiated, or even subverted through the use of the mobile phone.

If 130 million beeps circulate the mobile communication channels on a daily basis, to what extent can this figure be translated into empowered development? More specifically who constitutes these beeps, how are the beeps informed, executed, and perceived? In the process of answering these questions, this study draws on previous research that has identified social, contextual rules to the practice that go beyond cost-saving strategies (Donner, 2007). As such, if specific rules pervade the practice, additional questions such as how these rules are conceptualized and maintained will hopefully be answered. On another register, because majority of the accounts informing the gender and technology field are from Western experiences (see, Bray, 2007; Mellström, 2009), this article further contributes to the growing body of work emanating from developing regions, especially the meager contributions that come from the African continent.

Method

Limiting the study to Uganda, located in East Africa, this research is a result of a year of ethnographic observations, 23 conversational interviews, and a subsequent 53 qualitative interviews. The research group relating to beeping was further narrowed to young adults pursuing university education within close proximity to Uganda's capital city, Kampala because it is the young "urban users" who are "generally said to be the early adapters of newer communication technologies" (Berg, Mörtberg, & Jansson, 2005, p. 51; McGuigan, 2005; see also Katz & Sugiyama, 2006). Second, the practice of beeping is largely concentrated among university students (Castells et al., 2006; Donner, 2007) although the practice is known to also cut across generations (Castells et al., 2006). Third, more than 70% of Uganda's population is below the age of 30 as mentioned earlier.

I refer to the conversations from which some of the data presented henceforth were derived, as conversational interviews (see Dowson & McInerney, 2003) because all the informants were made aware that the thoughts shared were being gathered as data for a scholarly article. These conversations took place at informal gatherings, in public transport, with extended family members, with friends and acquaintances with whom rapport had been established (Aull Davis, 1999; Fetterman, 1998). The situations in which I engaged in these conversations, did not always provide an opportunity to digitally record the exchange because not only was the location inappropriate but the natural conversation flow would have been disturbed if I had stopped to request for permission to record (for a discussion on this method, see Fetterman, 1998). These conversations not only explained the beeping behavior but also helped give cultural meaning to the practice (Fetterman, 1998). On the occasions when I had the opportunity of discussing this research with the same informants twice, I asked for permission to record the exchange. These discussions lasted anywhere between 30 min to 2 hr.

Building on the observations and conversational interviews, I crafted a short questionnaire, which I personally administered to willing participants. Administering the questionnaire ensured clarity of the questions posed (Bryman & Cramer, 2001); it also provided an opportunity to follow-up on some responses with further questions (Nachmias & Nachmias, 1996). Meeting these students was done either in the libraries, outside their lecture halls, at the beginning or end of a lecture, or, as in one case, when I stationed myself outside a residence hall. The duration of the interviews with the students ranged from 10 to 30 min. I took notes as the respondent answered the questions, making sure to capture the nuances raised from question to question. The option not to record these encounters allowed the students to talk freely at times even read through my notes and provide more information. Majority of the students alluded to owning Nokia phones and as additional material, I also read through 20 Nokia mobile phone manuals—found online—of the more popular brands in 2008/2009 looking for any mention of instructions on how to perform beeping.

To establish whether beeping was also prevalent among young working recent graduates, ages ranging from 23 to 28, seven willing respondents were also interviewed from an advertising agency. These seven participants highlight an interesting class dimension to the practice of beeping that is intersected with gender and social norms. A qualitative presentation of data will be used although numerical mentions will be made liberally throughout the text.

Earlier Research

Castells et al. (2006) present a global perspective on mobile communication in society and rely on the term "mobile network society" to amplify social networks built on "wireless communication technology" (Castells et al., 2006, p. 6). Beeping receives mention as a last resort form of communication for mobile subscribers that beep the friends and relatives with the financial ability to return the call. Emphasis is also cast on the social rules to the practices, such as the social faux pas performed when a gentleman suitor beeps his love interest, suggesting gender rules to the practice (cf. Donner, 2007). Chango's (2005) contemporary review queries whether the desperate need to communicate by "the urban poor," who use beeping, does actually make them members of the information society. Mckemey et al. (2003) present a technical report that identifies patterns of telephony use unique to the African region, such as beeping and the potentials of using these practices as advertising campaigns to foster greater mobile phone use/integration. Horst and Miller (2006) in their anthropological study of the cell phone in Jamaica, encounter the practice as a cost-saving strategy. In Donner (2005, 2006, 2007), the author traces the practice across sub-Saharan Africa, and discovers a social etiquette to the practice. These articles pay the most social attention to beeping, and Donner concludes in his 2007 article that the practice is more than a "cost-saving practice," and that it can be categorized into three major beeps. He identifies a "call me back beep," a "prenegotiated beep" that involves prearranged messages such as "when I beep you I will have arrived at the agreed destination," and "relational beeps," which he notes cut across friendships, and romantic relationships, which this current article focuses on with an in-depth gender and technology analysis. Other sources that have informed this study include news reports such as a BBC article by Borzello (2001), titled "Uganda's 'beeping' nuisance." The article suggests an underlying financial impact to the practice, with the financially able users bearing the responsibility and cost of "the call backs."

Mutual Shaping of Gender and Technology

Beeping in this article will benefit from the social constructivist framework in gender and technology studies that sees "technology as both a source and a consequence of gender relations" (Wajcman, 2010, p. 149). This argument continues to inform feminist approaches to the field of Science, Technology Studies (STS), where it is suggested that science and technology do not "evolve in a vacuum rather they participate in the social world, being shaped *by* it and *shaping it*" (Law, 2004, p. 12). As such, feminist readings of STS seek to problematize the complex relationship women appear to have with technology as well as the obvious dominance of men in the field (Lagesen, 2008; Wajcman, 2004). Sociologist, Judy Wajcman (2010) notes that "gender relations can be thought of as materialized in technology, and masculinity and femininity in turn acquire their meaning and character through their enrolment and embeddedness in working machines" (Wajcman, 2010, p. 149). Among the youth in Uganda, one can see this mutual shaping emerging in beeping practices. This co-construction of gender and technology erupts in beeping as the youth engage with the mobile phone in specific and structured ways.

Performances of masculinity and femininity (cf. Butler, 1990, who looks upon gender as a performance) "are practices through which men and women engage in gender and effects of these practices [consequentially impact] practices on bodily experience, personality and culture" (Connell, 2005, p. 71). Masculine and feminine practices are varied, hence the emphasis on pluralized masculinities and expressions of femininity, and are learned and expressed individually, communally, or even culturally. Analyzing the dominance or subordinances within and among masculinities in particular, Connell (1987) notes a hegemonic form of masculinity that negotiates power relations in lieu of women and other men. The role of provider over a woman is one of the "prevailing hegemonic masculine ideal" [5] (Kiyimba, 2005; Nannyonga-Tamusuza, 2005; Nyanzi, Nyanzi-Wokholi, & Kalina, 2009; Sorensen, 1996; Wyrod, 2008, p. 809) and is a characteristic particularly developed through oral literature, language, music, dance, and folklore among Uganda's youth. Not only do these media create "near indelible impressions in matters of social organisation, political power relations, resource management and sharing, and gender relations, these impressions in turn lay down rules of the future social behaviour and contact of the young people" (Kiyimba, 2005, p. 254). Expected social behaviors are made visible in the games that young boys and girls engage in, the same are held in check by peers, relatives, the community "and recently through electronic mass media" (Nannyonga-Tamusuza, 2005, p. 116-119). I note that both boys and girls are constructed by and restricted by the local gender system and are groomed in opposition of each other, which helps establish relations of codependency (cf. Nannyonga-Tamusuza, 2005; Nyanzi et al., 2009). When asked if, how

often, why, and whom they beep, all the informants in this study alluded to being engaged in heterosexual relationships. As such beeping as a practice fortifies these cultural constructs but the same practice also makes visible the iterative quality of these expressions as well as their fragility.

Forms of Beeping: Results

The implicit economic argument put forth by (Borzello, 2001; Castells et al., 2006; Heavens, 2007; Horst & Miller, 2006; Mckemey et al., 2003) cannot be disputed regarding the practice of beeping. The empirical data provided henceforth are indicative of sociocultural practices among the young people, perhaps as a minion of economic inabilities. That said, it was also evident that even financially able female respondents (e.g., the working advertising firm female respondents), still beep their boyfriends, with a nonchalance that suggests an expectant "call me back" situation in most cases. Far from being a representative sample of the youth in Uganda, the subsequent numerical depictions are useful for the ensuing qualitative analyses as they provide general trends among the informants that provide the emphasis needed in detailing the qualitative data that follow this section.

From the 53 participants that informed this study, 46 of them are university students. Of the 46 students, 26 are female students and 20 are male students. The questions listed on the questionnaire first established if the respondent owned a mobile phone, after which demographic questions such as age, year of study, and name were also sought. Thereafter, the respondents' phone billing plan was queried. This question was essentially posed to establish if economic factors motivated the practice of beeping. All 53 respondents (including the working respondents) are on the prepaid phone billing plan, the choice plan made for the economically challenged who would want to monitor their expenditure (Skuse & Cousins, 2008; Waverman, Meschi, & Fuss, 2005). Further along the interview, the participants were asked to choose from a list of features available on their mobile phones, three of their favorite and most frequently used. Beeping is not a feature on any mobile phone, and scrutiny of the manuals sold with each mobile phone do not list this practice as a feature or supply instructions on how to go about the exercise.[6] Nonetheless, beeping was included (deliberately to establish if it was considered important) among the list of features to choose from, and with the exception of 1 female university student, who questioned whether beeping was a mobile phone feature, the rest of the informants either scorned at the feature, or jovially pointed at it as one of their favorite mobile phone features.

More specifically, 11 of the 26 female university students selected beeping as one of their favorite features and 8 men out of 20 male university students selected it. That this nonfeature is one of the most favored offered by the mobile phone by 19 of the university student respondents is indicative of just how prevalent the practice is among this group of users. The respondents were also asked how often and with whom they engaged in the practice of beeping. About 50% (13) of the female university students beep on a daily basis, and 35% (7) male university students beep on a daily basis. From the same group of university respondents, it was also established that 21 (81%) of the female respondents beep their boyfriends, while 8 (40%) of the male respondents alluded to beeping their girlfriends. With the 7 recently graduated respondents, all employed, the following responses were recorded. Of 4 female respondents, 3 chose beeping as one of their favorite mobile phone features, and 1 out of 3 male respondents selected beeping as a phone feature they favored. Two of the girls from the advertising agency claimed to beep on a regular basis, and were quick to point out that the beeping was to specific people such as to their boyfriend, or family or parents, never their peers, because "now that I am working I don't think it is good. When I was in college, it was very understandable. I only beep [my friends] if there is a prearranged message" (Daniella, 24, PR executive). None of the young working males beeped their girlfriends or buddies. But they did admit to beeping parents or older family members whom they knew would be able to call them back. Comments made especially by male respondents when asked if they beeped their girlfriends were noted. The strong reactions toward beeping their partners were mostly experienced from the male participants, while majority of female respondents almost seemed to consider the same practice as normal. To analyze these attitudes and further deconstruct the data provided here, the subcategories enlisted below relate to a relational beeping analysis further contextualized within the construction of gender in Uganda and among these participants.

Beeping Performances

To untangle the structures that order beeping performances as shown in the previous section, the empirical presentation in this section is enveloped in discussions on masculinity, and how this enterprise is produced through beeping practices. Seen as a process, masculinity is a relational concern within gender relations or as Connell (2005) puts it "'[m]asculinity' does not exist except in contrast with 'femininity'" (Connell, 2005, p. 68). As such bringing masculinity into focus further makes visible other constructions of gender categories (Carrigan, Connell, & Lee, 1985; Donaldson, 1993) that call various models of masculinity into play within beeping. The production of masculinity in beeping is assisted by class dynamics driven by economic resources as alluded to earlier. As the economic processes vary for informants in this study, the forms of masculinity are plural in their production. The strong affiliation between economic resources and expressions of masculinity provides another explanation for beeping. As a matter of course, Connell's

(1987) concept of hegemonic masculinity is taken as a point of departure and is maintained as a central focus in the process of analyzing the order in beeping practices. Hegemonic masculinity to recapitulate is when "particular groups of men" validate, maintain, and defend their positions of power over women and other men (Carrigan et al., 1985; Connell, 1987, 2005). In beeping practices, hegemonic masculinity is called into play through one's ability not to beep but also the ability to respond to a beep.

During one of my conversational interviews, I was made to realize how the relationship between masculinity and economic resources is the source of specific beeping practices:

> While I was at campus this was the case; a guy who is interested in a girl, will want to show her that he is capable of taking care of her. And guys like to show off, so the guy will start buying the girl airtime and he tells her that she should beep him, and he will call her back, and when her airtime runs out she must tell him, he will load more on her phone. So the girl gets used to this airtime. (Salome, November 2008)

The specific gender relations constructed in the interaction described in this excerpt can be better understood when the historicity of the gender order (cf. Donaldson, 1993) in Uganda is considered. If parallels are drawn between the breadwinning practice and the guy's ability to buy airtime, and respond to a girl's beep, hegemonic masculinity is conditioned upon and modeled, toward the guy not beeping, simultaneously encouraging his girlfriend to beep him. It is imperative for the guy in question to provide the call credit, which according to this excerpt must never be used to call the credit provider. This pattern of masculinity, may not necessarily be oppressive to the girl in question, in fact "some expressions of the hegemonic pattern [may be] more familiar and manageable" (Donaldson, 1993, p. 645) for women, who are actively engaged in its production. I discussed some of the patterns emerging in my data with Salome, whose insights were very useful in fleshing out the various reasons many young men in my data consciously chose not to beep.

Responsible Beeping

I came across 20-year-old Richard early October 2008, on his way to the library, and he mentioned that even though he did not have a girlfriend at present, he would never beep her because "I have to show an 'I can' attitude," and he followed this comment with "you know how it is for us guys." Richard spoke with hand gestures as he tried to get his opinion across. The historicity of the gender order referred to earlier is particularly evident in Richard's reference to "you know how it is for us guys." How society grooms and regulates gender relations can be reflected on how Richard interprets and adheres to these regimes. I also realize that I cannot be absolved of Richard's perception of me and the impact this might have had on the interview. The "I can" attitude alluded to may have also been directed at me as a woman. Shortly after meeting with Richard, I chanced upon 21-year-old Michael outside his residence hall. When asked who he practiced beeping with, his disapproval toward beeping his girlfriend was visible in his facial expression before he stated, "I would rather send her a SMS, beeping is too 'high school.' It's for immature boys." Michael, however, talked about beeping his parents claiming, "Oh yes, there is always money there." This was a clear theme among the informants. Namely, that the same young men do engage in the practice with their family displaying that construction of a good son, in relation to the familial hierarchies of parents caring for and protecting their children is unlikely to berate a beep from the son. Whereas the construction of a good boyfriend, within this category of respondents may be contingent on being responsive to a girlfriend's beep. Michael's reference to "there is always money there," with regard to his parents steers the discussion back to the economic resources that are central to Michael's beeping practices and depending on the social arrangements the same can both motivate and undermine who Michael beeps or does not beep.

On another occasion, stationed a few meters away from a roadside phone kiosk on one of the University campuses, I happened upon 21-year-old Steven. Steven was very chatty, and when asked who he beeps, his friends and his girlfriend were not in this category. As a way of explanation to not beeping his girlfriend, Steven supplied, "beeping doesn't show responsibility, I must show my girlfriend that I am responsible!" Steven gave some fascinating alternatives as opposed to beeping his girlfriend. We were standing a few meters from a roadside phone kiosk as mentioned already, and he stressed that rather than beep his girlfriend he would call her from the public phone, which he gestured toward. Call charges to mobile phones from one of these public phone booths are cheaper than direct mobile-to-mobile calls. This practice has social connotations to the effect, that one who calls another mobile phone from one of these phone booths communicates their financial inability to communicate directly from a mobile phone particularly if the call recipient is aware that the caller is in possession of a mobile phone. When Steven stressed that he would rather call from these phone booths he demonstrated his abhorrence toward beeping his girlfriend and also created hierarchies in communication practices that are worth pursuing in further research.

Ronald, a 22-year-old, had just come out of a lecture on a Tuesday afternoon when he agreed to an interview. We sat in the empty lecture hall and soft-spoken Ronald quietly responded to my questions with a conviction I had not seen with some of the other male informants. Ronald disapproved of beeping in general but with specific reference to his girlfriend he said, "I have to show that I care about her." I asked as a follow-up if beeping the girlfriend was an indication that he did not care for her, he simply nodded.

Beeping Behaviors

The context and the environments in which I met these informants further contributed to the material at least with regard to the expressions of masculinity. For example, John, an advertising agent, sat in an open office with several other colleagues and this interview could be heard by all who were present. When we reached the section in my question sheet that asked about his beeping practices, 25-year-old John exclaimed "I never beep my girlfriend, she is different." I asked him what was different about his girlfriend and he explained while looking around at his female colleagues "you know how girls are they expect you to call them." And to his credit, the female colleagues gestured and mentioned that it was important that a boyfriend calls his girlfriend as opposed to beeping. Seated a couple of desks away from John was 27-year-old Ivan. Ivan was a jolly boisterous interviewee who may have also wanted to show off for his colleagues in the same open office space as John. As he responded to the questions, he encouraged the participation of his colleagues by challenging them with questions or comments. Although Ivan did not mind his girlfriend beeping him, he took issue with friends or colleagues beeping him and when he mentioned this, one of his female colleagues calmly mentioned that sometimes a call is unnecessary when all you want to say is "hi, I am thinking of you." To this Ivan responded "you might as well call me and blurt out the 'hi' and then hang up, beeping is so unloving."

The environments were themselves highly gendered and produce varied enactments of masculinities. Alternatively, the context had an influence on or helped convey masculinity. Ronald John and Ivan put forward the same argument but expressed it differently. While John and Ivan fed off of their colleagues, Ronald's quiet conviction if paralleled exemplified different expressions of masculinity, that are still "centrally connected with the institutions of male dominance" (Donaldson, 1993, p. 645). Put another way, the girlfriends are a critical aspect in the production of hegemonic masculinity as long as they are the ones doing the beeping. For example, the female respondents (students and employed), who alluded to beeping their boyfriends, did not follow their response to this question with strong comments as noted above. A case in point is 24-year-old advertising executive, Daniella, who mentioned that she does not beep as much now that she is working, unless she is communicating a pre-arranged message. Daniella and her female colleagues suggest here that as working women, they have to adhere to specific codes of behavior such as only beeping when they have to. The same informants claimed to beep their boyfriends, however, with Daniella providing that "he expects me to beep him" (January 2009). Even though these ladies may be in a position to afford to call their boyfriends, they do not do so. As such when Wajcman (2010) talks about technology being a source and a consequence of gender relations, the performance of masculinity, emerging in this category of informants emphasizes the hegemony of these men in relation to women (Connell, 1987). Richard's comment about having an "I can" attitude, or Michael's comment about beeping being an "immature" trait for men, along with "beeping is so unloving," or "I have to show that I care about her," are comments that demonstrate that if, as a man, you love or care about your girl you cannot *devalue* what she means to you by beeping her (Chango, 2005) or risk the embarrassment of being labeled as "*cheap*" (Donner, 2007, p. 6).

Emphasis is made with words such as devalue and cheap, to highlight the socioeconomic implications of beeping at the same time as I problematize the exclusive focus on economic resources in previous research by making visible the gender structure that informs beeping. Put differently, the economic argument that feeds its way into specific sociotechnical practices is part of "large-scale social structures" (Carrigan et al., 1985, p. 577). In addition to the earlier research on beeping, this article argues that beyond the economic argument, the production of masculinity can be undermined or strengthened within specific conventions of gender practices. Each of the young men in this section have to show that they "care" about their girlfriends, and their inability to maintain the relationship could be seen as a loss of their masculine identification (see for comparison, Connell, 1987, pp. 183-187; Connell & Messerschmidt, 2005, p. 840; Kimmel, 1987, p. 266). As informed by another conversation interview,

> I had a chat with this guy who was complaining that his previous girlfriend who was from the [south] of Uganda was fond of beeping him. He said, "I used to buy her the airtime and she would beep me only, never called, so I started wondering—if I am buying her the airtime, who is she calling with my money? After a while we broke up and I am now seeing this lady from the [east] of Uganda, this lady is different, she will even go to the phone kiosks when she has no airtime, I don't understand, I have told her to beep me, but she always calls, is she more desperate than girls coming from the [south] or what?" (Mary, October 2008)

Several issues are raised in this excerpt, and as mentioned earlier, the various attributes that negotiate beeping such as ethnicity as inferred in this quote are topics for future research. But the idea promoted here illustrates the construction and expression of masculinity, around notions of power. This man, encourages his girlfriends to beep him and the girl who fails to adhere to his ability of assuming the role of provider, has him pondering whether or not she is desperate, a trait he clearly finds disconcerting toward the performance of masculinity. There is a lot of showing off, of being financially able, and for hegemonic masculinity to emerge successfully; women like Daniella have to be compliant in "accommodating the interests and desires of men," a process

that Connell (1987) labels as "emphasised femininity" (p. 183) meaning that the women are active participants in the process and production of hegemonic masculinity.

Expressing Resistance

Whereas Donner's (2007) and York's (2002) materials suggest that employees as well as students in this particular case can beep their employers or lecturers, my data does not agree with this assessment. The students and the advertising agents provided very emphatic nos to this question. Both male and female student respondents did not beep their lecturers and the employed respondents did not beep their employers, suggesting that social hierarchies can renegotiate the gendering of technology and why a situated context-specific analysis of sociotechnical practices is important.

Thinking back to Michael's comment when asked if he beeps his parents, where he reveals "oh yes, there is always money there," as opposed to him not beeping his lecturers or girlfriend, one can conclude that masculinity is expressed, both in situations of beeping and nonbeeping. Hence, within the family discourse, constructions of masculinity are still prevalent where the construction of "son" builds upon certain traits of masculinity at the same time as it dismisses others. Similarly the construction of "boyfriend" rests upon and is reinforced by certain traits of masculinity—caring for your woman—just as femininity when confronted by particular rankings in society, may not be an acceptable excuse for engaging with technology in specific ways.

Other tensions were registered with some of the subverted beeping performances as in the example of eight university male students who alluded to beeping their girlfriends, while five female university students did not beep their boyfriends. These young women considered the whole practice unfair to the recipient of the beep. Rebecca in particular was very vocal on her feelings toward the practice saying "it's a moral issue I don't consider it proper, like I am *detoothing* him." Detoothing is a form of dependency literally meaning to render someone toothless by deliberately enjoying financial favors from them with no intention of returning the goodwill. Hence young ladies who succeed in having men lavish money on them without a capitulation on their part are said to detooth guys. I contend the same can happen the other way around or even within same gender relations. For Rebecca, 22, a 3rd-year university student, she considers this reliance on a boyfriend to call her back as immoral and a practice only for the "cheap" girls.

To suggest that the forms of masculinities expressed among the category of men who beep as well as those who appear to have no qualms with receiving calls as opposed to beeps from girlfriends like Rebecca are subordinated forms in relation to hegemonic masculinity would be a rather simple conclusion. Rather the power structures that demotivate certain groups of men from beeping, and encourage others to beep, may be different but it also shows that the differences are not just about power—or power may assume a different character. Put differently, some of these masculinities resist the established assumptions at the same time as they stabilize the associated notions of power. The breadwinning position, shaped by social norms also shapes masculinity, and some of these norms are made visible through the young men who express the provider role by responding to a beep. The empirical material does not delve into how these groups of men negotiate alternative forms of masculinities, but it shows that not all men engage with technology in relation to conventional gender expressions, and certainly not women like Rebecca either for whom beeping the boyfriend may subjugate her boyfriend's masculinity rather than qualify it. I return to my conversation with Salome to help flesh out this section,

> Those guys who end up saying "my girlfriend is dethoothing me," have brought that upon themselves, you can't start buying airtime and change your mind along the way, she is not dethoothing, she is doing as you have showed her. (Salome, November 2008)

It would in fact appear that some men find the requirement to respond to a girlfriend's beep, disempowering especially if they refer to the process as one that *detooths* them. Even if it is the men who encourage the beeping according to Mary and Salome, the power structures that inform the exercise become subverted as women assume greater control and power over the process. The fragility and process-oriented quality in expressing masculinity, especially hegemonic masculinity, is evident in the men who fail to control their girlfriends beeping and as in the case of Mary's discussion also the girlfriends calling patterns. The economic incentives that appear to create the hegemonic pattern in beeping are not ahistorical. Their construction would benefit from a deeper examination of what practices sustain them (cf. Connell & Messerschmidt, 2005; Kimmel, 1987). But it is noteworthy that the five girls in this category do not engage in any form of beeping, and passionately dislike the practice. While on the flip side, the eight men who beep their girlfriends—an interesting contrast to the women they are being compared with in this category—noted how it was perfectly alright for their girlfriends to beep them, but they would also do (did) the same.

Peer Beeping Practices

Within the male homosocial/male–male circles, an unspoken tension was communicated on beeping one's male friends. The notion of responsibility discussed by the male respondents according to their reference of girlfriends was in this instance put forward as a boyish trait that is not popular even among their male friendships. As Donaldson has it, "[t]hrough hegemonic masculinity most men benefit from the control of women. For a very few men, it delivers control of other men" (Donaldson, 1993, p. 655). Held in check by

peers, as mentioned earlier, the gendered power relations among men (Connell, 2009) are amplified in relation to other men and manifest among the informants in this study in how they communicate through the mobile phone with each other. The male respondents were emphatic when they mentioned that beeping their friends almost always involved communicating a prenegotiated message, such as "I have arrived at the 'agreed meeting place.'" The girls on the other hand did not communicate the same friction regarding beeping each other. Beeping fellow girlfriends, just to communicate a "hello" message was a common practice mentioned by the female respondents. Some female respondents narrated the number of beeps required to communicate particular messages and that it was commonplace to beep friends on a regular basis.

Among the male respondents, 78% (18) in total mentioned beeping their friends, most of them following this with a comment to the effect that a beep had to be agreed upon prior to its emission. Among the female respondents, 83% (25) indulge in the practice liberally with fellow girlfriends, without the same tension as communicated by the guys. One female respondent noted that some of her friends did not appreciate being beeped so she did not beep them, but those that did not mind she beeped them every once in a while to say "hi." This demonstrates that the practice is not used within all female-to-female relationships, but it is more prevalent among this group than the male-to-male friendships. This shows that even among their male-to-male social bonds, there is the unspoken responsibility of elaborating masculinity. Because as Steven phrased it, "beeping doesn't show responsibility"; a trait that men are taught very early on in their childhood. The girls on the other hand, are taught to emphasize their dependence, (Kiyimba, 2005; Nannyonga-Tamusuza, 2005) and for them displaying these traits even among each other, may not always result in their femininity being compromised, to the level that masculinity is among the guys (Connell, 1987, p. 186f).

Conclusion

It is highly likely that beeping as a form of communication will wane as other features are introduced on the mobile phone. The smartphone revolution has afforded many students and Uganda's youth in general with cheaper affordable communication features powered by Wi-Fi or mobile Internet bundles. With Wi-Fi services spread liberally at many universities in Uganda, communication through these various applications can take place with little or no cost to the student. But even though the mobile phone may continue to provide affordable means of communication, this article has illustrated how to some extent expressions of gender are changing with the advent of ICT, but to a greater degree, the gender order at play is reinforced as made visible through mobile phone communication practices.

Beeping is ubiquitous within the Ugandan society (Borzello, 2001). The concept has been analyzed by researchers with the conclusion that it is not only a nuisance but is a practice deeply embedded in the socioeconomic status of Ugandans and Africans in general (Borzello, 2001). The current article develops this analysis by emphasizing an aspect of beeping that has received passing attention in the published works that of beeping being primarily informed by or negotiated by gender. This article has not dismissed the economic aspect put forth by previous research, but has sought to build on this enterprise by illustrating the conscious efforts applied by mobile phone users, to convey their gender identities when they communicate through the mobile phone. For example, the pressure, perhaps even pleasure, and power that young men face when they receive a beep from a love interest requires a cultural performance of the local understanding of what it means to be a responsible man. One also has to understand that the girls' practice of beeping does not preclude them as passive in the practice because their interactions with the mobile phone precipitates the enactment of hegemonic masculinity as made clear by Daniella's confession. Some of these ideas substantiate the hegemony of some groups of men that can be paralleled to the culturally idealized form of breadwinning, an enterprise that is actively sought by most men. Under the umbrella of socioeconomic processes, juggling the various positions of son, boyfriend, student, or employee, for the young men in this study requires various expressions of masculinity in beeping as regulated by the social arrangements. Not only does a mundane (in)significant practice such as beeping contribute to establishing these gendered norms, the practice through this technology also enables a glimpse of how gender is structured (Bray, 2007) in Ugandan society.

Cyberfeminists have alluded to a Cyberutopia that has the potential to redress inequalities in societies (Kirkup et al., 2000; Plant, 1998). A positive way of looking at the 130 million beeps is to consider the rapid uptake of the mobile phone by a region that has for the most part struggled to catch up with the rest of world with regard to ICT uptake (Polikanov & Abramova, 2003). But looking at the technology, and the social and cognitive structures in place, the way technology is used depends on the subject, and making the kind of jump alluded to by the cyberfeminists of a more equal society through technology (especially if a hype is created around the 130 million daily beeps), may in fact obscure more than illuminate interactions with ICT. Rather,

> both technology and gender are products of a moving relational process, emerging from collective and individual acts of interpretation. It follows from this that gendered conceptions of users are fluid, and that the same artefact is subject to a variety of interpretations and meanings. (Wajcman, 2010, p. 150)

Bray (2007) suggests that "[t]echnical skills and domains of expertise shape [and are shaped by local] masculinities and femininities" (p. 37).

As a social symbol of negotiating beeping practices, gender illustrates among this category of informants that the hegemonic masculine ideal of provider, prevails among majority of the young men who ascribe to this role by consciously choosing not to beep. In the same vein, majority of the young women practice beeping in line with local cultural emphasized femininity. These social signals are useful in "charting the coproduction of global and local culture" (Bray, 2007, p. 53) in the perpetual movements toward globalization. Masculinity and femininity expressions are tenuous within gender relations. They are confronted by or intersect with other social attributes such as class, kinship, cultural norms, age to name a few, which create a situated expression of the same, and beeping as a practice is particularly indicative of this. The socioeconomic aspect of beeping when shown to interact with the gender reveals a practice that extends beyond the economics of beeping. In principle, the social relationship to the beep recipient negotiates this practice to a greater extent than the economic resources. Africa's 130 million daily beeps are thus doing much more than merely informing recipients of a missed call. They are reinforcing and sometimes negotiating relationships and hierarchies across the continent.

Acknowledgments

I would like to thank the anonymous reviewers for their insightful valuable comments. I also thank Dr. Tanja Paulitz, editor of SAGE Open, for the great comments provided on the draft. To Jennie, Ericka, Ulf, and Petra, thank you for the thorough job you did in reviewing earlier drafts of the article.

Declaration of Conflicting Interests

The author(s) declared no potential conflicts of interest with respect to the research, authorship, and/or publication of this article.

Funding

The author(s) disclosed receipt of the following financial support for the research and/or authorship of this article: The research undertaken for this article was made possible with funding from the European Union FP7 Project—Networking for Communications Challenged Communities: Architecture, Test Beds, and Innovative Alliances. www.n4d.eu.

Notes

1. http://news.zdnet.com/2100-1035_22-166877.html
2. Users as active agents in thwarting what producers initially have in mind for a technology is an interesting thought worth pursuing, but it is beyond the scope of the current article. For other similar studies that denote similar arguments (cf. Cockburn & Ormrod, 1993; Oudshoorn & Pinch, 2005).
3. http://news.zdnet.com/2100-1035_22-166877.html
4. www.populationaction.org
5. I realize the debate surrounding the concept of hegemonic masculinity, where a number of arguments disagree with the

notion of hegemonic masculinity (cf. Donaldson, 1993) as first introduced by Connell (1987). Connell and Messerschmidt (2005) rethink the concept by advising that hegemonic masculinity as a concept be situated when utilized to avoid, homogenizing the notion.

6. For example, the Nokia 3310 manual mentions "beep" as the noise made by the phone when an incoming call interrupts an ongoing one. A "beep" sound can also be the signal selected for incoming calls. Another example in the Nokia 6300 manual indicates beeping as the sound made when an ongoing call is recorded by the recipient of a call. With the 6210, a beep is described as the noise made by the phone when a voice tag to a phone number is added. Even though beeping as a phrase appears in the Nokia manuals, it is not listed as a communication feature in the same sense that the informants of this study relate to the practice. None of the manuals studied supplied instructions on how to perform beeping as a communication process.

References

Aull Davis, C. (1999). *Reflexive ethnography: A guide to researching selves and others*. New York, NY: Routledge.

Berg, E., Mörtberg, C., & Jansson, M. (2005). Emphasizing technology: Socio-technical implications. *Information Technology & People, 18,* 343-358.

Borzello, A. (2001). Uganda's "beeping" nuisance. Retrieved from http://news.bbc.co.uk/2/hi/africa/1132926.stm

Bray, F. (2007). Gender and technology. *Annual Review of Anthropology, 36,* 37-53.

Bryman, A., & Cramer, D. (2001). *Quantitative data analysis with SPSS release 10 for Windows*. New York, NY: Routledge.

Butler, J. (1990). *Gender trouble: Feminism and the subversion of identity*. New York, NY: Routledge.

Carrigan, T., Connell, B., & Lee, J. (1985). Toward a new sociology of masculinity. *Theory and Society, 14,* 551-604.

Castells, M., Fernandez-Ardevol, M., Linchuan Qiu, J., & Sey, A. (2006). *Mobile communication and society: A global perspective*. Cambridge, MA: MIT Press.

Chango, M. (2005). Africa's information society and the culture of secrecy. *Contemporary Review, 286,* 79-81. Retrieved from http://www.bizcommunity.com/Article/416/78/13760.html

Cockburn, C., & Ormrod, S. (1993). *Gender and technology in the making*. London, England: SAGE.

Connell, R. W. (1987). *Gender and power*. Cambridge, UK: Polity.

Connell, R. W. (2005). *Masculinities* (2nd ed.). Berkeley: University of California Press.

Connell, R. W. (2009). *Gender* (2nd ed.). Cambridge, UK: Polity.

Connell, R. W., & Messerschmidt, J. W. (2005). Hegemonic masculinity: Rethinking the concept. *Gender & Society, 19,* 829-859.

Coyle, D. (2005). An overview, of in Africa: The impact of mobile phones. Retrieved from www.vodafone.com/etc/medialib/attachments/cr_downloads.Par.78351.File.tmp/GPP_SIM_paper_3.pdf

Donaldson, M. (1993). What is hegemonic masculinity? *Theory and Society, 22,* 643-657.

Donner, J. (2005, May). *The rules of beeping: Exchanging messages using missed calls on mobile phones in sub-Saharan Africa*. Paper presented at the 55th Annual Conference of the International Communication Association, New York, NY. Retrieved from https://blog.itu.dk/DMKS-E2008/files/2008/10/donner-rules-of-beeping-ica.pdf

Donner, J. (2006). The social and economic implications of mobile telephony in Rwanda: An ownership/access typology. *Knowledge, Technology, & Policy, 19,* 17-28.

Donner, J. (2007). The rules of beeping: Exchanging messages via intentional "Missed calls" on mobile phones. *Journal of Computer-Mediated Communication, 13.* Retrieved from http://jcmc.indiana.edu/vol13/issue1/donner.html

Dowson, M., & McInerney, M. D. (2003). What do students say about their motivational goals? Towards a more complex and dynamic perspective on student motivation. *Contemporary Education Psychology, 28,* 91-113.

Fetterman, M. D. (1998). *Ethnography step by step* (Applied Social Research Methods Series) (2nd ed.). Thousand Oaks, CA: SAGE.

Goodman, J. (2005). Linking mobile phone ownership and use to social capital in rural South Africa and Tanzania, Africa: The impact of mobile phones. Retrieved from www.vodafone.com/etc/medialib/attachments/cr_downloads.Par.78351.File.tmp/GPP_SIM_paper_3.pdf

Heavens, A. (2007). Phone credit low? Africans go for "beeping." Retrieved from http://www.reuters.com/article/worldNews/idUSHEA92325720070926

Horst, A. H., & Miller, D. (2006). *The cellphone: An anthropology of communication*. Oxford, UK: Berg.

Johnson, J. (1988). The mixing humans and nonhumans together: The sociology of a door-closer. *Society for the Study of Social Problems, 35,* 298-310.

Katz, J. E., & Sugiyama, S. (2006). Mobile phones as fashion statements: Evidence from student surveys in the US and Japan. *New Media & Society, 8,* 321-337.

Kimmel, M. (1987). Men's responses to feminism at the turn of the century. *Gender & Society, 1,* 261-283.

Kirkup, G., Janes, L., Woodward, K., & Hovenden, F. (2000). *The gendered cyborg: A reader*. London, England: Routledge.

Kiyimba, A. (2005). Gendering social destiny in the proverbs of the Baganda: Reflections on boys and girls becoming men and women. *Journal of African Cultural Studies, 17,* 253-270.

Lagesen, V. A. (2008). A cyberfeminist Utopia: Perceptions of gender and computer science among Malaysian women computer science students and faculty. *Science Technology & Human Values, 33,* 5-27.

Law, J. (2004). *After method: Mess in social science research*. New York, NY: Routledge.

Lie, M., & Sørensen, K. (1996). *Making technology our own? Domesticating technology into everyday life*. Oslo, Norway: Scandinavian University Press.

McGuigan, J. (2005). Towards a sociology of the mobile phone. *Interdisciplinary Journal of Humans in ICT Environments, 1,* 45-51.

Mckemey, K., Scott, N., Scouter, D., Afullo, T., Kibombo, R., & Sakyi-Dawson, O. (2003). Innovative demand models for telecommunications services: Final technical report. Retrieved from www.telafrica.org/pdfs/FinalReport.pdf

Mellström, U. (2009). The intersection of gender, race and cultural boundaries, or why is computer science in Malaysia dominated by women. *Social Studies of Science, 39*, 885-907.

Nachmias, F.-C., & Nachmias, D. (1996). *Research methods in the social sciences* (5th ed.). London, England: Arnold.

Nannyonga-Tamusuza, S. (2005). *Baakisimba: Gender in the music and dance of the Baganda people of Uganda.* New York, NY: Routledge.

Nyanzi, S., Nyanzi-Wokholi, B., & Kalina, B. (2009). Male promiscuity: The negotiation of masculinities by motorbike taxi-riders in Masaka, Uganda. *Men and Masculinities, 12*, 73-89. doi:10.1177/1097184X07309503

Oudshoorn, N., & Pinch, T. (Eds.). (2003). *How users matter: The co-construction of users and technology.* Cambridge, MA: MIT Press.

Oudshoorn, N., & Pinch, T. (2005). Introduction: How users and non-users matter. In N. Oudshoorn & T. Pinch (Eds.), *How users matter: The co-construction of users and technology* (pp. 1-29). Cambridge, MA: MIT Press.

Plant, S. (1998). *Zeros and ones: Digital women and the new technoculture.* London, England: Fourth Estate.

Polikanov, D., & Abramova, I. (2003). Africa and ICT: A chance for breakthrough? *Information, Communication & Society, 6*, 42-56.

Skuse, A., & Cousins, T. (2008). Getting connected: The social dynamics of urban telecommunications and use in Khayelitsha, Cape Town. *New Media & Society, 10*, 9-26.

Sorensen, P. (1996). Commercialization of food crops in Busoga, Uganda, and the renegotiation of gender. *Gender & Society, 10*, 608-628.

Wajcman, J. (2004). *Techno feminism.* Oxford, UK: Martson Book Services.

Wajcman, J. (2010). Feminist theories of technology. *Cambridge Journal of Economics, 34*, 143-152.

Waverman, L., Meschi, M., & Fuss, M. (2005). The impact of telecoms on economic growth in developing countries. In *Africa: The impact of mobile phones* (pp. 10-23). The Vodafone Policy Paper Series, Number 2. Retrieved from http://info.worldbank.org/etools/docs/library/152872/Vodafone%20Survey.pdf

Wyrod, R. (2008). Between women's rights and men's authority: Masculinity and shifting discourses of gender difference in Urban Uganda. *Gender & Society, 22*, 799-823.

York, T. (2002). Wireless taking African sub-continent by storm. Retrieved from http://web.archive.org/web/20030705211335/cooltown.com/mpulse/0602-africa.asp

Nokia user phone manuals reviewed (Retrieved through Google search: July 18, 2009, and June 2, 2012)

3310, 6300, 6210, 5230, 5140, 5110, 3330, 5210, 3410, 3210, 8810, 8210, 2100, 1650, 6085, 6280, 3220, 6111, 7280, 7373, 6288.

Bio

Caroline Victoria Wamala is a postdoctoral researcher at the Center for Gender Studies, Karlstad University, Sweden. Her research focuses on the innovative uses and access processes of Information Communication Technologies (ICT) adapted in developing regions. Located in the field of gender and technology, Caroline's research contributes to this field, a deeper appreciation of the cultural embeddedness of technology and gender.

4

Seeking *Personal Autonomy* Through the Use of Facebook in Iran

Mona Hajin[1]

Abstract

In Iran, where males and females are kept separated in different spheres, Facebook may be used as an opportunity to bridge this gap between the genders. However, this study showed that Facebook, as a *nonymous* platform in which people are in contact with their already-made *social ties*, didn't seem to be liberating from the existing norms and rules within society. Facebook was a *stage* that became restricted with the involvement of *social ties*. The study's analysis of interviews with six young Iranians showed that social meanings and norms of self-presentation on Facebook are defined to a large degree in terms of gender. The informants used a variety of strategies when presenting themselves on Facebook. They used Facebook simply for gaining *personal autonomy*. *Strategies* were adopted especially when one's *personal and community needs* were in conflict. Efforts made to apply strategies were gendered and were used mainly by females. Males conformed to and women resisted societal norms and expectations.

Keywords

Facebook, Iran, youth, strategic self-presentation, and social control

In Iran, where males and females are kept separated in different spheres, the Internet can be used as an opportunity to bridge this gap between the sexes. Based on Islamic law, both sexes in society should dress according to the Islamic dress code. Therefore there is not much room for choosing dress according to personal preferences. Different aspects of everyday life, *voices*, freedom of speech, and acts are regulated by the Islamic rules and gender segregation policies. In such a society where the choice of dress is strict, Facebook may provide a vocal opportunity to this muted population both textually and visually. It might also provide an environment for people to experience online the way of life they choose while living under the control of the authorities.

There are different scholarly approaches toward the Internet depending on whether it is a context-dependent or an independent platform from offline life. Given these approaches, there is tension among scholars over whether the Internet is an open virtual space independent from the offline world or an extension of the offline world (Bryant, 2008). Many scholars have considered the Internet as an independent sphere from offline life, its structures, discourses, ideologies, and so. They believe that online life is independent from offline life, and therefore it can make its own discourses, which are free from the offline discourses (Kendall, 2002; Mitra, 2005; Turkle, 1995). Thus, people can behave independently in regard to their offline life and experience a new way of life on the Internet. Accordingly, those who haven't had a *voice* before get a chance to be heard and might be able to present themselves as they wish. In this regard, the

Internet is viewed as a free space in which people can portray their identity freed from the structures within society. The Internet allows them to experience different aspects of themselves, evaluate their identities, and play different selves (Bryant, 2008; Turkle, 1995). Other scholars consider the Internet as parallel to offline life, representing and reflecting the structures of the offline world (Agger, 2004; Fornas, Klein, Ladenfork, Suden, & Sveningsson, 2002). With respect to this view, the Internet can act like a mirror and can show how people behave and live in the offline world. However, it doesn't need to reflect the offline world perfectly, but can be along the same lines with its structures and norms. Considering the Internet as either *mirror* of the offline life (see Agger, 2004, Fornas et al., 2002) or *independent sphere* from the offline life (Kendall, 2002; Mitra, 2005; Turkle, 1995) seems mutually exclusive. To make a conclusion to what extent the Internet is dependent on or independent from offline norms, structures, and son, one should focus on a specific context, for instance a sociocultural situation in which the Internet (generally) and social media (specifically) are used. The approach presented here is inspired by Markham's (2004) definition of the Internet as "a

---

[1]Stockholm University, Sweden

Corresponding Author:
Mona Hajin, Stockholm University, SE-106 91 Stockholm, Sweden
Email: mona.hajin@gmail.com

communication medium, a global network of connections, and a sense of social construction" (p. 119). According to Markham, "the shape and nature of internet communication is defined in context, negotiated by users that may adapt hardware and software to suit their individual or community needs" (p. 119). This approach encourages a refreshing view of the Internet that goes beyond online and offline segregation and motivates studying the Internet within the context in which it is used.

When studying the Internet (generally) and social media (specifically), it is important to make sure not to fall into a *naïve belief and cyber-utopianism* approach (Morozov, 2011) that considers them as liberating and emancipatory per se, without considering their downsides, especially when it comes to authoritarian states in which social media can be used as a tool for *controlling, tracking*, and *harassing* people based on their acts (Morozov, 2011). It is no longer possible to segregate between online and offline when people get *arrested, tracked*, and *harassed* due to their acts online (Morozov, 2011). According to this approach, the Internet and social media are an extension of one's life, rather than parallel to or separate from it, and people use them to fulfill their *individual or community needs* (Markham, 2004).

In symbolic interactionism tradition, all actions have *social meaning* (Farquhar, 2009). However, as Farquhar (2009) points out, "in the tradition of Mead, social meanings and norms are fluid" (p. 26). Accordingly, even though social meanings and norms are socially constructed, they are negotiated by the users and therefore are all *fluid*, rather than being fixed and static. As Farquhar writes, drawing on Boyd (2004, p. 9) "it is not that we live separate lives, one in the technology, one in the 'real' world. Rather, it is that we live lives with a mix of interaction types" (p. 17). Therefore, actions would be based on "the norms and meanings of society in general and the Facebook community specifically" (Farquhar, 2009, p. 23). That is, Facebook, due to both the limitations and the possibilities that are part of the technology, might make possible the construction of new *social meanings* in conjunction with the users' own individual and community norms. The self-concept, according to Farquhar, drawing on Mead (1932, 1936, 1938) and Goffman (1959) "comes through experiences with those around us" (p. 27). Accordingly, the self is not constructed in isolation, rather is constantly *constructed socially* and in relation and in interaction with others and their perceptions, expectations, and attitudes. Accordingly, any self-presentation is basically made to be performed before others—who are called *Friends*—on Facebook. As Farquhar points out "Facebookers would learn how to perform their own identities from their interpretations of other Facebookers" (p. 31).

Inspired by the definition of the Internet provided by Markham (2004) as fulfilling one's *individual or community needs*, and also *fluid*ity of *social meanings* and *norms*, this study investigates how a small number of young Facebook users in Iran reflect on their own self-presentation on Facebook (on a daily basis). In particular, this study investigates whether there are any gender differences between these reflections, and also how these reflections, claims, and self-presentational acts are in line with or different from the social meanings and norms in the sociocultural context of the society (Iran) in which they live and use Facebook.

As Farquhar (2009) notes, computer-mediated communications (CMC) "offer greater control over presentation of self" (p. 30; Boyd, 2004; Turkle, 1995), compared with unmediated communication. In a similar sense, social networking sites, as Papacharissi (2011) writes, "enable individuals to construct a member profile, connect to known and potential friends, and view other members' connections. Their appeal derives from providing a stage for self-presentation and social connection" (p. 304). Facebook members can communicate with other members and engage in interaction on an everyday basis. Users can create a personal profile, add personal information, exchange information and messages, and get in touch with other people. However, on the Internet in general, and Facebook specifically as Farquhar points out, we can see "the influence of institutions . . . the influence of societal norms and group affiliations; and individual ability to make decisions" (p. 33).

Facebook in the Middle East and Iran

Less "Fictional" Self-Presentation and Social Control on the "Nonymous" Platform of Facebook

According to Grasmuck, Martin, and Zhao (2009) people tend to be more *realistic and honest* in more *nonymous* websites. This is in contrast to the anonymous websites in which people tend to "act as somebody else" (Grasmuck et al., 2009, p. 158). The *nonymous* setting refers to an environment in which "identity claims appear to be grounded in offline realities" (Grasmuck et al., 2009, p. 158). It is assumed that people are more realistic, are honest, and act as themselves, and they, for instance, reveal their real name and identity. However, the *nonymity* of environment doesn't guarantee that there would be no fake identities.

Even though on Facebook, users themselves construct the self they want to present before others, self-presentations seem to be less *fictional* compared with previous means of using the Internet in which self-presentations were generally anonymous, *fictional*, and *fabricated*. However, it doesn't mean that Facebook users do not exaggerate some aspects of their identity plays using the platform. On Facebook, users often interact with their "pre-existing networks of social ties" (Boyd, 2004; Ellison, Steinfield, & Lampe, 2006, in Farquhar, 2009, p. 31) of *friends* who might be generally familiar with who they in fact are.

When talking about social meanings and norms in this study, the focus is not only on the rules and codes imposed by the Islamic Republic but also on cultural meanings and norms within the socioculture context of Iran.

Possibly, one of the most significant cultural concepts in the Middle East including Iran, according to Moghadam (2003), is *social control*, especially *control over women.*

The social network opportunities that Facebook offers increase drastically in importance when Facebook is utilized by a society like that in Iran, in which all media are state-run and the media contents are controlled by the authorities. Iran has a relatively young population. In 2010, "about 66 percent of Iranians are age thirty and younger" (Hashem & Najjar, 2010, p. 126). Many of the youth own computers and mobile phones and are adept at using them for a variety of purposes. In addition, "Iran is a well-wired and networked nation, with more than twenty million people who use the Internet on a regular basis" (Hashem & Najjar, 2010, p. 126).

Facebook can be considered as a place where gender segregation is harder to enforce and it also allows "for more playfulness about identity, and the possibility of adding a transnational context to it" (Khosravi, 2008, p. 101). Accordingly, Facebook can be considered as a *stage* for "practice of defiance" (Khosravi, 2008, p. 3). However, this doesn't mean that the downsides of social media use are neglected here. On the contrary, their use has turned into a *tool* for authoritarian states to *control, track*, and identify opposition (Morozov, 2011). Morozov's (2011) insights regarding social media is most accurate for defined political acts, but for instance, when it comes to decreasing the gender gap within the society, providing a platform for people to have dialogue with one another, and also for banal practices, social media may be more liberating compared with what is practiced within the society and between people.

"Cultural invasion" (*tahajom-e farhangi*) has always been one of the state's main concerns over the introduction of new communication technologies and media to the country and according to Khosravi (2003, p. 32) "is perceived to be conducted by the Great Satan, i.e. The US, and its 'indigenous agents'… [and is considered to be] more dangerous than the military ones. According to Esfandiari (2011), drawing on the Islamic authorities, social media are basically made to "lead young Iranians astray and to distance them from the ideas and principles of the republic" (Esfandiari, 2011).

As York (2012) writes "for months . . . another story about the Islamic Republic's ambitions has been gaining ground. that story is about the Iranian government's attempt to create its own 'halal' internet, cut off from the outside world" (York, 2012). According to her, "Iran's intent, it would seem, is to create an Internet where Iranians are 'safe' from the pornography, hate speech and cultural influence that the World Wide Web provides" (York, 2012). As she further writes, "despite an official ban on the site, Facebook is widely used (through the use of proxies and VPNs, which a reported half of Iran's population uses) and Twitter is increasing in popularity."

Access to the Facebook website since the 2009 election onwards has been blocked and its usage is considered as illegal in Iran. Its use might be considered as a "cultural crime" (*jorm-e farhangi*). "Cultural crime", as Khosravi writes (2003, p. 42), "appeared in the post-revolutionary Penal law as a new term for breaking Islamic rules.

According to Sharif (2012), "officials of the Iranian regime consider the new media and social networks on the Internet [as] serious threats to the existence of [the] regime" and that is the reason why they use any means to limit the access of social media to the Iranian people. The Islamic republic authorities, as Sharif writes, "have just recently announced that Iranian Cyber Police has begun operation in Tehran arresting a number of social activists on Facebook."

Accordingly, Facebook users are "branded as law-breakers in their trivial everyday life" (Khosravi, 2008, p. 125), as mundane practices like "showing more hair or skin than allowed" (Khosravi, 2008, p. 125) can violate the "'collective sentiment' of the Muslim community" (Khosravi, 2008, p. 17). According to JakartaGlobe (2012), citing an Iranian ayatollah, "the social networking service Facebook was un-Islamic and being a member of it a sin." The growth of Facebook has increased in a way that has caused concern and led to Islamic clerics issuing a *Fatwa* on Facebook, which introduces its use as *Haram*. According to one of the highest clerics in Egypt, Facebook use is considered absolutely immoral, as it threatens family cohesion, causes more break-ups, and increases the number of divorces (Haaretz Service, 2010). This *Fatwa* was issued a little while before the political unrest in Egypt.

When talking about Facebook, it is not all about popularity and growing usage. There are many people who leave Facebook every day for different reasons such as privacy concerns, or seeing it as a waste of time or too superficial. Although Facebook is in decline in certain parts of the world, it is still popular and peaking in the Middle East (Itameri, 2011). The number of Facebook users in the Middle East has increased by 30% from the time when political unrest began in the region (Itameri, 2011).

Facebook has been one of the few social networking sites that has experienced such rapid growth in popularity among Iranians. Despite Facebook being blocked during the Iranian presidential election in 2009 and afterwards, its usage is still increasing among Iranians. There is however, no clear statistic showing the exact number of Facebook users in Iran as the website is blocked and people access it through the VPNs and proxies and therefore the real number of users cannot be specified. However, as the Green voice of Freedom (2011) reports referring to an official website, "17 million Iranians are Facebook members and in a way, live on this website, despite the fact that this website has been censored in Iran."

Aims

This study relies on the individual Facebook users' own experiences and reflections, and instead of assuming Facebook as a political or a social platform, takes an inductive approach by which individuals themselves lead the researcher to come to a conclusion in this sense.

This study doesn't claim to be representative of Iranian youth on Facebook, rather each informant is a case here. Each informant informs the researcher, not only about his or her own experiences of self-presentation on Facebook but also about a larger network of social meanings and norms. This study also explores whether social meanings and norms are opposed or reinforced by the informants' use of Facebook in the context of Iran.

This qualitative study focuses particularly on how young Iranians themselves reflect on their own self-presentation on Facebook. To do so, a limited number of young Facebook users (five informants) who currently live in Iran and were interested in participating in this study were recruited and then interviewed in person.

The main questions that are explored in this study are as follows:

- How do young Iranians (informants of this study) reflect on their own presentation of self on Facebook?
- Are there any gender differences in the informants' reflections and if so, which differences?
- To what extent are the existing social norms and meanings of the society seen as being reinforced or opposed through the informants' reflections?

Strategies

In the following, the term *strategy* will be defined using definitions provided by Wodak, de Cillia, Reisigl, and Liebhart (2009) and Erving Goffman (1959). These definitions will then be combined and modified in relation to self-presentation on Facebook.

As Wodak et al. (2009) write, "the concept of strategy derives from Greek *strategia* and, since the nineteenth century at the latest, has meant 'the art of a commander-in-chief; the art of projecting and directing the larger military movements and operations of a campaign'" (p. 31). According to Wodak et al., drawing on Oxford English Dictionary definition (1998, p. 852, point 2a) "a strategist is someone who is skilled in leading an army."

As Wodak et al. (2009) further write, "abstracted from a purely military context, the concept of strategy generally denotes a more or less accurate plan adopted to achieve a certain political, psychological or other kind of objective" (p. 31). Moreover, they write that "the strategist attempts to anticipate all those factors which may have an impact on his or her actions" (Wodak et al., 2009, p. 31). In fact, the use of the term *strategy* in this study is particularly derived from the

latter definition in which the *strategy* refers to anticipations that one makes in order to control factors which might have impact on his or her actions. As Wodak et al. write, drawing on Bourdieu's definition, "the significance of strategies cannot be associated with a simplistic, finalistic and voluntary perspective" (p. 31). According to Wodak et al., "strategic action is oriented toward a goal but not necessarily planned to the last detail or strictly instrumentalist; strategies can also be applied automatically" (p. 31). In addition to the definition provided by Wodak et al., Goffman (1959) writes that "when an individual appears before others his actions will influence the definition of the situation which they come to have" (p. 6). He further argues that "sometimes an individual will act in a thoroughly calculating manner, expressing himself in a given way solely in order to give the kind of impression to others that is likely to evoke from them a specific response he is concerned to obtain" (Goffman, 1959, p. 6), and sometimes he acts in "a calculating manner" because of "the tradition of his group or social status require this kind of expression and not because of any particular response" (Goffman, 1959, p. 6). However, the definition provided by Goffman seems more applicable for self-presentational situations in which people's use of strategies/calculating manners might be modified due to "the tradition of his group or social status" which necessitates those types of expressions. According to both of these definitions, there are some anticipations or calculating manners considered to be able to make some aims or needs met. Accordingly, the term *strategy* here will refer to the anticipation and calculating manner that one considers to *denote a plan* to achieve a/some particular aim(s) or an "attempt to anticipate . . . factors which may have an impact on his or her action" (Wodak et al., 2009, p. 31).

In the following paragraph, self-presentation is defined and discussed in relation to online self-presentation and also in relation to Facebook and young Iranians.

Self-Presentation

Self-presentation in Goffman's (1959) view is compared with a theatrical performance. In relation to any presentation, there is an *actor*, a *performance, setting, viewer(s), front stage*, and *backstage*. What is important in this definition is to see how actors perform before people on the *front stage* where they are separated by the *setting*. Goffman uses the word *performance* to refer to "all the activity of an individual which occurs during a period . . . before a particular set of observers and which has some influence on the observation" (p. 22).

In relation to the importance of self-presentation and impression management, Leary et al. (1994) write that "because people's outcomes in life depend, in part, on others' perceptions and evaluation of them, people sometimes try to convey impressions that will help them obtain valued goals" (p. 664). Leary et al. further write that "as a result, they often monitor and attempt to control the impressions they are making, a process known as self-presentation or impression management" (p. 664).

According to Zarghooni (2007), drawing on Walther, Slovacek, and Tidwell (2001, p. 110), "one of the reasons why self-presentation on social networking sites may be different from face-to-face is that online one may 'inspect, edit and revise'" (p. 4). So, one can present himself or herself before the presentation is made available to others. According to Zarghooni, "the self-presenters are the same people as before, but they have got a new self-presentational tool and a new arena for social interaction" (p. 6). Therefore, it seems that the two main concepts of this study: *strategy* and *self-presentation online* are quite interconnected. Self-presentation online provides a situation in which one can *inspect, edit, and revise* for managing it suitable for some particular aims.

Goffman (1959) uses the term *front* to refer to "that part of the individual's performance which regularly functions in a general and fixed fashion to define the situation for those who observe the performance" (p. 22). Therefore, in relation to the *front* itself, it would be interesting to see "what seem to be standard parts of front" (Goffman, 1959, p. 22). I use the word *standard* not only referring to standard and visible factors in *front* such as *furniture, décor*, and *physical layout* which Goffman calls *setting* but also to standard behaviors, acts, norms, manners, and activities. As Goffman writes, "those who would use a particular setting as a part of their performance, cannot begin their act until they have brought themselves to the appropriate place and must terminate their performance when they leave it" (p. 22). But more interesting than the term *setting*, Goffman writes about "the scenic parts of expressive equipment," or *personal front* which is "the expressive equipment . . . [which] we most intimately identify with the performer himself," such as "clothing, sex, age . . . size and looks; posture; speech patterns; facial expressions; bodily gestures" (p. 24). As he carries on, sometimes "the signs are relatively fixed and over a span of time . . . [and some] are relatively mobile or transitory" (p. 24).

Drawing on Goffman's (1959) definition of *backstage*, Courtney, Britt, and Mckibben (2011) write that *backstage* is "an area as a place where individuals can drop their guards, expose their flaws, and seek to determine how they will later portray themselves" (p. 19). Goffman writes that "very commonly the back region of a performance is located at one end of the place where the performance is presented, being cut off from it by a partition and guarded passageway" (p. 113). He further writes, "the back region will be the place where the performer can reliably expect that no member of the audience will intrude" (p. 113).

Middle Stage and the "Detached Self"

Zarghooni (2007) uses the term "detached self-presentation" (p. 17) to refer to the self, shaped somewhere in between the backstage and *front stage* self (offline), that modifies his self-presentation (online) in a way to distance himself from his own self and from the *current state*. What he refers to is more applicable in relation to the online and the offline selves.

Similar to the difficulty of separating between *front* and *backstage*, it seems that it is quite difficult to specify where public and private spheres meet on Facebook. These two spheres seem quite overlapping and in many ways the characteristics of both are observed, for instance, in *private groups* where access can vary from public to private/closed members. Moreover, people can make private groups in which the access is limited to those addressed. Private group access however varies between private and public.

Conforming to the Norms and Expectations of Society?

According to Goffman (1959), "when the individual presents himself before others, his performance will tend to incorporate and exemplify the officially accredited values of the society, more so, in fact, than does his behavior as a whole" (p. 35). I assume that this statement might be correct in the offline (in person) interactions, and in situations in which people are in contact with those they know. On Facebook, however, it may be a bit complicated. Depending on people's intentions for using Facebook, their audiences might vary from closer friends to newly met people and even strangers. Therefore their actions might not necessarily conform to the expectations and norms of the society in which they live. People might be there to experience something new. Moreover, they might manage to perform different selves in different spheres.

Goffman (1959) writes that "in thinking about a performance, it is easy to assume that the content of the presentation is merely an expressive extension of the character of the performer and to see the function of the performance in these personal terms" (p. 77). According to him, "this is a limited view and can obscure important differences in the function of the performance for the interaction as a whole" (p. 77). As he writes, "it often happens that the performance serves mainly to express the characteristics of the task that is performed and not the characteristics of the performer" (p. 77). Accordingly, if young Iranian Facebook users try to perform different roles in different situations simultaneously, it shouldn't be read simply as their own characteristics but rather as the characteristics of the task which can be a self-presentation on Facebook.

Leary et al. (1994) write that "the kinds of impression people try to create are affected" by different factors such as "norms and roles, the characteristics and values of the people whom the person want to express, others' existing impressions of the person, the person's own self-concept, and his or her desired images of self" (p. 664). Therefore, to get a better image about self-presentation, in addition to presentation itself, one should give particular attention to the cultural, social, and religious context/situation in which the performance is made.

Self-Presentation in Different Cultures

Jungsik, Seongsoo, and Wansuk (2011) investigate "how the length and strength of an expected social relationship are related to different self-presentation" (p. 63). They reflect on Eastern cultures such as Korean, Japanese, and Chinese. They write that in Korea, "being a collectivist culture, the Korean participants might feel a greater need for humble self-presentations" [and there is more pressure] "when an individual is in a long and intensive relationship" to get socially accepted (p. 71).

Moores (2000) argues that there is a need for empirical case studies which consider the *double consciousness* of people who inhabit plural or multiple cultures. In relation to *double consciousness*, which was brought up in Moores's study, Khosravi's (2008) study on young Iranian youth also shows that they perform different public and private selves simultaneously. As Khosravi writes "everybody knows the rules of the game. They know where and when they should switch from one to the other. The youth of today have grown up in this between-ness" (p. 123). However, one might discuss that it may not be unique only to Iran. Jungsik et al. (2011), explain "how Japanese people manage two opposite selves, 'honne' (inner self) and 'tatemae' (self-expressed to others) to coexist in an interdependent structure" (p. 71). They write about "dual selves" and conclude that their study has shown empirically that "Koreans seem to flexibly handle such dual needs for individual competence and being socially humble in response to the opportunities" (p. 71). They also criticize "the research on self-view and self-presentation [since they] have neglected various social situations as influencing self-presentation behaviors" (p. 71). They write that, "self-presentation and the motivation behind it must be understood within a context of social relationships, rather than within isolated individuals" (p. 72).

Chiung-Wen, Ching-Chan, and Yi-Ting (2011) in a similar study on Taiwanese users of Facebook discuss that the users might "interact differently with different types of friends when choosing Facebook tools" (p. 474). However, that cannot be limited to the Eastern culture, rather to everyone who wants to make sure which types of image one gives of oneself. They also discuss the differences in interaction, and write that "users might act differently with their new friends than their close friends" (p. 474). Their study's findings show that the users "tend to choose less time- and effort-utilizing tools to interact with newly met friends and more privacy-controlled tools when interacting with closer friends" (p. 476). They don't discuss their findings further, but one can interpret that in relation to the *social control* by which one might feel more concerned about the impression

they give to close friends and family members rather than newly met people (Moghadam, 2003). As mentioned by Leary et al. (1994), as people's outcome in life depends on how other people perceive and evaluate one, people try to transfer impressions that help them attain appreciated aims.

When discussing different cultures and how self-presentational acts are performed, one should refer to some cultural factors within the Iranian culture that seem important to be mentioned. In addition to the *social control* which was mentioned earlier, one of the main cultural factors in Iran is one's own and family *reputation* (Moghadam, 2003). It is important to mention that, as Moghadam (2003) writes, "a family's honor and reputation rest most heavily on the conduct of women . . . sex segregation and veiling, legitimated on the basis of the Shari'a, is part of the Islamic gender system" (p. 123). Women usually, according Moghadam "is seen as a direct threat to manhood, community and family" (p. 165). And that can be appeared in forms of "the control of extended families over women" (Moghadam, 2003, p. 60). Compulsory veiling, according to her "has been a mechanism of social control: the regulation of women" (p. 160).

It is worth mentioning that veiling is not only limited to physical coverage but also to voice, ways of conduct, looking, and so on (Khosravi, 2008). However, meanings attached to veiling should be explored in any particular sociocultural context, rather than generalization. According to Naficy (2000, p. 562) "veiling of the voice includes using formal language with unrelated males and females, and decorous tone of voice, avoidance of singing, boisterous laugher, generally any emotional outburst in public than the expression of grief or anger" (cited in Khosravi, 2003, p. 73). "The patriarchal father's attention to the female virtue (namous) of his family is now part of how the state manages space" (cf. Focault 2000: 207, cited in Khosravi, 2003, p. 74).

Veiling may not be all about limiting women; different studies have shown that within closed and fundamentalists situations, it has provided some sort of freedom of acts for women to be participants within the society and live in public (Mohanty, 1991). Hourani (1991) gives a cultural background about the Middle East and explains how public and private spheres had been separated along the gender lines by which public spheres were in control of men, business, and trades, and private spheres of home and family were in control of women in which women could have "maintained a culture of their own" (p. 120). One of the factors he mentions in relation to separation of public and private is "hijab" (women's head to toe veiling), which was actually meant to separate public from private, so women could still keep their private spheres even while attending in public.

According to Goffman (1959), as it also mentioned earlier people's presentation of self might differ depending on the audience's gender, age, facial expressions, and so on. This is particularly accurate in Iranian culture in which there is a wide range of different norms and social meanings enacted differently for men and women. Iran is a gendered segregated society in which males and females are expected to present themselves differently.

In relation to the theoretical discussions above, it is important to clarify how principal issues such as *needs, strategies*, and *resistance* are interrelated with each other, both on Facebook and within the context of Iranian society. There are *needs* when one decides to use Facebook. Needs can range from *personal* to *community* and involve a variety of social, cultural, and political expectations. However, *needs* are not always compatible with one another, and sometimes are in conflict. In these types of situations, in order to fulfill conflicting *needs*, one might apply *strategies. Strategies* are useful as long as one tries to maintain two or more expectations, but when one prioritizes satisfying one *need*—for instance, the *personal* over the *community*—a form of *resistance* can appear. This *resistance* doesn't necessarily appear with the larger aim of transforming cultural and political restrictions, but rather can be simply for gaining one's own *personal autonomy*. Thus, *resistance* is closely related to the acts of fulfilling one's personal *needs* and striving for *autonomy*.

Method

Focus Group Interviews

The overall approach to the methodological perspective of this study is *symbolic interactionism* (Blumer, 1969).

The main method used in this study is focus group. After focus group interviews were done, the interview transcriptions were further analyzed using qualitative *thematic analysis*.

Focus groups, as Flick (2006) writes, "start from an interactionist point of view and want to show how an issue is constructed and changed in a group discussing the issue" (p. 197). It started from a critical point of view toward regular interviews in which the interviewees are "separated from all everyday relations during the interview" (Flick, 2006, p. 189). Focus groups aid "extending the scope of data collection . . . to create a situation of interaction that comes closer to everyday than the (often one-off) encounter of interviewer and interviewee" (Flick, 2006, p. 189). They "correspond to the way in which opinions are produced, expressed, and exchanged in everyday life" (Flick, 2006, p. 191) and the stress is "laid on the interactive aspect of data collection" (Flick, 2006, p. 197). Accordingly, focus group interviews aim to provide a social context similar to that of everyday life in which people's opinions are produced, expressed, and exchanged in relation to others. Interviewees are allowed to reflect, respond, and comment, not only on their own opinions and attitudes but also on those of other interviewees. Thus, focus groups appear to be a proper method for capturing the dynamic, interactive, and vibrant nature of self-presentation, as people's self-presentation is constantly constructed, negotiated, and even changed within the group and

in relation to others. Therefore, it should be analyzed in relation to others as it is presented and performed.

Analytical Method: Thematic Analysis

The analytical method used for the focus group interviews' transcriptions is qualitative *thematic analysis* (Yin, 2003). This analytical method is based on a *pattern matching* procedure which starts with readings of the transcribed material over and over again and then classifying all informants' answers according to the questions posed. It aids the process of analysis with finding similar or different patterns in regard to the theoretical discussions in the empirical materials to construct a descriptive approach (Melián, 2012; Yin, 2003).

Data Gathering Strategy and the Informants

The informants of this study were gathered through a *friendship network*. To recruit the informants, the researcher informed her own *friends* on Facebook (those who have a large number on their friend lists, that is, about 500 and more) about the study. Those "friends of friends" who accepted to be the informants of this study were then sent a friendship invitation on Facebook by the researcher and were recruited for the interviews in person in Tehran. After a 2-week period, 10 informants (five males and five females) accepted to be participants of this study and attend two focus group sessions during a 10-day period. The number of informants, considering the qualitative approaches of this study, as well as the illegality of using Facebook in Iran and it being a "cultural crime," seemed appropriate.

Even though the strategy of *friendship network* was used to gain the informants' trusts, and they had been reminded of the interview the day before, five of the interviewees never attended on the day of the interviews. The study continued with six informants present. The absent informants were contacted by telephone the same day to explore the reasons for their absence. Two of them never answered their mobile phones. The remaining three provided different explanations that implied they were concerned about attending interviews about their Facebook use. One of them explained on the telephone that "well . . . You know . . . when signing my new job contract (in a bank), I was asked to sign a commitment letter beside that too, not to join Facebook . . . I signed the paper, while I was active on Facebook and I continued using it having a pseudonym . . . honestly, I got afraid of participating in the interviews, after I thought more about it . . . I am so sorry, I should have told you in advance . . . I don't want to risk losing my job after a long time [of unemployment]" (Female, age 28).

The result of this study was gathered through the use of focus group interviews that were conducted with six young Iranian Facebook users (remaining informants) in Tehran in late 2011.

The empirical material of the study includes the complete transcription of two session focus group interviews with six informants (including notes on their *perspective* and *point of view* about their own self-presentation as well as their reflection on each others'; Silverman, 2006). All six informants showed their profiles to the researcher on Facebook and the relevant information was considered during the interviews.

The informants all were between 18 and 25 years old. Three of the interviewees were male and three were female. They were again assured that they would remain anonymous. All of them were living in Tehran at the time of the interviews and claimed to use Facebook almost every day. All of the informants either had just started studying at university or had completed their education at the bachelor level.

The interviews were conducted twice with the same set of users. The order of questions was not followed strictly as planned, and the interviewees were given the chance to lead the interviews to a certain extent to grasp their main issues and concerns. Interviews were held in Persian and were conducted at the researcher's home in Tehran. Each interview lasted for about 60 min and they were recorded by a mobile phone's recording device and were transcribed immediately afterwards.

I introduced myself as a doctoral student and explained my doctoral project as well as the aim of these interviews. I assume that I was perceived as relatively neutral. The informants reported afterwards that they had had a pleasant session.

Analysis

The interview analysis revealed that the informants' everyday self-presentational acts and experiences on Facebook appeared to be more important and negotiable than the political ones. Informants of the study themselves declared that *everyday issues* are what they are *engaged with* constantly, while political self-presentations through the use of links, pictures, and videos seem to be *short-lived* and limited to specific moments and events which are forgotten as quickly as they start.

A number of themes were identified through the discussions in the focus group interviews which can be categorized into *Strategies: New and old users and public/private distinctions* and *need for confirmation (Likes)*. These themes might not be specific only to Iranian Facebook users, as separating spheres between public and private might be seen elsewhere as well. It is common for people to want to have *control* over their self-presentation which is performed before others.

Analysis of the interviews in the context of Iran revealed that there is a large gender difference between how men and women in this study tended to present themselves as well as reflecting on each other's self-presentation. *Self-presentational strategy* was a theme which was heavily gendered and was connected to cultural factors such as *social control* and one's own as well as *family reputation*. Female

informants claimed to experience a tension between taking responsibility for their own or their *community*/family *needs* and expectations, as they were usually in contradiction.

Strategies: New and Older Users and Public/Private Distinctions. Duration of being on Facebook seemed to play an important role in how the informants tended to present themselves. There appeared to be a relatively significant difference in usage between newly joined Facebook users versus those users who had been using Facebook for 3 to 4 months or more. New users wanted to be seen and presented themselves quite openly. They easily shared their personal information and pictures, and declared their political and religious views on their profiles while placing political links and pictures in their statuses. These six Facebook users added people as friends even though they hardly knew them in real life. They were explicit when they described where they had been and with whom, what they had done recently, and tended to tag (Facebook Help Center, 2013) their friends and relatives in their albums and also be tagged in pictures. However, it seemed that after a while when their *friend lists* had become extensive (not only with close friends and relatives, but also with colleagues), they started to be more cautious in the way they shared their information. They thought that it may not be safe to present their views so openly. Therefore, they used different *strategies* for controlling what they shared, how they shared, as well as with whom. Some of these users presented strategies they had used to separate closer friends from their family members. In this way, they wanted to keep their *voice* heard by the right audience. One of the users said that,

> In real life, I am used to considering the age, gender, level of respect, and intimacy when I talk to people, but on Facebook, it is difficult . . . I need to keep using strategies to make sure that my words don't reach the wrong audience . . . I also found out that it is becoming harder for me to update statuses if they are visible to everyone. (Female, age 21)

The informants mentioned that they use the Facebook privacy-control tools to control how they share dissimilar content to different audiences (*friends* on Facebook). This strategy however didn't appear to be so popular. They assumed that their information had leaked out, and their audiences realized that they were using strategies to hide some information from them. According to Zarghooni (2007) "a person who self-presents very differently from audience to audience may have difficulties to maintain . . . [favorite] impressions over a long time" (p. 10). This is particularly relevant here; one of the informants mentioned the use of privacy settings (which are applied differently to different audiences) as "disgraceful and tiresome," as the audiences may "discover [what you are doing] eventually" (Female, age 21).

Some other users mentioned that they use multiple accounts for different purposes and for addressing different audiences (e.g., separating family and friends). That was one of their main concerns on a platform in which they were supposed to be presented *nonymously* with their real names (Grasmuck et al., 2009). On the *nonymous* platform of Facebook, users can be found easily by the members of their *social ties* (families and relatives). Using multiple accounts allows them to separate between closer and less close Friends and more importantly, they could draw a line between family and friends. The use of *private groups* and creating different accounts (*selves*) for diverse purposes and audiences were, so far, their most reliable strategies.

The informants also showed concerns about how extensive their *Friends lists* had become, to include family members and relatives with whom communication had become uncensored and *front stage*. This issue recalls two cultural aspects, which might not be specific only to Iran, but also relatively prominent in Iranian culture: *social* and *family control*. The matter of *reputation* and *control* makes people concerned about how society (their own *social ties*) evaluates them. The closer *social ties*, the more tensions, considerations, and controls should be applied. It is worth mentioning that controls, tensions, and considerations were gendered and addressing women.

As one of the informants mentioned,

I have some male friends with whom I keep in contact (on Facebook) as well. My boyfriend is also on Facebook who keeps writing to me and tagging me in pictures . . . this is while my relatives think that I have no boyfriend . . . Now they are on my friends lists because I couldn't reject their friendship requests . . . So you see, I need to use *strategies* to keep some information invisible to some people, otherwise it doesn't work. (Female, age 24)

She claimed (24 years old) that she couldn't reject the extended family friendship requests on Facebook as she could have been assumed to "be doing something improper, otherwise there was no reason [for her] not to accept their requests on Facebook." However, if she does accept, "they will find *something* to talk about." If her family and relatives realize that she is in contact with male friends, specifically having a boyfriend, it threatens not only her own but also her family *reputation*. According to her, "personal and family reputation for us are not two different things, but rather connected" (Female, age 24). Thus, analysis of the interviews in this study displayed a quite similar result to the one provided by Chiung-Wen et al. (2011) in their study. The informants in this study, similar to the Taiwanese Facebook users, were more concerned about how their family members and close relatives judge them and what *impression* they give to them, rather than to newly met people.

It is important to mention that *reputation*, according to the informants, seemed to be a gender issue and didn't seem to be as significant for the male informants. Male informants said that they are not concerned about it, and are not cautious about what other people might think about them, and more

importantly, there is no connection between their own and their family *reputation* (Moghadam, 2003, p. 123), though that was the case for women. Interviews showed that female informants experienced a *double pressure* and tension on Facebook in taking responsibility for their own *individual* as well as their *community needs* (Markham, 2004), because usually "these *needs* and expectations are in contradiction with one other" (Female, age 24). "Accepting or not accepting an insignificant friendship request on Facebook might raise an issue" (Female, age 21). Possibly, using strategies in self-presentation enables women to fulfill their own *needs*, being true to themselves as well as being responsible for their family *reputation* (Moghadam, 2003). However, as one of the other informants mentioned,

Using different strategies for controlling what we share on Facebook is exhausting and at the same time ironic . . . we are using Facebook to be separated from the limitations of our daily lives, but we take all of those limitations along [on Facebook]. (Female, age 21)

Need for Confirmation (Like). The informants seemed to be concerned with receiving *like* from their *friends* on Facebook. It seemed that they were altering the way they presented themselves according to other people's tastes and approval. One of the male informants (22 years old) mentioned that "I don't update statuses which I know no one will *like*." The informants seemed to tend to upload pictures in which they looked perfect. This can also be interpreted in relation to *self-play* and people's effort in making the best *impression* of themselves before the social circle they are in contact with to receive approval.

Considering *double consciousness* for receiving approval by family, friends, and society in general was more of a concern for women than men, and women seemed to be more adept at that. The use of strategies for three of the informants seemed to be *automatic* (Gilroy, 1993, cited in Moores, 2000), while the other three found using *calculating manners* to be a challenging task, which was forced by the cultural tradition of the family and society in which they were living and using the platform. However, one might note that *double consciousness* in relation to one's self-presentation might not be specific only to Iranian society and can be seen elsewhere as well.

The informants' self-presentation on Facebook in relation to different *stages* seemed more complicated than the one presented by Goffman (1959) in everyday life. Self-presentation of the Facebook users in this study wasn't limited to separating only between the *front* and *back* stage, but also the *middle stage* in which they could experience the quality of both stages. The use of private groups by which they could go *backstage* and be with their own *team members* (Goffman, 1959) is one of the examples of using the *middle stage*. In these private groups, they could enjoy having a relatively *public sphere* and be in contact with those they trust and with whom they experience less concern over how they present themselves and interact with one another.

Analysis of the interviews revealed that the informants use *strategies* to control what they share on Facebook and with whom. The use of Facebook for the informants wasn't *liberating* per se (from the cultural norms and codes in the society), because they needed to make sure they were not hurting the family *reputation* through their acts on Facebook.

According to one of the informants, "the use of Facebook for political acts gets limited to some specific time when political tensions are high, but it is temporarily, while everyday use of Facebook is constant and permanent" (Male, age 25). Accordingly, even though the use of Facebook for political acts was limited to some specific time periods, the use of strategies was constantly a method for *resistance*. For women, the use of strategies seemed to facilitate how they could play with the self they present, being fair to their own *needs* by using Facebook, while keeping cultural norms and codes in consideration.

Accordingly as mentioned earlier, *needs, strategies*, and *resistance* on Facebook are interrelated and should be understood within this context. *Strategies* are useful as long as one wants to fulfill contradictory *needs* (*personal* and *community*) at the same time. *Strategies* are not applied anymore when one stops trying to fulfill two or more contradictory *needs* and expectations, and instead prioritizes the *personal* over the *community* one, for instance. In this type of situation, a form of *resistance* appears which might not be necessarily aggressive and directed at changing cultural and political frameworks, but rather at gaining one's own *personal autonomy*.

Final Words

The study's analysis of interviews with a limited number of young Iranian informants showed that *social meanings* and *norms* of self-presentation on Facebook are defined to a large degree in terms of gender. There appeared to be no specific definition of what is right and what is wrong to do on the platform in terms of self-presentation, but every act finds a specific meaning when gender of the users is considered. Gender played an important role in how informants presented themselves and interacted with one another on Facebook and even within the group discussion.

The analysis of interviews showed that social and everyday means of using Facebook seem to be more important, controversial, and debatable than the political ones. However, it depends on what one means by politics, and how one defines political acts on Facebook. If one means leading to direct political changes through the acts on Facebook, that might not happen. This is because in many ways, male informants at least appeared comfortable with maintaining the status quo, especially in terms of gender issues, *social control*, and women's responsibility for family *reputation* through one's acts on Facebook. However, if one defines

political acts in terms of any type of *resistance* toward the existing rules, codes, and guidelines, the informants are doing some sort of *resistance. Strategic self-presentation* was a form of *resistance* used to enable the users to get some sort of *control* through the medium of Facebook over their own self-presentation while dealing with existing expectations. However, it is worth noting that *resistance* was not shown particularly in relation to the Islamic Republic rules and norms, but rather to the existing norms of family *reputation* and gender issues within the socioculture situation of Iran. *Strategies* were helpful for the female informants to keep up with the norms and rules of *reputation*, while also fulfilling their own *needs*, for instance, wearing clothes according to their own preferences as well as using Facebook itself, despite it being illegal.

The informants' attitudes and self-presentational acts didn't seem directed toward any social or political movements, instead they sought "individual autonomy rather than political freedom" (Khosravi, 2008, p. 128). Their reflections and sentiments on their own as well as others' self-presentation within the group provided a refreshing contradiction to how both transnational and Iranian state media depict them and their lives. However, the lack of observation of any clear-cut political acts in this study could be challenged due to the self-selection of the informants. The possibility of having politically active informants has decreased in this study, as political activists might be reluctant to present their political actions within a group of semistrangers or even participate in these types of studies.

According to the informants, the use of Facebook for political acts seems to be limited to specific periods of time in which Facebook users may want to present their political views, but everyday, social and cultural reasons for using Facebook seem to be predominant and permanent. It is worth mentioning that the use of Facebook itself is considered illegal and is a form of *cultural crime* (Khosravi, 2003, p. 42). let alone direct political acts on it.

The primary aim of this study was to grasp how young Iranians reflect on their own self-presentation on Facebook. Interviews were initially planned to be conducted with 11 young Facebook users. Even though the data gathering strategy was based on a *friendship network* and a *mutual trust* between the informants and the researcher, and they were assured that they would remain anonymous, 5 informants who originally accepted didn't attend on the day of the interview; 2 of them didn't answer when contacted by telephone, 1 claimed that they were busy, and 2 said that they were concerned about attending the interview. The study nevertheless continued with 6 remaining informants in attendance.

The use of strategies appeared to be an important part of their self-presentation, especially for female informants. The strategies used varied from the use of privacy-control tools, through the use of private groups in which they could address particular audiences, to the use of different accounts for different purposes. The informants appeared to be adept at performing differently while being on *stage* before different audiences simultaneously. The informants used different strategies for different audiences and knew how to play *the game*. However, this might not be the case only for Iranian Facebook users, as it can be practiced to varying degrees all over the world.

The analysis of focus group interviews and interaction between the informants within the context of Iranian society showed some themes to be important such as *reputation, social control*, and *gender* issues (Moghadam, 2003). Facebook, as a *nonymous* platform in which people are in contact with their already-made *social ties*, didn't seem to be as *liberating* from the existing norms and rules within society compared with older ways of using the Internet where people could enjoy being completely anonymous and experience different selves simultaneously. Facebook in this study showed that it is a *stage* on which to be observed but this stage becomes restricted with the involvement of close *social ties* in the context of social control within Iranian society. Accordingly, it seemed that it is no longer conceivable to draw clear lines between *back* and *front stage* concerning how people present themselves on Facebook, as one can play between different *stages* and go back and forth between them simultaneously. *Strategies* were therefore used to link different *stages* and give some sort of control to the informants *over* self-presentational *impressions*. The informants seemed to draw a clear line between the way they presented themselves before their family members/family relatives, friends, and newly met people. With family relatives, they were more concerned and cautious about how they presented themselves, while they felt more comfortable before their friends and newly met people. Accordingly, the closer the *social ties*, the more concern was experienced regarding how they are perceived and which types of *impressions* are made. *Strategies* used and informants' approaches toward them were very gendered, though. Women seemed to need to use *strategies* to be able to enjoy relative freedom: being fair to their own personal *need* to be connected through Facebook and at the same time taking responsibility for the family's *reputation need*. The use of *strategies* seemed both natural and *automatic*, and also *challenging* and *ironic*. *Challenging and ironic* because they weren't intentionally using Facebook to conform to the existing social norms within society, but were conforming indirectly.

Self-presentational claims were different in isolation, compared with when they were analyzed in relation to each other. Analysis of the profile information alone without considering people's reflections on their own as well as others' acts could be misleading, because what informants said about themselves seemed to be slightly different than how they reflected upon others' reflections, as well as how they expected others to present themselves. For instance, even though male informants claimed that they are *modern* and *open-minded*, their expectations from women and their self-presentation on Facebook were in line with the existing traditional values of the *patriarchal* society in which different aspects of acts and experiences are biased based on gender and are expressed as *social control* under the shadow of family *reputation*. In other words, male informants' *front* and *backstage* seemed to be different. Thus, the traditional norms and meanings were reinforced by males through their acts and claims. *Facebook community meanings* and norms appeared to force them to appear *modern and open-minded*. However, the larger society's norms and social meanings seemed to be in contradiction with the platform's meanings and norms.

Self-presentation as a dynamic and interactive phenomenon, which is constantly *constructed socially* and in relation to others cannot be analyzed in a one-sided way without considering in which contexts it is performed. Therefore, it is important to use methods that keep track of both self-presentation as well as interaction, as they complete each other and each self-presentation is basically made to initiate an interaction.

Both male and female informants in this study were engaged in the process of reproducing the *social norms* and *meanings* of the Iranian society. However, that was slightly different for men and women. While male informants were reproducing the norms and codes of the society (such as *reputation, social control*, and *gender issues*) in a natural way, women were more forced to do that, due to having to take responsibility for their own as well as family *reputation*. However, in many ways, female informants appeared to be more *resisting* toward the contemporary norms and rules within the society compared with men.

The result of this qualitative study is about a very limited number of informants and their reflections on their own presentation of self on Facebook. It is not representative and cannot be generalized to all Iranian Facebook users, nor Iranian culture, codes, and norms. Instead, this study had an inductive approach, within the social context of focus groups to understanding self-presentation on Facebook and how it is *negotiated* within the sociocultural, political situation of the Iranian society in which they live and use Facebook.

The study's analysis confirmed that self-presentation is an interactive phenomenon and is *constantly constructed* in the society and in relation to others. However, a larger number of users may be needed to strengthen and pursue further the findings of the study. Other qualitative studies could include a wider range of people from different cities and sociocultural backgrounds. Other qualitative studies could explore for instance how one individual manages to fulfill one's own *personal needs* which are contradictory to family and community expectations and *needs*, while preserving the *reputation* being socially acceptable within the society.

Acknowledgments

I am so grateful to the Department of Media Studies, JMK at Stockholm University for their overall support as well as Iran Media

Program at University of Pennsylvania whose support enabled me pursue this project.

Declaration of Conflicting Interests

The author(s) declared no potential conflicts of interest with respect to the research, authorship, and/or publication of this article.

Funding

The author(s) disclosed receipt of the following financial support for the research and/or authorship of this article: This study was supported by IMP grant for Independent Research Projects from the Iran Media Program at the University of Pennsylvania in 2012.

References

Agger, B. (2004). *The virtual self*. Malden, MA: Blackwell.

Blumer, H. (1969). *Symbolic interactionism: Perspective and method*. Englewood Cliffs, NJ: Prentice Hall.

Boyd, D. (2004). Friendster and publicly articulated social networks. *Proceedings of ACM Conference on Human Factors in Computing Systems* (pp. 1279-1282). New York, NY: ACM Press.

Bryant, E. (2008). *A critical examination of gender representation on Facebook Profiles* (Conference paper). Washington, DC: National Communication Association. pp. 1-30.

Chaney, D. (1996). *Lifestyles*. London, England: Routledge.

Chiung-Wen, J., Ching-Chan, W., & Yi-Ting, T. (2011). The closer the relationship, the more interaction on Facebook? Investigating the case of Taiwan users. *Cyberpsychology, Behavior, and Social Networking, 14*, 473-476.

Courtney, L. G., Britt, T. W., & Mckibben, E. S. (2011). Self-presentation in everyday life: Effort, closeness, and satisfaction. *Self and Identity, 10*, 18-31.

Ellison, N., Steinfield, C., & Lampe, C. (2006, June). *Spatially bounded online social networks and social capital: The role of Facebook*. Paper presented at the Annual Conference of the International Communication Association, Dresen, Germany.

Esfandiari, G. (2011). Iran's Basij Head Turns Bard To Dismiss Twitter, Facebook. *Radio Free Europe/Radio Liberty*. Retrieved from http://www.rferl.org/articleprintview/24273009.html

Facebook Help Center. (2013). *Tag*. Retrieved from http://www.facebook.com/help/?faq=124970597582337

Farquhar, L. K. (2009). *Identity negotiation on Facebook.com* (Unpublished doctoral dissertation). University of Iowa, IA. Retrieved from http://ir.uiowa.edu/etd/289

Flick, U. (2006). *An introduction to qualitative research*. London, England: Sage.

Fornas, J., Klein, K., Ladenfork, M., Suden, J., & Sveningsson, M. (2002). *Digital borderlands: Cultural studies of identity and interactivity on the internet*. New York, NY: Peter Lang.

Goffman, E. (1959). *The presentation of self in everyday life*. London, England: Penguin.

Grasmuck, S., Martin, J., & Zhao, S. (2009). Ethno-racial identity displays on Facebook. *Journal of Computer-Mediated Communication, 15*, 158-188.

The Green Voice of Freedom. (2011). "17 million Iranians live on Facebook" says Basij official. *The Green Voice of Freedom*. Retrieved from http://en.irangreenvoice.com/article/2011/oct/06/3294

Haaretz Service. (2010). Top Egyptian cleric issues Facebook fatwa, Social networking site "breaks up families," says Cairo Sheikh after survey links online surfing to divorce. *Haaretz*. Retrieved from http://www.haaretz.com/news/top-egyptian-cleric-issues-facebook-fatwa-*1*.262860

Hashem, M., & Najjar, A. (2010). The role and impact of new information technology. Application in disseminating news about the recent Iran presidential election and uprisings. In Y. R. Kamalipour (Ed.), *Media, power and politics in the digital age: The 2009 presidential election uprising in Iran* (chap. 12, p. 126). Lanham, MD: Rowman & Littlefield.

Hourani, A. (1991). *A history of the Arab peoples*. London, England: Faber and Faber.

Itameri, K. (2011). Arabs Facebook users increase by 30 percent since 2011, report. *AlmasryAlyoum*. Retrieved from http://www.almasryalyoum.com/en/node/466131

JakartaGlobe. (2012). Facebook Un-Islamic, Membership a Sin, Iranian Ayatollah Says. *JakartaGlobe Tech* [online]. Retrieved from http://www.thejakartaglobe.com/tech/facebook-un-islamic-membership-a-sin-iranian-ayatollah-says/489718

Jungsik, K., Seongsoo, L., & Wansuk, G. (2011). Culture and Self-presentation: Influence of social interactions in an expected social relationship. *Asian Journal of Social Psychology, 14*, 63-74.

Kendall, L. (2002). *Hanging out in the virtual pub: Masculinities and relationships online*. Berkeley: University of California Press.

Khosravi, S. 2003. The Third Generation: The Islamic Order of Things and Cultural Defiance among the Youth of Tehran. PhD diss., Stockholm University.

Khosravi, S. (2008). *Young and defiant in Tehran*. Philadelphia: University of Pennsylvania Press.

Leary, M., Nezlek, J., Downs, D., Radford-Davenport, J., Martin, J., & McMullan A. (1994). Self-presentation in everyday interactions: Effects of target familiarity and gender composition. *Journal of Personality and Social Psychology, 67*, 664-673.

Markham, A. (2004). Internet communication as a tool for qualitative research. In D. Silverman (Ed.), *Qualitative research: Theory, method, and practices* (2nd ed., p. 119). London, England: Sage.

Mead, G. H. (1932). *Philosophy of the present*. Chicago, IL: Open Court Publishing.

Mead, G. H. (1936). *Movements of thoughts in the 19th century*. Chicago, IL: University of Chicago Press.

Mead, G. H. (1938). *The philosophy of the act* (C. W. Morris, Ed.). Chicago, IL: University of Chicago Press.

Melián, V. (2012). *Bridging the blocked river: A study on the Internet and mobile phone practices within an environmental movement between 2005 and 2008 in Argentina and Uruguay*. Doctoral dissertation, Stockholm University, Sweden.

Mitra, A. (2005). Creating immigrant identities in cybernetic space: Examples from a non-resident Indian website. *Media, Culture & Society, 27*, 371-390.

Moghadam, M. V. (2003). *Modernizing women: Gender and social change in the Middle East*. Boulder, CO: Lynne Reinner Publishers.

Mohanty, C. T. (1991). Under Western eyes: Feminism scholarship and colonial discourses. In C. T. Mohanty, A. Russo, &

L. Torres (Eds.), *Third world women and the politics of feminism* (pp. 51-80). Bloomington: Indiana University Press.

Moores, S. (2000). *Media and everyday life in modern society.* Edinburgh, Scotland: Edinburgh University Press.

Morozov, E. (2011). *The net delusion: How not to liberate the world.* London, England: Allen Lane.

Papacharissi, Z. (2011). A networked self. In Z. Papacharissi (Ed.), *A networked self-identity, community, and culture on social network sites* (pp. 304-318). New York, NY: Routledge.

Sharif, N. (2012). Mullahs fear of 6 million Iranian Facebook users. *Stop Fundamentalism in Iran.* Retrieved from http://blogs .stopfundamentalism.com/index.php?option=com_content&; view=article&id=83:mullahs-fear-of-6-million-iranian-face-book-users&catid=34:posts

Silverman, D. (2006). *Interpreting qualitative data: Methods for analyzing talk, text and interaction* (3rd ed.). London, England: SAGE.

Turkle, S. (1995). *Life on the screen: Identity in the age of the Internet.* London, England: Phoenix.

Walther, J. B., Slovacek, C. L., & Tidwell, L. C. (2001). Is a picture worth a thousand words? Photographic images in long-term and short-term computer-mediated communication. *Communication Research, 28,* 105-134.

Wodak, R., de Cillia, R., Reisigl, M., & Liebhart, K. (2009). *The discursive construction of national identity* (2nd ed.). Edinburgh, Scotland: Edinburgh University Press.

Yin, R. K. (2003). *Case study research: Design and methods* (3rd ed.). Thousand Oaks, CA: Sage.

York, J. (2012). Is Iran's halal internet possible? *Aljazeera.* Retrieved from http://www.aljazeera.com/indepth/opinion/2012/10/2012-10263735487349.html

Zarghooni, S. (2007). *A study of self-presentation in light of Facebook* (Master's thesis). Institute of Psychology, University of Oslo, Norway.

Author Biography

Mona Hajin is a PhD fellow in media and communication studies at Stockholm University and is currently researching visual culture in social media. Her broader research interests encompass social media, everyday life, Iranian media and culture, digital media, visual culture, visual communication, migration, and ethnicity.

5

Research Impact Unpacked?
A Social Science Agenda for Critically Analyzing the Discourse of Impact and Informing Practice

Simon Pardoe[1]

Abstract

U.K. policy is to embed "knowledge transfer as a permanent core activity in universities." In this article, I propose an agenda for analyzing and informing the evolving practices of communicating university research insight across institutions, and for analyzing critically the ways in which "research impact" is being demanded, represented, and guided in current policy discourse. As an example, I analyze the U.K. Economic and Social Research Council (ESRC) "Step-by-step guide to maximising impact," using the concept of "recontextualization." The analysis illustrates that the availability of relevant social science research does not ensure its use, even by the research funding council. It suggests that while a rationalist, *common sense* representation of communication may be functional in making research impact appear achievable within existing funding, it may be potentially counter-productive in terms of ensuring that research contributes to society. A recent project is cited to illustrate some of the intellectual challenges that are not indicated in the ESRC guide, and which demonstrate the value of social science insight.

Keywords

research impact, critical discourse analysis, recontextualization, knowledge transfer, knowledge exchange, research communication, research dissemination guide, ESRC impact toolkit

"Maximizing Research Impact": Discourse and Practice

A prominent issue for academic researchers across disciplines is the increasing demand to generate "impact" from their research beyond academic debate. The U.K. Government "Science and Innovation Investment Framework 2004-2014" states,

> The Government's aim for future policy is to create a funding regime that promotes and rewards high quality knowledge transfer, . . . and further *embeds knowledge transfer as a permanent core activity in universities* alongside teaching and research. (HM Treasury, 2004, p. 76, emphasis added)

Whether this will actually increase the use of research in industry and society will ultimately depend on the detailed practices of that cross-institutional communication. If it is ill-informed, routinely mishandled, and/or embedded in underlying assumptions that separate knowledge from practice, or view practitioners merely as *users* and *audiences* for academic research, then it may be counter-productive. Far from enabling research to be used, it could affirm the often negative (U.K.) practitioner preconceptions about research as arrogant, remote, and impractical.

There are well-established fields of research on knowledge transfer, research dissemination, diffusion of innovation, research utilization and science communication. So, it is useful to ask whether current policy and guidance is already sufficiently informed (and critiqued) by those fields of research. Or, whether there is still an important role for areas of social science with relevant insight and expertise, such as applied linguistics and social studies of science.

Relevant insight and expertise includes, for example, the analysis of networks of contingent elements within the development and the use of research claims (e.g., Latour, 1987), so challenging any simplistic causal expectation of "research impact." It includes analysis of the cross-institutional communication of research (e.g., Roberts & Sarangi, 2003), and of the relations and positioning of research participants, potential users, and stakeholders within the processes of research (e.g., Cameron, Frazer, Harvey, Rampton, & Richardson, 1992/2006). It includes the analysis

[1]PublicSpace Ltd., Research Dissemination, Lancaster, UK

Corresponding Author:
Simon Pardoe, Director, PublicSpace Ltd., Bletherbeck House, Ulverston, Lancaster, LA12 8DB, UK.
Email: simonpardoe2@publicspace.ac.uk

of institutional discourses and genres both as social practice and as a focus and mechanism of social and professional change (e.g., Fairclough, 1992b).

More broadly, social science offers expertise in juxtaposing theoretical perspectives, empirical evidence, and reflections on experience. For many social scientists, there is a fundamental belief in a need to put theoretical, "logical" and "common sense" accounts of the world up against empirical evidence, and vice versa, so that our understanding is neither *rationalist* (where socially located assumptions and logical reasoning replace the need for evidence) nor dogmatically *empiricist* (where assumptions within research questions and within observations of evidence can remain unchallenged).

In fields across social science, researchers have usefully challenged rationalist and common sense accounts, including of communication (e.g., Ivanič et al., 2009; Roberts, 1997), the management of professional change (e.g., Tengblad, 2012), and the interactions between science and other spheres (e.g., Jasanoff, 2005). They use research evidence to critique and cumulatively inform theoretical and practical insight.

The intention of this article is to identify the urgent need for such research to analyze, inform, and potentially challenge (i) the evolving practices of communicating publicly funded research insight across institutions, and (ii) the ways in which this communicative work is being demanded, represented, and informed within current policy and guidance.

The urgency is not just because this communicative work is central to making university research available and useable. It is also because current representations and guidance are problematic in familiar ways, which have long been analyzed and critiqued. Left unchallenged, they may have serious consequences for the future integrity, value, and use of publicly funded research.

Specifically, I propose an agenda of three inter-related contributions:

1. **Critical analysis** of the policy and guidance discourse from research funding bodies, including their representations of (a) research *knowledge*, (b) *impact*, (c) *researchers* and *users*, and (d) the *strategies* and *processes* of this cross-institutional communication.
2. **Case study analysis** of the communication of research in practice, including the intellectual challenges, potential pitfalls, and strategies; (with the intention to understand not only how research can be made useable for, and with, potential users and stakeholders, but also how it is selectively taken up or ignored within policy, debate, and professional practice).
3. **Conceptual guidance** to inform professional academic practice and reflection on practice in research communication, to take it beyond the problematic emulation of product marketing, and beyond the often banal and problematic notions of *publicity,*

communication skills, and *getting your message across*; the goal must be for this wider communication of research to become both productive for society *and* intellectually valued by researchers, rather than the reverse.

In this article, I focus primarily on the first. As an example, I analyze the "Step-by-step guide to maximising impact" (SSG) offered by the U.K. Economic and Social Research Council (ESRC) within its online "Impact toolkit" for university researchers. With the ESRC as its author, this toolkit is arguably the most authoritative and significant guidance on generating research impact for U.K.-funded social science researchers. Analysis of this guidance is necessary, first to show the need for informed input and critique, and second to show just how fundamental (even basic) the social science insights need to be to challenge the current representations of this professional activity.

To support this analysis, I also briefly illustrate the value of (2) and (3). I include observations from the experience of a recent European Commission (EC) FP7-funded project to illustrate some of the intellectual challenges in practice. I also briefly refer to some concepts that inform my own practice, to illustrate the kinds of conceptual understanding that it is possible to offer. Together, they help support the core argument in the analysis, that the current ESRC guidance is not merely simple, but simplistic.

The title "Research impact unpacked?" is a dual question. First, it is about whether current guidance usefully *unpacks* the challenges, pitfalls, and strategies involved in communicating research findings to those who might use them. Second, it is about the need to *unpack* a problematic rationalist discourse within current policy and guidance on "maximizing research impact." For the latter, the question mark keeps open the question of whether we have ever fully identified the ways in which it is problematic.

Method

"Unsuitable Terminology": Constructing the Object of Interest

The term *knowledge transfer* is the most familiar and established term used by research funders and others to refer to this area of communicative activity. Yet the term is also severely critiqued by authors involved in informing policy. In the context of applied social research, Davies, Nutley, and Walter (2008) describe it as one of "a plethora of unsuitable terminology," which "misrepresent the tasks that they seek to support" (p. 188). They argue,

> The metaphor . . . is, at best, one of gathering and integrating evidence from research, condensing this into convergent knowledge, and neatly packaging this knowledge for transfer elsewhere. . . . In other words, knowledge parcels for grateful

recipients. Such a view belies the inherent and, we would argue, largely insurmountable challenges of doing so for any but the most simple and incontrovertible of findings. Moreover, . . . the subtlety and complexity of research use in context further militate against simple models of "translate and transfer." (Davies et al., 2008, p. 189)

In this way, they suggest that the term and its use misrepresent each element within this communicative practice: the knowledge or findings, the actors, the challenges, and the processes of communication. Partly in response to such critiques, the ESRC now emphasizes the term *knowledge exchange* (KE; discussed below) as a means to "maximize your impact":

Knowledge exchange (KE) is about opening a dialogue between researchers and research users so that they can share ideas, research evidence, experiences and skills. This can involve a range of activities; from seminars and workshops to placements and collaborative research. By creating this dialogue, research can more effectively influence policy and practice, thereby maximising its potential impact on the economy and wider society. (ESRC, n.d.-b: Knowledge Exchange)

A further common term is *research dissemination*, which invokes the metaphor of sowing seeds. While it is problematic in implying a one-way flow of insight, if the metaphor is pursued it can also remind us that research ideas need to *germinate* in context, within professional practice, and will therefore develop in ways that cannot be fully predicted.

Bourdieu and Wacquant (1992) argue that such terms, along with their associated literatures, policy documents, and practices, "pre-construct" the object of our interest (p. 229). They therefore have to be viewed as part of the policy discourse to be analyzed. Put simply, a research *problem* may lie partly in the current ways of representing and understanding that *problem* and its *solutions*. Analysis should therefore "shift away from simply *using* socially pre-constructed categories or objects, towards exploring the practices involved in their construction and maintenance" (Fairclough, Pardoe, & Szerszynski, 2010, p. 414). In this case, the ESRC's discourse of "impact" and "KE" must be a focus for critical reflection and analysis.

Nevertheless, any analysis requires a term with which to refer to, and potentially re-conceptualize, the set of professional practices that are of interest. No term is neutral, so it needs to reflect our intellectual and practical interest.

Here I shall use "research communication" as a deliberately general term to describe the diverse set of communicative practices by which academic researchers engage with people in the domains of policy, professional practice, and public debate beyond academia, with the intention to develop and communicate significant insights from their research. The term has been used similarly, for example, by Scott (2000) within his scoping report for the European Environment Agency. In addition, the EC (2010) guide for

researchers in socio-economic sciences and humanities is called "Communicating research . . . "

The term *research communication* can reflect a social and linguistic interest in the cross-institutional communication of research insight. In contrast to knowledge *transfer*, it can keep in view the two-way and contingent nature of communication. In contrast to KE, the issue of whether the communication constitutes an exchange is not presupposed, but retained as a focus of analytic interest. In contrast to both terms, research communication avoids reifying *knowledge*, and so can keep in view the potentially negotiated, contextual and contingent nature of what is ultimately communicated. It can describe a process in which insight may be developed and refined through that communication, rather than just *transferred* or *exchanged*.

I deliberately use the same term to embrace both dialogue and produced artifacts (such as introductory leaflets, research articles and reports, online demonstrations, video documentary, broadcast interviews, and media publications). This helps to keep in view a core insight from Bakhtin (1986), that even such texts and artifacts are dialogic: They are constructed in anticipation of encountering a response.

Analyzing the Representation of Professional Practice in Discourse

As a framework to analyze the representation of research communication in policy and guidance, I draw on the concept of "recontextualization." Originally from Bernstein (1990), it has been developed by van Leeuwen (2008) to analyze the representation of social and professional practices in discourse.

Echoing van Leeuwen (2008), Fairclough (1992b), and others, I am using the term *discourse* to refer to language-in-use that represents (and so also constructs) the world in socially and institutionally established ways. That includes representing knowledge and things, as well as identities, social relations, actions, and social practices. (For example, van Leeuwen, 2008, analyzes the representations of a child's first day at school within expert advice to parents, and of immigration in an Australian newspaper article.) Texts, parts of texts, and individual utterances may instantiate and perpetuate a particular discourse, yet we can also draw on competing discourses unwittingly, deliberately, and creatively to negotiate and contest established representations (Fairclough, 1992b).

Bernstein's (1990) sociological concept of recontextualization was prompted by observing what he described as the "overwhelming and staggering uniformity" of classroom activity and teaching practices across cultures and across the curriculum (p. 169). He argued that social and professional practices such as physics and woodwork (his examples) are recontextualized by a "pedagogic discourse" which "fundamentally transforms" them into classroom physics and woodwork. Physics is "delocated" from the original social relations and purposes of research and industry, and "relocated" within the social relations of a pedagogic discourse, with its own purposes, sequencing, and systems of evaluation.

Van Leeuwen (2008) usefully develops the concept beyond pedagogy, as an analytical framework with which to "analyse all texts for the way they draw on, and transform, social practices" (p. 5). He describes his work as starting from the view "that all discourses recontextualize social practices, and that all knowledge is, therefore, ultimately grounded in practice, however slender that link may seem at times" (van Leeuwen, 2008, p. vii). He uses the term *discourse* explicitly in the Foucauldian sense of

> "a socially constructed knowledge of some social practice" developed in specific social contexts. (van Leeuwen, 2008, p. 6)

By empirically analyzing Australian news reports and written guidance for parents on their child's first day at school, van Leeuwen (2008) observes that discourses

> not only represent what is going on, they also evaluate it, ascribe purposes to it, justify it, and so on, and in many texts these aspects of representation become far more important than the representation of the social practice itself. (p. 6)

Specifically, he argues that recontextualization can involve the substitution, deletion, rearrangement, re-legitimation, and re-evaluation of elements of the social practice. As these "transformations" can be achieved by diverse linguistic means, his inventory is "sociosemantic" rather than traditionally linguistic (van Leeuwen, 2008, p. 23; Halliday, 1978). He argues that in a "recontextualization chain," such transformations "can happen over and over again, removing us further and further from the starting point" (van Leeuwen, 2008, p. 13).

Two important notes are necessary. First, recontextualization is not itself pejorative: Any representation involves recontextualization. Rather, it is an analytical concept with which to investigate and critically analyze representations of social and professional practice. Second, it is important not to reify a *pure* or *true* practice (such as physics, woodwork, or parenting) prior to its recontextualization (Pardoe, 1997); such practices are usually heterogeneous and contested. Moreover, a social or professional practice may be informed and transformed by the discourses that selectively represent it.

Drawing Insight From Case Studies

In advocating case study analysis (agenda item 2, above), it is important to acknowledge that the ESRC offers a growing number of "impact case studies" online. However, these are primarily retrospective reports of successes, rather than analyses of the intellectual, practical, interpersonal, and organizational challenges within the communicative process. I

suggest it is useful to draw a parallel with Latour's (1987) classic methodological distinction between studying "ready made science" and "science in the making." Rephrasing his core principle (p. 4), I would argue that

> our entry into understanding research impact should be through the back door of *impact-in-the making*, not through the more grandiose entrance of *ready-made-impact*.

As decades of research in social studies of science and elsewhere have shown, issues and challenges can be most explicit at moments when the opportunities are uncertain, strategies are not yet decided, and success is still unclear. Arguably, these moments are as important for developing guidance as they are for research, because the challenges are often forgotten once they are resolved. Studies of *impact-in-the-making* are likely to reveal complex networks of contingent factors that together generate, co-produce (and counter) impacts in the short and long term. Only some will be within the control of the researcher.

When proposing case studies, it is vital to be clear about what insight can be drawn from them. As many have argued in qualitative and case study research, a case cannot simply be generalized as truth or as a guide for future work. Instead, its value is to offer a situated and potentially "telling case" that contributes to building deeper conceptual understanding (discussed in Platt, 1988). It may enable us to refine or refute universal claims produced from *common sense* or from less detailed research.

It is with that aim that I include brief observations from one recent inter-disciplinary project below. The intention is simply to show that the intellectual challenges of research communication in practice are of a different intellectual order from those assumed within the ESRC guidance. This contrast helps to

- refute the general categorical assertions within the guidance,
- foreground the ESRC assumptions and problematize the discourse,
- illustrate why such guidance needs to be informed by social science.

The example project. As the U.K. partner for the EC FP7-funded ORCHESTRA project (2009-12: No. 226521) our collaborative focus was on generating understanding of recent research on computer-based (in silico) methods for assessing chemical toxicity. The project was timely because the 2007 EU REACH regulations demand that industry assess many thousands of existing chemicals in the coming years. This is predicted to cost billions and "consume" many millions of animals in traditional animal testing, despite EU policy to approve animal testing only "as a last resort."

The project therefore set out to inform professional thinking among regulators, industry, and toxicologists about in silico methods as a means to reduce in vivo testing. One of many dissemination strategies was to conduct detailed interviews with researchers, regulators, potential industry users, and other stakeholders to investigate and analyze their priorities and concerns, and the issues around take-up. These were communicated in a video documentary (Pardoe, Cazzato, Golding, Benfenati, & Mays, 2011) as well as online (www.in-silico-methods.eu) and in open access articles (e.g., Benfenati et al., 2011).

As a final ORCHESTRA output, we reviewed the experience of the project to help inform science and technology researchers in future EC projects. An updatable e-guide (Pardoe & Mays, 2012) identifies some of the intellectual and practical challenges for researchers of making science and technology research engaging, accessible, and useable, while also retaining the scientific rigor and integrity of the publicly funded research. The project was an affirmation that even when research lies firmly within the natural sciences, social science insight can offer a vital contribution to understanding the challenges and potential pitfalls involved in a wider communication, and informing strategies.

The U.K. ESRC Guidance as a Focus for Analysis

Why Analyze Guidance From the U.K. ESRC?

The ESRC is "the UK's largest organisation for funding research on economic and social issues," and describes itself as "an international leader in the social sciences" (ESRC, n.d.-a). It financially supports more than 4,000 researchers and postgraduate students at any one time. Its "role is to"

- promote and support . . . high-quality basic, strategic, and applied research . . . in the social sciences

- advance knowledge and provide trained social scientists . . . , thereby contributing to the economic competitiveness of the United Kingdom, the effectiveness of public services and policy, and the quality of life

- provide advice on, disseminate knowledge of, and promote public understanding of, the social sciences. (ESRC, n.d.-a)

With decades of social science research funded by it, the ESRC is in a unique position among the U.K. funding bodies to be able to do (in its guidance) exactly what it now demands from its researchers: *to show the value of social science research for informing policy and professional practice.* The ESRC funds much of the U.K. research that could usefully inform (and critique) our understanding of research communication and impact, including in education, linguistics, management, and science and technology studies.

After many years of development and refinement, the *impact toolkit* is the ESRC's online guide for funded researchers on how to maximize the impact from their research. Before analyzing part of it, it is useful to look briefly at the

discourse around it that justifies and explains the demand on researchers.

From U.K. "Global Economic Performance" to "Succinct Messages"

Consistent with its umbrella organization, the Research Councils UK (RCUK), the ESRC defines *research impact* as "the demonstrable contribution that excellent research makes to society and the economy." The ESRC elaborates this as

- fostering global economic performance, and specifically the economic competitiveness of the United Kingdom

- increasing the effectiveness of public services and policy

- enhancing quality of life, health and creative output. (ESRC, n.d.-b: What is impact?)

These societal and economic goals are articulated to explain and justify the U.K. "funding regime that . . . embeds knowledge transfer as a permanent core activity in universities" (HM Treasury, 2004, p. 76). The required shift in professional practice for university researchers is represented in terms of "the behaviour and attitudes that RCUK wishes to foster" (RCUK, n.d.).

At the level of implementation, the policy becomes a demand for each funded project to produce a "pathways to impact" plan, and to generate and demonstrate impact. The ESRC's role includes evaluating impact plans and giving the advice analyzed below.

In this chain of recontextualization, the policy goals and the potential ways forward are progressively transformed or narrowed by decisions about how they can be achieved through specific demands and procedures. By viewing these policy statements in a sequence, from global economic priorities, to researcher "behavior and attitudes," to calls for "succinct messages" (below), social scientists and discourse analysts can usefully recognize and question each successive narrowing.

While such analysis is not the focus of this article, such successive transformations and narrowing can be observed within single texts. The extract below, from a page within the impact toolkit titled "Why is KE important?" consists of a series of categorical and potentially non-controversial statements. The policy and rhetorical action is achieved through making them sequential. As is often observed in critical discourse analysis (e.g., Fairclough, 1992b), the most ideological and problematic steps are those which are omitted, and which the reader has to presuppose to make sense of the text.

The UK has a strong science base, but performs less well in capitalising on new research to generate innovation. Effective KE is vital in ensuring research is translated into policy and practice. As a funding body, the ESRC spends over £211 million a year on research, training and knowledge exchange. We want to ensure that our funded research is not only of the highest quality, but also has a positive impact on society. KE is therefore fundamental to the way we work.

(*)However good your research, there is little point in doing it if nobody knows about it. If your research is to make a difference to policy or practice it must be accessible to potential users and other interested parties. Thinking about who these might be and how to actively engage with them through the lifespan of your research will help you to:

- Gain a better understanding of the needs of potential users, their expertise and their perspectives on your chosen topics

- Inform and improve the quality and focus of your research

- Gain valuable new skills

- Increase the prospects of your research being applied. (ESRC, n.d.-b: Why is KE important?)

For example, the statement at (*) would appear to be utterly obvious in a profession so focused on peer review and publication. Yet juxtaposing it with the subsequent statements potentially redefines "nobody" as "nobody outside academia," which has fundamental implications for research.

The extract ends by bringing the U.K. societal and economic goals right down to the personal "gains" and skills for "you" as a researcher. The role of the impact toolkit is then to guide social science researchers in how to achieve those societal, economic, and personal goals, by "ensuring research is translated into policy and practice."

The ESRC's Impact Toolkit and "Step-by-Step Guide to Maximising Impact" (SSG)

Our impact toolkit *gives you everything you need* to achieve the maximum impact for your work. The toolkit includes information on developing an impact strategy, promoting knowledge exchange, public engagement and communicating effectively with your key stakeholders. (ESRC, n.d.-b, emphasis added)

The impact toolkit is the core of the ESRC's written advice for its funded researchers. Launched in January 2011, it is described by the ESRC as "a practical tool" which "draws on best practice from investments" (ESRC, 2011).

It consists of more than a hundred separate web pages. Its *size* is relevant to this analysis because any omissions are not due to brevity. In some areas, it has surprising levels of detail. For example, it offers advice on networking, including "creating rapport" and "making small talk," including advice to "ask an open rather than a closed question" in a coffee queue and to "match and mirror body language" of the person you are talking to (impact toolkit: tools: networking: creating rapport). A section, called "tools," offers extensive *to do* lists on "branding," "digital communications," "events," "media

relations," "networking," "publications," and "public engagement." So from brief observation, it might appear almost encyclopedic.

Within that toolkit, I analyze the "SSG" because it is one of the most explicit and practical guidance sections. Elsewhere in the toolkit, the researchers are directed to the SSG for "practical guidance on planning research impact." It is a section in which the ESRC goes beyond merely reiterating demands and policy statements, to offer *step-by-step* advice and understanding on what research communication involves *in practice*. (I have also reviewed pages across the impact toolkit to check that an observed omission in the SSG is not simply located elsewhere.)

The ESRC claims that the SSG addresses the readers' need, both practically and intellectually. The introductory page "Developing a strategy" claims that "This part of the toolkit gives guidance on how to maximises impact" [*sic*], it "takes you through each stage of the process." A page directing researchers to the SSG states,

This takes you through *the issues you will need to think about* and gives advice on planning activities to help you generate impact. (ESRC, n.d.-b: What the ESRC expects, emphasis added)

The page headings within the SSG identify the steps and approach. The introductory page (1) is followed by

2. Setting objectives
3. Developing messages
4. Targeting audiences
5. Choosing channels
6. Planning activities
7. Allocating resources
8. Measuring success. (ESRC, n.d.-b: Developing a strategy: Step-by-step guide)

Pages 2, 3, 4, and 5 are most specifically about communication, so they are my focus in the analysis and the frequency tables below. But first, it is worth simply quoting from pages 3 and 5 to offer a further sense of the conceptual understanding of communication offered by the SSG, and of the style of the toolkit as a whole:

(3) Developing messages

Drafting your messages An effective strategy needs to have clear, succinct messages that summarise your research . . .

When drafting your key messages, avoid using overly complex statements . . . Remember that key audiences such as journalists and policymakers are overloaded with information and may not remember your messages if they are too complex.

Ensure that the language you use is appropriate for the audience . . .

Using different formats . . . It's useful to try out your messages in different formats, for example: a media release; a report; a research briefing; a newspaper article; a website page . . .

Creating a brand In order to convey your messages more effectively you need to think about branding . . .

(5) Choosing channels

It is important to consider the most appropriate channels to reach your target audience . . . for example: . . . why an email bulletin rather than face-to-face contact?

Researching your audience Find out how your target audience prefers to receive information . . . (ESRC, n.d.-b: Developing a strategy: Step-by-step guide)

Comment: The SSG as an Example of Informing Professional Practice

Across the SSG, and more widely in the impact toolkit, the guidance consists of simple and categorical statements of this kind. The simplicity and conceptual repetition (and the often circular linking of pages) make the SSG and wider toolkit far less comprehensive than it first appears.

It is perhaps necessary to remember that it is written for use by social science researchers. Even if the reader is not involved in research specifically related to communication or knowledge claims, they are nevertheless involved daily in complex communicative practices ranging from engaging students to managing teams and institutional politics, and in making subtle decisions about their communication strategies in meetings, emails, reports, and academic articles.

Many of the guidance statements above could be applied equally to those familiar communicative practices. If they were, they might be criticized for merely stating the obvious. In other words, such statements are *generalized* and *decontextualized* communication advice that does not capture what is different, new or challenging about communicating research insight across institutions to inform policy and professional practice.

Reading such guidance can prompt us to ask the basic questions that we should ask of any text that aims to inform professional practice. Is it really informing the readers (here social scientists) of things they do not already know? Does it really address the challenges they face (in communicating their research to people in other contexts and professional worlds)? Does it offer the kind of conceptual understanding and practical detail they need to achieve it? Moreover, from all that we know from research and practice, is this really the most useful information that we are currently able to offer?

When involved in a research communication project, I raise such basic questions and concerns with the team, and suggest strategies for finding out the answers. I would raise very serious concern if a research team produced a guide for professionals that diverged so markedly from the discourse practices of those professionals, was so devoid of explicit research insight or evidence to support the guidance, and was so universal, directive, and categorical. I would be concerned about whether it could create a productive relationship with

the user, whether it would inspire or actually deter use of the research, and whether the team had really investigated or understood the challenges and the varied contexts of use. Those questions seem highly relevant in this case.

I would argue that while the *authorship* of the SSG and the ESRC's decision-making around its production might be interesting, that is not the primary issue. The authority of this guide and the claims made for it (see previous section), make the text itself the focus of interest. Moreover, it is no maverick text. Its consistency with other texts across the impact toolkit and beyond, and its durability over the years, mean that the text itself and the policy discourse that it instantiates and perpetuates are the interesting and legitimate focus for analysis.

Analysis

Van Leeuwen identifies several ways in which a social or professional practice may be recontextualized and so transformed by representations of it. To structure this analysis, I explore six potential transformations:

1. Relocating purposes
2. Rearrangements and (re)ordering of actions
3. Evaluations: the concepts offered to guide and review communication
4. Substitutions: the representations of policy makers, practitioners, and publics
5. Substitutions: the representations of communicative actions
6. Legitimating the guide and the demand to maximize impact

Relocating Purposes

Questions like "why are we doing this?" or "what for?" are as central to research communication as they are to any activity. Van Leeuwen (2008) observes that representations of practice in discourse can involve adding, articulating, and/or potentially transforming the purpose(s) of that practice, including the participants' own sense of "what for" (p. 20).

A first observation of the purposes articulated within the SSG is that these are almost all *internal* to the ESRC's funding procedures; they are delocated from the kinds of societal goals and personal goals in the policy statements. A second observation is that the purposes are all nevertheless apparently unique to "you," the reader:

. . . Every strategy is different, but you can use these steps as a template to develop your own. Your strategy takes you from where you are now to where you want to be. (ESRC, n.d.-c)

In this way, the text simultaneously locates the purposes within ESRC procedures, yet also claims they are individual and come from "you." This is a familiar feature in advertizing, in which demands are both personalized and mitigated by invoking notions of freedom and choice (Fairclough, 1992b). It also echoes a pervasive and problematic discourse in education (critiqued in various fields including in Hyland, 2003; Mercer, 1995; Swales, 1990) in which individuals are required to generate *their own purpose* and unique text, without the institution recognizing or making explicit the institutionalized practices, constraints, and purposes within which that individual purpose has to develop and will be judged.

The guide presents "you" with the demand to be (already) clear about "your objectives," but without any useful indication of how those objectives might be developed and clarified (see *reordering* below). Only one example is offered: "A typical set of objectives might be to:"

- Build awareness of the project among a defined audience

- Secure the commitment of a defined group of stakeholders to the project aims

- Influence specific policies or policy makers on key aspects

- Encourage participation among researchers or partner bodies (ESRC, n.d.-b: Step-by-step guide: Setting objectives)

Arguably, these objectives are merely an extension of the policy demands rather than useful intentions informed by practice. They appear irrefutable because in each case it would be difficult to advocate the opposite. In van Leeuwen's (2008) terms they are "moralized actions": abstractions that merely "trigger intertextual references to the familiar discourses and values that underpin them" (p. 126). For the research team, each objective only begs the kind of practical (what, who, how, and why) questions that the team already face. They are about impacts, rather than about the people, ideas, and communicative processes that might one day achieve those impacts. They contrast markedly, for example, with the respectful, self-questioning, experience-based "tips" for engaging with policy makers offered by Goodwin (2013).

The SSG offers no conceptual basis, and no investigative strategies, from which readers can develop their own objectives, or review them. There is no link to case studies that could usefully show, for example, (a) that it is necessary to *find out* what the "key aspects" of the research are for policy makers or (b) that for stakeholders to "commit" to the project aims, the research team may need to engage in interactions that develop trust and credibility over time. Similarly, it offers no downstream illustration of how such objectives might inform the communication strategies.

Instead, the ESRC merely tells researchers to "make sure that you set SMART objectives": "specific," "measurable," "achievable," "relevant," and "time-bound." This is a framework available on many project management and business advice websites. Its representation as a checklist of demands, with these particular elaborations of each letter (cf. Jisc InfoNet, 2008), reinforces the ESRC demand that "impact must be demonstrable" (impact toolkit: what is impact?).

This representation reflects a current managerial policy focus on "performance indicators"; it is an interesting shift from its original use. Morrison (2011) and others attribute SMART to Doran (1981), and observe that as an industry consultant, Doran made it explicit that "in certain situations it is not realistic to attempt quantification" (cf. "measurable"), and that it "can lose the benefit of a more abstract objective" (quoted in Morrison, 2011). Doran's caution is vitally important in research communication, given the temptation to pursue measurable and achievable performance indicators in place of the less tangible and more challenging long-term societal impacts desired in U.K. policy.

In the FP7 ORCHESTRA project (see "Drawing Insight From Case Studies", above) the team started by investigating the current policy commitments, professional practices, and concerns and priorities of policy makers and potential users. As in many projects, it became clear from interviews and discussions that knowing how to use the research in practice was only one small part of the challenge faced by practitioners. Merely "building awareness" and providing information from the research was not the route to professional uptake. Instead, the project needed to investigate and address wider issues, including current user experience, confidence and concerns, reliability in practice, and issues around the regulatory demands debates and acceptance.

Second, it became clear that to "influence policy makers" or "secure the commitment" of some users, as the ESRC suggests, would merely increase the concerns of others, and so position the project on one side of an unproductive debate. To retain the integrity and value of the research, we needed to critically inform the debate, rather than just join it. Social and historical studies of both science and education have shown that it has been the trajectory of too many new methods and technologies to be overstated initially, and then discredited when they are misapplied or used too widely or used with insufficient scrutiny. So in this case, instead of promoting use of the technology, our objective was to promote a critical understanding of it and of its limitations to inform a wise and appropriate use.

In other words, a credible, effective, and professionally responsible approach required a long-term view of impact, and required objectives and strategies directly counter to simply "maximizing impact" in the short term. In each way, the intellectual challenge of developing and clarifying objectives is of a different order from the simple self-determined process advocated by the ESRC guide.

Rearrangements and (Re)ordering of Actions

Questions like "where do we start?" "what do we do next?" and "where are we heading?" are likely to emerge during any unfamiliar communicative endeavor. Addressing them can be a potentially vital part of any guide. Yet van Leeuwen (2008) argues that in representations of practice, "elements of the social practice, insofar as they have a necessary order, may be rearranged" (p. 18). Representations can reverse action and reaction, as well as the chronological or rhetorical sequencing.

As its name suggests, the *step-by-step guide* offers an ordered or sequenced account of research communication. It states that the starting point is "your research," "your strategy," and "your objectives."

> The first stage of developing your strategy is to set out a clear statement of your objectives. This should link to your goals and how you will evaluate the success. (ESRC, n.d.-b: Step-by-step guide: Setting objectives)

The subsequent *steps* further represent research communication as a self-determined process of the researcher "setting objectives," "developing messages," "targeting audiences," and "choosing channels." The decisions and rationale for the actions of research communication all derive from the research and the researchers, rather than from the users and the contexts of use (further analysis below).

Arguably, this functions as a "regulatory discourse" (Bernstein, 1990; Chouliaraki & Fairclough, 1999) as it constructs the social relations in such a way that the researcher appears as the main (or sole) actor, and so appears responsible for the success or failure of the impact. However, it also represents and potentially encourages a kind of arrogance that researchers and research policy would be wise to avoid if U.K. research is to engage other professionals.

By contrast, I would argue that it is only through investigations and dialogue that a researcher can develop a sense of (a) what the research might contribute to professional practice and debate, (b) what may be seen as significant by practitioners, and (c) what may be needed to make it useable in practice. As in the ORCHESTRA example (see previous section), a primary element in almost any research communication process will be to *investigate* what practitioners are already doing, the understandings and implicit or explicit theories that inform their practice, and the current issues, concerns, and debates.

Contrary to the guide, I regard it as a primary organizing principle for research communication, that *interaction* inform the planning, and come before *production*. That interaction is not merely to research the "audience" to inform the presentation. It is actually about content: that interaction may generate further understanding, may identify new areas of significance, and may present new challenges for the research evidence. This interactive and investigative process can become "a different kind of *peer review*" (Pardoe & Mays, 2012). It can vitally inform the objectives and strategies.

Case study observations of research communication practice can usefully help us to understand alternative, but no less rigorous, orderings. For example, in the ORCHESTRA project the dissemination actions and outputs had to be proposed by partners prior to the contract (as is usual in EC proposals) with no opportunity for face-to-face discussion. Within the later discussions, I realized that our agreed actions and outputs functioned as "boundary objects" (Leigh Star & Griesemer, 1989), with partners from different disciplines holding very different assumptions about what an output would be and what purpose it would serve. It was only when actually organizing the event or collaboratively writing the output, that the different assumptions became apparent. At that stage, making good decisions required explicit and reasoned discussion about the function and form of each output: the leaflet, the workshop, the video, etc.

Such an ordering of actions is actually the opposite of the simple ordering assumed by the ESRC. Yet it has a clear rationale: Each media and genre ("channel") is effectively a "resource for making meanings" (Halliday, 1978, p. 192) or "semiotic resource" (van Leeuwen, 2005, p. 3) that brings possibilities for what can be communicated and to whom. Having proposed a leaflet or video, the team is able to imagine and discuss together what could be communicated and potentially achieved by them. In other words, the planned "output" or "channel" provides a frame for then collaboratively "developing messages," and developing and refining the "objectives." This reverse ordering does not displace the ESRC ordering; both may be evident and useful. But understanding it is vital to enable researchers to see the value and the potential for creativity and intellectual rigor within their own ordering. If researchers only have the ESRC model in mind as the goal, they may view their own collaborative process merely in frustration, as being a result of cross-institutional misunderstanding and a failure to define their objectives and messages first.

Evaluations: The Concepts Offered to Guide and Review Communication

In the various literatures around research communication, there is an often-cited and valuable distinction between the "instrumental utilisation of research" and the "conceptual utilisation of research" (Caplan, Morrison, & Stambaugh, 1975, cited in Scott, 2000, pp. 5, 11; ESRC, 2009; Nutley, Walter, & Davies, 2007).

An often quoted development of this point from Weiss (1980), is that

> Instrumental use is often restricted to relatively low-level decisions, where the stakes are small and users' interest relatively unaffected. Conceptual use . . . can gradually bring about major shifts in awareness and reorientation of basic perspectives. (quoted in Scott, 2000, p. 11)

Davies et al. (2008) further emphasize the value of the using research in conceptual ways

> on the ground, research and other forms of knowledge are often used in more subtle, indirect and conceptual ways: bringing about changes in knowledge and understanding, or shifts in

perceptions, attitudes and beliefs, perhaps altering the ways in which policy-makers and practitioners think about what they do, how they do it, and why. (p. 189)

This conceptual/instrumental distinction is itself a good example of a concept that offers practical value. Having heard it, it becomes a way to think about formulating and communicating research ideas, and about what conceptual understanding may be necessary for any *impact*.

Given the diversity of U.K. social research, and the diversity of potential users, stakeholders, and contexts of use, it is legitimate to ask what conceptual understanding of communication is offered by the SSG. Table 1 therefore lists the concepts offered to researchers to inform and evaluate their "strategies," "messages," "language," and "channels."

Table I. Concepts to Inform the Communication.

Concepts to inform strategies, messages and language use		Concepts to inform the channels of communication	
Complex	4	Large	I
Clear	4	Small	2
Effective	3	Regular/quarterly	2
Succinct	2	Occasional	I
Appropriate	2	Audience preference/ audience needs	2
Simple	2	Personal	I
Long	2	Direct	2
Specific	I	Well placed (in media)	2
Over-arching	I	Time and money	I
Targeted	I	Active membership	I
Right	I	Two-way communication	I
Accessible	I	Building relationships	I
Capture attention	I		
Formats	I		
Stories	I		
Case studies	I		
Packages of info	I		

Van Leeuwen (2008) observes that representations of social and professional practices involve such representations of what is "good" or "bad" or "useful" or "interesting," and why (pp. 18-21). In this case, the SSG reader already knows that "clear," "effective," and "accessible" are good, because the opposites are clearly not good. These are *common sense* goals. What is absent is any basis for achieving them. For example, the need to be "succinct" is obvious; the challenge for researchers is how to achieve it for a particular text and a particular audience. What is "succinct" in a report is not succinct in a leaflet. What may be a succinct report for an experienced practitioner may be inaccessible to a manager or policy maker. Yet each term is offered without any

conceptual understanding of how to achieve it or how to judge that it has been achieved.

In practice, to produce a "succinct" text for an unfamiliar readership is likely to involve investigating the professional expectations of the genre, and more specifically, what information can be assumed and what needs to be articulated for those readers. That kind of interactive investigation is vital if the research outputs are to have professional credibility, and not be dismissed as, say, abstract, patronizing, useless, or unreadable.

The concept of "appropriate" texts and language has been a focus of critique for many years (e.g., Fairclough, 1992a). What it actually *means* when applied to a particular communication is highly cultural, institutional, and genre specific. The advice to make a text "appropriate" raises familiar questions, but answers none.

In this way, the guide appears to draw on a familiar and much-critiqued discourse of communication as "decontextualised skills" (Ivanič et al., 2009), as if one can engage in "appropriate" and "effective" communication simply guided by a universalized notion of "clarity," without knowing or investigating the specialist genres and discourses of the professionals you want to communicate with.

The SSG does offer one example of how to "ensure that the language you use is appropriate for the audience":

For example, the Institute for Social and Economic Research published a report called *The Impact of Atypical Employment on Individual Wellbeing*. The press release had the more accessible title of "What Kind of Work is Bad for Your Health?" The first line of the release summed up the research finding: "Temporary jobs and part-time employment do not have adverse consequences for people's health."

While the original press release may have been wise, its inclusion as a single example of "appropriate" language is highly problematic, especially without any concepts with which to review it. It is likely to reinforce the common notion that researchers need to adopt a more "tabloid" discourse to communicate beyond academia and generate impact. There are no warnings about the consequences.

I would argue that if publicly funded university research is to have impact, then it needs to retain the integrity and rigor that is its strength and core value. So any advice on making research engaging and accessible must, above all, include advice on monitoring whether the original research claims have shifted in the process. In her classic article, Fahnestock (1986) analyzed "the fate of scientific observations as they passed from original research reports intended for scientific peers into popular accounts aimed at a general audience" (p. 275). She identified significant transformations in the apparent focus of the research and the scientific claim (e.g., here from "atypical employment" to "kind of work," and from "well being" to "bad for you"). She observed that scientific claims can become more certain within popularizations, as the theoretical perspectives, experimental constraints, and elements of context are put aside in favor of simply reporting a finding.

The ORCHESTRA project showed repeatedly the fundamental socio-linguistic point that changing the language changes the meaning. To increase the "clarity," "relevance," and "accessibility" of research findings often involves requesting further conceptual clarity from the researchers. To ensure that every statement is *wise* as well as *true* then requires an iterative process that is (a) grounded in the research and (b) highly attentive to the implications of lexical and grammatical changes. Without this, the pursuit of more engaging outputs can mislead potential users and risk fatally undermining the integrity of the research.

Substitutions: The Representation of Policy Makers, Practitioners, and Publics

Substitution is described by van Leeuwen (2008) as "the most fundamental transformation" (p. 17). Participants and actions can be particularized or generalized, aggregated or nominated in ways that can transform their apparent relations and identities.

In the SSG, the researcher is individualized and activated as "you" with responsibility for all actions. At the same time, the policy makers, industries, practitioners, and publics who may use the research (and/or benefit or lose by its use) are all aggregated and passivated. Of 58 references to other people, 29 represent them as an "audience." Indeed, 65% of all references to other people define them solely in their relation to "your research" (Table 2, left column).

Table 2. Representations of Those To/With Whom the Researcher Communicates.

Audience	29	Organizations/individuals/ everyone	4
Who / those who may have an interest	3	MP; Backbench MPs; MPs' researchers	3
Your contacts	3	Researchers	2
User groups	1	Journalists	2
potential beneficiaries	1	Policy makers	2
Gatekeepers to your audiences	1	Stakeholders	2
		CEOs and personal assistants	2
		Umbrella body and members	2
		Partner bodies	1
Total	38	Total	20

The right column shows the only groups identified specifically by profession. (An additional linked pdf page simply lists sector categories such as "Large business," "SMEs," "Senior Civil Servants," and "Trade unions.")

Of these, MPs and journalists are two professions that can create visibility at government level for ESRC-funded research, and so may influence policy. A new section appeared in the toolkit in early 2012 on "Taking the research to Westminster" and "Contact[ing] government organisations." Yet a focus on them represents a potentially centrist, top-down view of innovation and professional change. It is also highly problematic in terms of achieving informed and wise impact for the U.K., since MPs and journalists are at least one step removed from the detail of most professional practice.

Taking U.K. university research directly to MPs and journalists risks bypassing vital peer review from experienced practitioners. From experience, I would argue that professionals in fields relevant to the use of the research are a vital source of insight—about whether and how the research could be useful in practice, what problems, concerns, and questions there will be, and what might need to be communicated for people to understand it. Practitioners are potentially the most able to trial it, challenge it, refine it, and initiate its use.

The SSG contains no reference to such dialogue with practitioners. There is nothing on what the researcher might gain or learn from them, or on the ways in which different "audiences" might use different aspects of the research in different ways. This omission in the SSG contrasts with the ESRC demand, for example, that

by considering impact from the outset, we expect you to

- Explore who could potentially benefit from your work

- Look at how you can increase the chances of potential beneficiaries benefiting from your work. (ESRC, n.d.-b: What the ESRC expects)

These two *investigative* actions are fundamental to any process of research communication. The first is arguably the starting point that should inform any impact objectives. So one might expect a step-by-step guide to offer strategies for such investigations, and for drawing insight from them.

Substitutions: The Representation of Communicative Actions

The page headings of the SSG represent the communicative process as "setting objectives" to deliver "messages" to "audiences" through "channels." In the detail, the communicative process is similarly represented as material actions: *doing things*. Table 3 lists the frequency of finite processes that represent what a researcher does or says to (or with or for) the "audiences" and other people.

There are no interactive processes like "discuss" or "ask" or even "explain." Communication is represented as a mechanical process of information management, or "shunting information" (Smith, 1985), rather than generating interaction and understanding. (The single reference to "consider" is the only finite process that explicitly represents a response to what the other people may say; "two-way communication" is mentioned once as a noun.)

Table 3. Actions by "You," To/With/For Others.
According to the guide, you:

Material action		Semiotic action	
Target [them]	4	Influence [policy makers/ audiences]	2
Gain [them]	1	Know [key audiences]	1
Select [them]	1	Concentrate on the most influential	1
List [them]	1	Consider their timescales	1
Prioritize [them]	1	Research [your audience]	1
Rank [them]	1	Demonstrate your relevance to them	1
Focus on [them]	1	Build [their] awareness	1
Keep track of [them]	2	Communicate (one way)	1
Manage [them]	1	Communicate with	1
Reach [them]	1	Understand their needs	1
Mail [them]	1	Encourage their participation	1
Contact [them]	2		
Capture the attention of [them]	1		
Secure their commitment	1	Cognitive reaction	
Obtain feedback from [them]	1	Consider [what messages emerge]	1
Share contacts [within team]	1		
Cultivate [them]	1		

Table 4. All Actions by Others.
According to the guide, other people:

Benefit	2	Take notice (of your work)	1
Prefer	2	Read (your work)	1
May not remember	1	Tell you	1
Suffer from info overload	1	Determine whether your material ever reaches …	1
Are overloaded	1	Capture attention (of others)	1

Consistent with this, the actions also passivate others as recipients of activity or as being subjected to it. For example, "journalists are worth cultivating too." The only actions (finite processes) for which other people are the agents are shown in Table 4, column 2, yet in each case their actions are reactions, and solely in relation to the research. (The activated process in the right column is associated only with MPs, MPs' researchers, and CEO's assistants, discussed above.)

There are pages elsewhere in the toolkit that can at first appear to counter this passivation of others. For example, a page alongside the SSG, titled "How to maximise impact," calls for "establishing networks and relationships with research users," for "involving users at all stages of the research," and for "well planned public engagement and knowledge exchange strategies" as "key factors that are vital." This sounds like it is advocating dialogue. However, it remains consistent with the SSG: first, it gives all agency to "you", the researcher, and second, it offers no indication of what you may learn from doing it.

The reality of any research communication process is that the potential users and other key "audiences" are likely to be already informed by extensive professional experience, and possibly by other research. They will have well-established institutional practices, professional debates, current controversies, and concerns, and they may have commitments to existing policies. All of that forms the context and the real challenge of research communication.

Yet there is nothing in the SSG to prepare you for this. It implicitly reproduces a view of research communication, and more broadly of science and society, that has been much critiqued (e.g., Irwin, 1995; Myers, 2003; Wynne, 2005), in which the professional and public worlds are unknowing, separate from the research world, and a blank slate waiting to receive the research.

Legitimating the Guide and the Demand to "Maximize Impact"

Finally, I turn to the issue of legitimation. Van Leeuwen (2008) argues that "texts not only represent social practices, they also explain and legitimate (or delegitimate, critique) them." Texts explain (or assume) why a social or professional practice "must take place in the way that it does" (p. 20). In analyzing the SSG, I suggest it is useful to explore how it legitimates (a) itself and (b) the demand for every research project to "maximize impact."

For the guide itself, one source of legitimation is the use of an all-knowing expert voice: The categorical statements appear to know the problems "you" will face, and how to address them. It is self-evidently in "your" interest to follow the advice:

An effective strategy needs to have … ;

it is worth … ; it is useful to … ; it is important to … ;

In order to … you need to …

A further source of legitimacy is simply that the advice offered is so basic that, in van Leeuwen's (2008) terms, it appears to be "common sense and in little need of legitimation" (p. 20).

Yet categorical claims and imperatives (like "avoid using … ," "ensure that … " "find out … ," and "don't assume . . . ") usually need a further source of legitimation. In this case, it is provided implicitly by the role and status of the institutional author, the ESRC. This author has the ultimate power of judging "your" funding applications. So their advice is an indication of what will get research proposals funded. Indeed, in procedural terms, it can be useful for the ESRC-funded researcher if the advice is singular and categorical.

It is therefore vital to recognize that this guide is *not* a model of the "clear" and "succinct" communication that a social science researcher might use. Instead, it relies on a relationship of authority; one that is in marked contrast to the relationship between researchers and potential users of their research. When seeking to generate impact from research, researchers usually have no decision-making role: like a consultant, they can only advise and suggest. In fact, they are likely to be in the difficult role of an *uninvited* consultant, who first needs to explain why other professionals should even want to hear about the research. That challenge can be a source of considerable anxiety for researchers, yet it is not addressed here.

Perhaps the most interesting observation in terms of legitimacy is the ESRC's departure from academic and scientific practice. Within research, a traditional source of legitimacy is to cite and build on what others have said and done before. In social science, legitimacy also involves offering empirical evidence of some kind. Those are the professional practices of the intended readers of this guide. As a research funder, the ESRC usually insists on such practices. Moreover, there is a huge amount of research available to inform this guide.

Yet the ESRC breaks with academic practice, to offer no research insights or evidence. There are no social science references. Even in a policy document outside academia, we could expect some support and referencing. In this case it appears that the institutional power of the author, and the common sense nature of the advice, make research and evidence unnecessary.

The ESRC guide and toolkit thereby place the practice of research communication, and the ESRC's process of guiding it, *outside* usual academic practice. Research communication is represented as not being a focus for intellectual interest or debate. It is not something to be informed collectively by building research insight. Instead, the guide can just describe what "you" need to do. In this way, the guide is more similar to a "You can do it too" manual from a DIY store, than a realistic or informed account of the cross-institutional communication of research.

An alternative approach could have been to make research communication intellectually attractive to researchers. The guidance could have concisely summarized at least some of the academic insights which point to the challenges of research communication and which reveal it as an intellectually interesting activity. The equivalent guide from the EC (2010) does so. The ESRC guide could build on the traditions of participatory and action research in the social sciences, to inform a process of creating productive relationships with stakeholders and potential users. Yet the ESRC chooses not to do so.

Finally, I turn to our second question of how the guide legitimates the policy demand for every project to "maximize impact." This is the demand for impact generation to become "a permanent core activity in universities" (HM Treasury, 2004, p. 76). The answer would appear to be precisely in simplistic nature of the advice. The representation of research communication as a series of self-evident and easy tasks, needing only common sense strategies, makes impact *appear* achievable in practice within existing funding. The misleading deletion of complexity is potentially functional.

If the ESRC had drawn on the available literature and offered a fuller understanding of research communication and impact generation in practice, it would risk pointing to the "largely insurmountable challenges" (Davies et al., 2008, p. 189). Instead, the ESRC discourse echoes the commercial discourse analyzed by Fairclough (e.g., 2002, p. 115) where companies minimize the apparent task: You "just" do this.

As if anticipating this guide, van Leeuwen (2008) observes that "some texts are almost entirely about legitimation . . . and make only rudimentary reference to the social practices they legitimise" (p. 20).

Concluding Comments

In this article, I have suggested a three-point agenda for social science to inform the cross-institutional communication of research as a rapidly growing area of professional academic practice. I have illustrated the need for a fundamental critique of the current guidance discourse by analyzing the ESRC's *step-by-step guide*. By contrasting the ESRC guide with brief examples from a single project, I have shown that the practical and conceptual challenges in research communication are of a different intellectual order to those in the guide.

Ironically, the guide provides a good example of why the uptake of research, and the connecting of research and practice, is not just about making research accessible or available. The guide is written by the U.K. organization whose function is to promote the use of social science research, and which has unique access to the body of research and to scholarly advice. It nevertheless chooses to put all this aside in favor of informing a U.K. shift in professional practice on the basis of a naively rationalist common sense discourse. It can even seem functional to do so.

A major challenge when communicating with policy makers and practitioners is precisely to make the case that research insight may be useful in practice. It often involves struggling with a dominant professional ideology that practice is simply common sense, and that it is better guided by the common sense of managers without reference to research. This is a particular challenge for social scientists, where the findings are not technologies or *facts* with self-evident value, but may instead involve questioning the ways in which things are currently understood. For example, generating impact can be a challenge for researchers in applied linguistics, because common sense views of language and communication are so powerful and problematic. It is therefore both ironic and dispiriting to observe it within the ESRC guidance.

So does the guide matter? As always, practitioners may realize the guidance is inadequate, and may draw on other insights and develop their understanding from reflective practice to achieve some success. That is very likely in the case of communicating research and generating impact. However, I suggest the inadequacy does matter in practice.

First, it matters within the culture of universities. Separating this vital activity from academic debate, and representing it as simple common sense, only serves to give it low status in relation to research. Focusing the purposes and actions on bureaucratic and procedural demands separates and undermines it further. There is an already pervasive assumption that research communication happens after the intellectual work of research, rather than constituting part of that work. This guide appears to confirm it. It therefore undervalues the intellectual work carried out by anyone who genuinely wants to engage with practitioners and policy makers to communicate their research. It is an example of how to undermine and *disengage* the key actors on whom the wider use of U.K. research depends.

Second, it matters outside the university, in industry, policy, and public arenas. In each part of the analysis above, I have suggested that if the guide were actually followed, it would risk producing uninformed actions that lack credibility among the intended "audiences," and risk undermining the integrity and rigor of the research. Moreover, the lack of status of the activity, the focus solely on the researchers' own objectives, the lack of any recognition of the value of practitioner review, and the view of professionals simply as receiving "audiences" will not go unnoticed by those "audiences." The focus on quick visibility with politicians and journalists, rather than engaging with the complexities of practice, can only undermine it further. In each way, the guidance can actually undermine the policy of maximizing the impact of social science research.

Third, it matters within the funding councils themselves and the wider U.K. policy, because it institutionalizes an attractive and reductive understanding of research communication among those who influence funding. The evident danger of institutionalizing such a simplistic view is that the ESRC reviewers of research proposals may not recognize the dangers and the sheer inadequacy of proposals that merely promise to write to MPs and "get the existing information to everyone for very little cost." Consequently, they may reject as "over-evaluated" and too costly, those proposals that recognize the challenges and that realistically cost the intellectual and practical work needed to distill the research in new ways, engage with practitioners, and investigate current practices, concerns, and debates. If so, such funding decisions will run directly counter to achieving the goal of enabling social science research to contribute to society and the economy.

Author's Note

Earlier versions of this analysis were presented at the ESRC conference "Bridging the Gap Between Research, Policy and Practice" (Pardoe, 2011a), and at the interdisciplinary conference on "Applied Linguistics and Professional Practice" (Pardoe, 2011b). I would like to thank Dr. Sherilyn MacGregor, the participants in the two conferences, and the anonymous peer reviewers for their helpful, critical and encouraging comments.

Declaration of Conflicting Interests

The author(s) declared no potential conflicts of interest with respect to the research, authorship, and/or publication of this article.

Funding

The author(s) disclosed receipt of the following financial support for the research and/or authorship of this article: European Commission (FP7 project 226521, 2009-12).

References

Bakhtin, M. M. (1986). *Speech genres and other late essays* (V. W. McGee, Trans., C. Emerson & M. Holquist, Eds.). Austin, Texas: University of Texas Press.

Benfenati, E., Diaza Gonella, R., Cassano, A., Pardoe, S., Gini, G., Mays, C., . . .Benighaus, L. (2011). The acceptance of in silico models for REACH: Requirements, barriers, and perspectives. *Chemistry Central Journal, 5*, Article 58. Retrieved from http://journal.chemistrycentral.com/content/5/1/58. doi:10.1186/1752-153X-5-58

Bernstein, B. (1990). *The structuring of pedagogic discourse.* London, UK: Routledge.

Bourdieu, P., & Wacquant, L. (1992). *An invitation to reflexive sociology.* Cambridge, UK: Polity Press.

Cameron, D., Frazer, E., Harvey, P., Rampton, B., & Richardson, K. (2006). Power/knowledge: The politics of social science. In A. Jaworski & N. Coupland (Eds.), *The discourse reader.* Abingdon, UK: Routledge. (Original work published 1992)

Caplan, N., Morrison, A., & Stambaugh, R. (1975). *The use of social science knowledge in policy decisions at the national level.* Ann Arbor, MI: The Institute for Social Research.

Chouliaraki, L., & Fairclough, N. (1999). *Discourse in late modernity: Rethinking critical discourse analysis.* Edinburgh, UK: Edinburgh University Press.

Davies, H., Nutley, S., & Walter, I. (2008). Why "knowledge transfer" is misconceived for applied social research. *Journal of Health Services Research & Policy, 13*, 188-190. doi: 10.1258/jhsrp.2008.008055

Doran, G. T. (1981). There's a S.M.A.R.T. way to write management's goals and objectives. *Management Review, 70*(11), 35-36.

Economic and Social Research Council. (2009). *Taking stock: A summary of ESRC's work to evaluate the impact of research on policy and practice.* Retrieved from http://www.esrc.ac.uk/_images/Taking%20Stock_tcm8-4545.pdf

Economic and Social Research Council. (2011, February 2). *Key Priorities & Future Strategy: ESRC Draft Delivery Plan 2011-15* (PowerPoint presentation by Adrian Alsop, Director of Research and International Strategy). Retrieved February 17, 2014 from www.rhul.ac.uk

Economic and Social Research Council. (n.d.-a). *About us: What we do.* Retrieved February 17, 2014 from http://www.esrc.ac.uk/about-esrc/what-we-do/index.aspx

Economic and Social Research Council. (n.d.-b). *Impact toolkit* (Impacts and findings: Impact toolkit). Retrieved February 17, 2014 from http://www.esrc.ac.uk/funding-and-guidance/impact-toolkit/

Economic and Social Research Council. (n.d.-c). *Step-by-step guide* (Impacts and findings: Impact toolkit: Developing a strategy:

Step-by-step guide). Retrieved February 17, 2014 from http://www.esrc.ac.uk/funding-and-guidance/impact-toolkit/developing-plan/index.aspx

European Commission. (2010). *Communicating research for evidence-based policymaking: A practical guide for researchers in socio-economic sciences and humanities.* Brussels, Belgium: European Commission Directorate-General for Research Communication.

Fahnestock, J. (1986). Accommodating science: The rhetorical life of scientific facts. *Written Communication, 3,* 275-296. doi: 10.1177/0741088386003003001

Fairclough, N. L. (1992a). The appropriacy of "appropriateness." In N. L. Fairclough (Ed.), *Critical language awareness* (pp. 33-56). London, UK: Longman.

Fairclough, N. L. (1992b). *Discourse and social change.* Cambridge, UK: Polity Press.

Fairclough, N. L., Pardoe, S., & Szerszynski, B. (2010). Critical discourse analysis and citizenship. In N. L. Fairclough, *Critical discourse analysis: The critical study of language* (pp. 412-436). Harlow, UK: Longman.

Goodwin, M. (2013, March 25). How academics can engage with policy: 10 tips for a better conversation (Higher Education Network, UK Blog). *The Guardian.* Retrieved February 17, 2014 from http://www.theguardian.com/higher-education-network/blog/2013/mar/25/academics-policy-engagement-ten-tips

Halliday, M. A. K. (1978). *Language as a social semiotic.* London, UK: Edward Arnold.

HM Treasury. (2004). *Science and innovation investment framework 2004-2014.* Norwich, UK: HMSO.

Hyland, K. (2003). Genre-based pedagogies: A social response to process. *Journal of Second Language Writing, 12,* 17-29.

Irwin, A. (1995). *Citizen science: A study of people, expertise and sustainable development.* London, UK: Routledge.

Ivaníc, R., Edwards, R., Barton, D., Martin-Jones, M., Fowler, Z., Hughes, B., Mannion, G., Miller, K., Satchwell, C., & Smith, J. (2009). *Improving learning in college: Rethinking literacies across the curriculum.* London, UK: Routledge.

Jasanoff, S. (2005). *Designs on nature: Science and democracy in Europe and the United States.* Princeton, NJ: Princeton University Press.

Jisc InfoNet. (2008). *SMART targets.* Retrieved from www.jiscinfonet.ac.uk/tools/smart-targets/

Latour, B. (1987). *Science in action: How to follow scientists and engineers through society.* Cambridge, MA: Harvard University Press.

Leigh Star, S., & Griesemer, J. R. (1989). Institutional ecology, "translations" and boundary objects: Amateurs and professionals in Berkeley's Museum of Vertebrate Zoology, 1907-39. *Social Studies of Science, 19,* 387-420. doi: 10.1177/030631289019003001

Mercer, N. (1995). *The guided construction of knowledge: Talk amongst teachers and learners.* Clevedon, UK: Multilingual Matters.

Morrison, M. (2011). *Why SMART Objectives don't work* (Online article for Business & Organizational Development tools, training & services—Human Resources, OD & Leadership [RAPIDBI]). Retrieved February 17, 2014 from http://rapidbi.com/why-smart-objectives-dont-work/

Myers, G. (2003). Discourse studies of scientific popularisation: Questioning the boundaries. *Discourse Studies, 5,* 265-279. doi: 10.1177/1461445603005002006

Nutley, S., Walter, I., & Davies, H. (2007). *Using evidence. How research can inform public services.* Bristol, UK: Policy Press.

Pardoe, S. (1997). "We're talking about reality": Vocational claims and the recontextualization of professional practice. In *Writing professional science: Genre, recontextualization and empiricism in the learning of professional and scientific writing, within an MSc course in Environmental Impact Assessment* (Unpublished PhD thesis). Lancaster University, UK.

Pardoe, S. (2011a, December). *Bridging the gap or just making a splash? Why funders need to recognise that generating impact beyond academia is as intellectually challenging as research.* Platform presentation at the ESRC conference Bridging the Gap Between Research, Policy and Practice: The Importance of Intermediaries in Producing Research Impact, RIBA, London, UK.

Pardoe, S. (2011b, June). *Collaborating in the cross-institutional communication of research insight: The challenge of informing reflective practice.* Platform presentation at the 2011 conference of Applied Linguistics and Professional Practice (ALAPP), The Health Communication Research Centre, Cardiff University, UK.

Pardoe, S., Cazzato, L., Golding, A., Benfenati, E., & Mays, C. (2011). *QSARs in REACH? Uses, issues and priorities* (Video documentary). Lancaster, UK: PublicSpace. Retrieved February 17, 2014 from http://www.ORCHESTRA-QSAR.eu/

Pardoe, S., & Mays, C. (2012). *Disseminating science and technology research: A practical e-guide for researchers.* Lancaster, UK: PublicSpace. Retrieved February 17, 2014 from www.researchdissemination.eu/guide/

Platt, J. (1988). What can case studies do? In R. G. Burgess (Ed.), *Studies in Qualitative Methodology* (Vol. 1, pp. 1–23). Greenwood, CT: JAI Press.

Research Councils UK. (n.d.). *Excellence with impact: Statement of expectation on economic and societal impact.* Retrieved February 17, 2014 from http://www.rcuk.ac.uk/Publications/archive/StatementofExpectationon/

Roberts, C. (1997). There's nothing so practical as some good theories. *International Journal of Applied Linguistics, 7,* 66-78. doi: 10.1111/j.1473-4192.1997.tb00105.x

Roberts, C., & Sarangi, S. (2003). Uptake of discourse research in interprofessional settings: Reporting from medical consultancy [Special issue]. *Applied Linguistics, 24,* 338-359. doi: 10.1093/applin/24.3.338

Scott, A. (2000). *The dissemination of the results of environmental research: A scoping report for the European Environment Agency.* (European Environment Agency Experts' corner report.) Copenhagen, Denmark: European Environment Agency.

Smith, F. (1985). A metaphor for literacy: Creating worlds or shunting information. In D. R. Olson, N. Torrance, & A. Hildyard (Eds.), *Literacy, language, and learning: The nature and consequences of reading and writing* (pp. 195-213). Cambridge, UK: Cambridge University Press.

Swales, J. (1990). *Genre analysis: English in academic and research settings.* Cambridge, UK: Cambridge University Press.

Tengblad, S. (2012). Overcoming the rationalistic fallacy in management research. In S. Tengblad (Ed.), *The work of* managers: Towards a practice theory of management (pp. 3-17). New York: Oxford University Press.

Van Leeuwen, T. (2005). Introducing social semiotics. London, UK: Routledge.

Van Leeuwen, T. (2008). Discourse and practice: New tools for critical discourse analysis. Oxford studies in sociolinguistics. New York, NY: Oxford University Press.

Weiss, C. H. (1980). Knowledge creep and decision accretion. Science Communication, 1, 381-404. doi: 10.1177/107554708000100303

Wynne, B. (2005). Risk as globalizing "democratic" discourse? Framing subjects and citizens. In M. Leach, I. Scoones, & B. Wynne (Eds.), Science and citizens: Globalization and the Challenge of Engagement (pp. 66-82). London, UK: Zed Books.

Author Biography

Simon Pardoe is Director of PublicSpace.ac.uk, a not-for-profit social enterprise specialising in the distillation and communication of publicly-funded research. He works with university research teams (as a project partner, co-author and video producer) to help to communicate the significance of their research to inform professional practice, policy and public debate. He was a teaching fellow, research associate, consultant and honorary research fellow at Lancaster University, U.K., following a PhD in applied linguistics with science and technology studies. He was previously a senior lecturer in business studies in inner London. He has taught, advised and researched in communication and academic and professional writing for 25 years.

6

Meaning Making Through Minimal Linguistic Forms in Computer-Mediated Communication

Muhammad Shaban Rafi[1]

Abstract

The purpose of this study was to investigate the linguistic forms, which commonly constitute meanings in the digital environment. The data were sampled from 200 Bachelor of Science (BS) students (who had Urdu as their primary language of communication and English as one of the academic languages or the most prestigious second language) of five universities situated in Lahore, Pakistan. The procedure for analysis was conceived within much related theoretical work on text analysis. The study reveals that cyber-language is organized through patterns of use, which can be broadly classified into minimal linguistic forms constituting a meaning-making resource. In addition, the expression of syntactic mood, and discourse roles the participants technically assume tend to contribute to the theory of meaning in the digital environment. It is hoped that the study would make some contribution to the growing literature on multilingual computer-mediated communication (CMC).

Keywords

computer-mediated communication, meaning, morphemic reduction, syntactic reduction, mood

Introduction

Computer-mediated communication (CMC) is proliferating in the lives of most people today. The ubiquity of CMC invites interesting debate on the phenomenon of meaning making or success of communication in the digital environment. As a result, communication theorists (Baron, 2008; Crystal, 2001, 2006; Herring, 1996; Thurlow, Lengel, & Tomic, 2004) have been investing considerable time to investigate the emergence of digital meaning. This scenario seems to be shaping if not determining many aspects of our real life, as has been argued by Turkle (2011). Herring, Stein, and Virtanen (2013) further assert that the Internet mirrors co-construction of the self in an ad hoc manner. The transformation of "the self" from real to virtual began with the invention of electronically mediated communication, such as through the telegram, telephone, fax, and so on, phenomenon that is almost a century and half old. However, the virtual self has attracted much attention only in recent years. This transformation manifests an evolution of meaning-making self, which is expressed through such mediated communication.

One assumes that there is a good reason to believe that the expression of meaning is fundamental to language. The present study hypothesizes that certain mechanisms, primarily morphemic and syntactic reduction, are one of the ways used to achieve this end. We also claim that the use of such minimal mechanisms, more obviously in the context of CMC, is not random, but rather, it is necessitated by the expression of meaning under the special circumstances in question. It may sound as though we hold deterministic attitude toward the relationship between form and meaning, but our intension is simply to claim the priority of meaning over form.

Much work by the researchers of CMC (Danet & Herring, 2007; Segerstad, 2002; Tagg, 2009; Thurlow & Poff, 2013, as well as those referred previously) shows a split approach, either taking formal reduction or minimization to be an act of linguistic transgression or considering this kind of variation to be an inherent property of language, as had always been evident in the nature of historically older forms of reduced language, for instance, diary register (Weir, 2012, and the references cited therein). Crystal (2006) asserts that CMC heralds novel manifestations of this property mirroring linguistic behavior, in ways primarily different, from the traditional modes of communication as has been noted by Bodomo (2009). Rafi (2010), arguing on the same lines, states that Internet users are habitually and increasingly customizing language to capture their experiences and to express their e-identity through various linguistic innovations.

While discussing grammatical features of English, Ko (1996) finds that electronically governed communication typically involves relatively short or shortened words. While

[1]University of Management and Technology, Lahore, Pakistan

Corresponding Author:
Muhammad Shaban Rafi, University of Management and Technology, C-II, Johar Town, Lahore 10033, Pakistan.
Email: shabanrafi@hotmail.com

explaining these features, he shows that the use of present tense, coordinations, adverbials, demonstrating pronouns, and intensifiers is used more frequently in real-time or online communication than in any other type of discourse. However, grammatical forms such as prepositional phrases, relative clauses, and perhaps subordinate clause in general are comparatively infrequent in CMC discourse. Furthermore, he claims that discourse markers and hedges are, likewise, less common and the patterns of turn taking are far from neat and orderly. He also mentions this proportionately the huge role that icons play in CMC. Kalman and Gergle (2009, 2010) assert that repetition of punctuation and letters indicates the stretching of a word, emulating a stretched out syllable like in spoken conversation. Furthermore, they explain that these repetitions tend to communicate tempo, pitch, prosody, and other paralinguistic elements for achieving visual emphasis. Reporting similar results, Herring (2011) additionally points out the prevalence of "nonstandard" typography and orthography practices. Danet and Herring (2007) note that research on other languages has exhibited similar tendencies.

Probably, it is no longer the case that the Internet communication is predominantly in English. Bodomo (2009) asserts that bilingualism/multilingualism has now become the norm in CMC. As the Internet has been increasingly becoming multilingual, researchers have explored new patterns of use and language combination in bilingualism/multilingualism communities (e.g., Androutsopoulos, 2007; Axelsson, Abelin, & Schroeder, 2007; Barasa, 2010; Durham, 2007; Paolillo, 2007; Seargeant, Tagg, & Ngampramuan, 2012; Warschauer, El Said, & Zohry, 2007). The Urdu spoken in Pakistan, largely adopting the linguistic ecology of English (Rafi, 2013), is case in point. Thus, the use of Urdu on the Internet is always embedded within a larger Anglophone context. Not infrequently involving switches between the two languages or substitution in writing of an English letter like *b* for the whole corresponding semi-homophonous Urdu word more specifically یھب /b ʰ iː/ which means "also." The substitution noted above does not lead to any change in meaning; it does, though, just help the users make meaning by means of minimal linguistic forms.

In general, the present study intends to investigate forms that are reduced or have unique configurations, and which are exploited to project the same range of meaning as corresponding full forms. In section "Method," the methodology and theoretical as well as the linguistic backgrounds are outlined. The remaining sections except the concluding one present an analysis of the data.

Method

Much like researchers dealing with any kind of data, researchers who deal with digital data are often confronted with a variety of non-trivial questions. Most of them relate to the size and representativeness of data samples, data processing techniques, limitations of genres, kind and amount of

necessary contextual information, and ethical issues, such as anonymity and privacy protection. Keeping such questions in mind, this section describes the procedures of data collection, ethical considerations of collecting and handling data, data analysis, and the theoretical underpinnings.

Data Collection

The first stage of data collection consisted of the selection of a sample of appropriate participants. The sample included both male and female students who were between 18 and 24 years of age and were registered in the Bachelor of Science (BS) program in five different universities situated in Lahore, Pakistan. A pilot study suggested that this cohort was indeed suitable for the investigation under discussion (see, for example, Ilyas & Khushi, 2012; Leppänen, 2007; Rafi, 2008, 2010, 2013). This study assumes that CMC is typical of this age group almost anywhere in the world.

According to Grinter and Eldridge (2001), young people bring with them to college a well-developed practice of e-styles. As social networks found relatively young people its potential users, the present study assumes that they know how to accommodate and appropriate the English language in CMC. Jørgensen (2001, as cited in Leppänen, 2007) argues that youth can take language in their own possession and typically active in generating more general language change. Researchers (e.g., Leppänen, 2007; Paolillo, 2011; Peuronen, 2008) assert that there is always association between being young and using new patterns of English along with the native language.

Facebook being popular forum for conversations among young people was assumed to be a good source of the data (cf. Boyd & Ellison, 2007; Raskin, 2006; Sengupta & Rusli, 2012). The data collection was confined to Facebook wall, which gives easy access to enormous subject pool, without violating anyone's privacy. Furthermore, it is characterized by the use of a broad range of traits at the very informal end of the linguistic spectrum like in a natural setting, which is the primary focus of the present study.

Sample. The sample was drawn from the five private (by and large not funded by the State) institutions of higher learning listed in Table 1. The reason to choose these institutions was to make the sample as representative of BS students as possible. In terms of academic, linguistic, and cultural background, the sample was homogeneous. All the participants were from BS program but strictly speaking in different disciplines (e.g., Business of Administration, Engineering, Computer Sciences, and English) and academic years (e.g., covering first year to fourth year). As the participants were between 18 and 24 years of age, they might be considered young people. Youth is defined here as a chronological age—number of years since birth (cf. Leppänen, 2007). Also, communication on Facebook was more or less within the same group. Notwithstanding the possibility of this forum to

Table 1. Distribution of Sample.

Institution	Number of participants
COMSATS Institute of Information and Technology	50
National University of Computer and Emerging Sciences	21
Superior University	50
University of Lahore	50
University of Management and Technology	29
Total	200

Table 2. Demographics and Nature of Data.

I. Total number of participants	200
a. Average age	21 years
b. Average time spent on Facebook in a week	6.25 hr
c. Average number of followers of each participant	150
II. Total number of linguistic postings in the data	
a. Roman Urdu	588
b. English	1,135
c. Mixed Roman Urdu and English	793
III. Total number of words in the data	27,476
a. Range of words in a posting	1-612
b. Average number of words in a posting	137

connect people from different linguistic backgrounds, the participants in this study were evidently Urdu/English bilinguals. It should be added that, strictly speaking, the first language of most of them was Punjabi but Urdu was their primary language of communication, both at home and elsewhere, and English was their most important academic language and the most prestigious second language. They acquired Punjabi, however, through informal contact with, for example, lower or middle lower working class—who speaks it maybe to mark solidarity and identity. They used to meet with each other both in offline and online contexts. Moreover, their communication was not only limited to Pakistani contexts but it could be referred to international settings also. The participants linked and shared activities covering information exchange, debate, problem solving, exchanging picture, jokes, and video related to varying themes, such as greeting, politics, showbiz, sports, and sex. Peuronen (2008) argues that these activities "best describe their being" (p. 104). Their communication in online context comprised new creative ways. Importantly, they identified themselves mostly either through Urdu or English or both. The data were mainly in Romanized Urdu, English, or a mixture of Urdu and English. The corpus used was simply in multiple languages, all written in the same (Roman) script. Tone used in the message threads was typically informal; however, a number of snippets seemed to be formal too. A lot of words that the participants used were clearly from their specific academic context.

Five volunteers who were also students at these institutions were engaged for help during the process of data collection. Each one of them coordinated the data gathering process for one of the five institutions. The researcher shared with them the purpose and ethical boundaries of the study. Each of them managed to add on average 375 students over a period of 2 months. Thus, the researcher had access to all the students through these volunteers. The data from only 50 participants from each institution were analyzed on the basis of the quantity of their posting output. Table 1 shows the distribution of the sample over the five institutions. The study investigated linguistic postings of each participant transmitted over the period of a week. The logic behind collecting the whole week's data was to

observe most of the linguistic features that the participants used in their communication. Each posting consisted of usually more than one utterance. Thus, the data collected were naturalistic and observational, with minimum interference from the researcher. However, the participants were asked questions to seek clarifications where it was absolutely necessary.

Nature of the data. As shown in Table 2, the data were collected from 200 participants whose average age was 21 years. The data indicate that each participant spends on average 6.25 hr on Facebook per week and hosts on average 150 followers, thus creating the likelihood of feedback on his or her posting. Of 2,516 linguistic postings, 588 are in Romanized Urdu; 1,135 are in English; and the remaining 793 postings are a blend of both Urdu and English. This shows an increasing trend of English along with mixed Urdu and English in digital discourse. As is evident from Table 2, the participants have shown less and marginal need for Urdu as a medium of communication. The most plausible reason behind the marginal use of Urdu is perhaps the importance given to English in academia. Importantly, English is their most prestigious second language. They come across relatively fewer opportunities to use written Urdu in their academic context, for example, examination, project, assignment, and so on. On average, each posting consisted of 137 words ranging between 1 and 612 words across the data. As many as 27,476 words are accumulated for the analysis.

Much like face-to-face conversation, communication between the participants reflected physical approximation. The illustrations in [1] confirm that the distant or even near future marking system is marginal in the context of CMC. Overall, the data included the postings, which demonstrated participants' relationships regarding how close they were with their followers.

[1]

a. <*tum kahn gum ho bhai . . . ?*> (Where are you brother?)

b. *<Guys kia plan hai iss Sunday ka ??>* (Guys, what is plan on this Sunday?)

c. *<wat an awsome dance . . . !!!>* (What a dance it is!)

d. *<girl . . . is an innocent nd beautiful creature . . . which only deserves to be luvd . . . ♥>* (A girl is an innocent and beautiful being who deserves only to be loved.)

e. *<Gud Morng To all my chintoo mintoo frndzzZZzzzz:)>* (Good morning to all my dear friends.)

f. *<2mrw morning em going near xxx on xxx mangni . . . yahuuuuuuuu>* (I am going to attend xxx's engagement near xxx tomorrow.)

Types of exchanges. Interactivity is a defining characteristics of CMC, which is organized around topics or threads. Barnes (2003) defines that "a thread is a chain of interrelated messages that respond to each other" (p. 20). A linguistic posting on the wall can be generally described as fragments, phrases, and clauses. The postings on the wall evolved in a coherent whole around topics of varying themes. Mostly, a topic was initiated by a single person followed by his or her followers' comments, which was either closed with words of thanks or reflection from the originator or left without proper closing. The exchanges on the wall can be classified into a three-part structure of Initiation, Response, and Reflection (IRR). This IRR structure can be seen in the following example:

[2]
A: <newly made my own pic>
B: <its awsome bro>
A: <thanks>

Unlike IRR structure, instances of dialogue structures, two-part adjacency pairs (e.g., question–answer), and non-linear flow of conversation were also prevalent. The dialogue structure and adjacency pair can be seen in [3] and [4]. The postings in the data were extrovert in nature and each provided a framework within which the next was formulated.

[3]
A: <sick of formatting now . . . (>
B: <m sick of dissertation now>
C: <Ahhhh!!>
D: <aww!!! my poor baby>
E. <O hoo . . .)>
A: <there is nothing more killing than formatting more than 50 times line spacing . . . margins . . . pagination . . . ToC, LoT, LoF . . . wot not and wot not . . . there is nothing more mechanical and boring than this . . . > :(
F: <do not be fed up now . . . its a continuous process of accomplishing task.>
A: <have been taking breaks and delaying it . . . first deadline is over now wanna get it over and forever>

A: <i hate this document now . . . dont wanna even open it up>
[4]
A: <wt r u doing now a days?>
B: <BS social sciences 4m xxx.>
A: <how abt xxx?>
B: <he/she too with me.>

Although Facebook is thought to be an asynchronous technology and it often transmits messages in near-real time, many participants were observed replying instantly to postings, rendering the technologically asynchronous medium effectively synchronous. Thurlow and Poff (2013) argue that with the convergence of new (and old) media, technological boundaries and generic distinctiveness of instant messaging, texting, and emailing are becoming blurred. Notable examples of this are found in micro-blogging (e.g., Twitter and status updates on Facebook), the multi-functionality of smart-phones (e.g., BlackBerry) and to some extent, Apple's iPhone. This study assumes that dividing Facebook conversations on the wall between synchronous and asynchronous modes may be misleading because the speed of communication mostly inhibits us to draw a clear line between synchronous and asynchronous communication. As there is a range of possibilities, we may consider the participants' conversation on the wall as falling between synchronous and asynchronous modes. But strictly speaking, this study considered Facebook communication on the wall analogous to asynchronous communication.

The participants used to grid a kind of coherent discourse because most of their postings were in response to a single individual's reflections. It was found that communication on the wall usually occurred between one-to-one or one-to-many—a kind of interpersonal communication. Barnes (2003) reinforces the fact that the Internet can be used to distribute messages in any number of directions, that is, one-to-one or one-to-many.

Ethical considerations. Ensuring that pre-existing ethical standards were properly met was crucial. In this study, we adhered to guidelines suggested by Mann and Stewart (2000) while collecting data.

The participants were informed about the nature of the study. They were given assurances regarding confidentiality, security of information, and unauthorized eavesdropping; that is, information that might identify the participants, places, institutions, and times was never to be disclosed. The participants were identified by means of cryptonyms in reporting research. Access to the database was restricted to the participants and the researcher. However, the researcher could not forbid the use of racist and sexist language, and other contentious and provocative material. Mann and Stewart (2000) argue that Internet research does not have to conform to these restrictions.

Data Analysis

To attempt the research question, minimal linguistic forms and what they communicate in terms of mood were analyzed. Jørgensen, Karrebæk, Madsen, and Møller (2011) remark that linguistic features are best suited as the basis for the analysis of language in polylanguaging context. These forms were classified into lexical and syntactic tiers. Word reduction was further subclassified into abbreviations (e.g., *gf* for "girlfriend"), clippings (e.g., *pic* for "picture"), and logograms (e.g., *some1* for "someone"). For clarity, logograms were further subclassified into phonetic spelling (*u* for "you"), lexo-numeric (*gr8* for great for "great"), digito-lexeme (*2morow* for "tomorrow"), and digit word semi-homophone (4 for "for"). The purpose of these classifications and subclassifications was to tabulate precisely the lexical features for measure of frequency. Frequency was calculated to gauge how many times a particular feature occurred and to suggest its permanence in CMC. The frequency of occurrence was determined if at least a word was repeated twice within a conversation and a minimum of 5 times in the whole data.

What CMC-unique structures are, and how they uncover functional roles, I examined (a) types of structures, (b) deletion of grammatical features, and (c) syntactic expression of mood. Types of structures were further classified into fragmented structure, simple structure, compound structure, and complex structure. To analyze whether or not the proportions of structure type are different across the languages, I calculated chi-square test of independence. As the characteristic of linguistic reduction was assumed as one of the key features across the corpus, grammatical properties, which were commonly deleted, were measured. This measure provided us with further insight to assess how far the participants compromised on structural rules when communicating in digital environment. Furthermore, a careful analysis of structures constituting mood (e.g., declarative, interrogative, imperative, and exclamatory) was examined to uncover functional roles and negotiation common to Facebook.

In addition to analysis of quantitative data, I also drew on the message threads to elaborate and support my verdict concerning the research questions, which helped bring triangulation to increase the credibility and validity of the results. These snippets were in English, Romanized Urdu, and mixed English and Urdu. They were demonstrated within mathematical symbols (such as < >) along with their transliteration in the parentheses. In relation to paralinguistic features, I considered their visual aspects only as a meaning-making resource in a message thread. Thus, the analysis was backed up by a large corpus covering both quantitative and qualitative data sets, which assisted in attempting the research questions with more extensive analysis, with far more participants and more rigorous sampling procedures.

Theoretical underpinning. The procedure for analysis was conceived within much related theoretical work on text analysis. Sinclair (1991, as cited in Stubbs, 1996) argues that intuitive judgments are particularly untrustworthy with respect to frequency and distribution of different forms and meanings of words, and to the interaction of lexis, grammar, and meaning. In the present study, against the reliance on invented data, example utterances were explicitly taken from the data. The accountability to data was thus taken into account. Unlike contemporary studies, which are based on a small data set, a large corpus based study was planned to address the underlying research question. The actual language text duly recorded was the main concern in the present study. In addition, interpretation of cyber-linguistic features was based on the whole data rather than relying consistently on a few examples.

There is reason to believe that a few linguistic features of a text are distributed evenly throughout. To encompass the whole data, the study recorded the frequency of commonly occurring minimal lexical as well as syntactic features across the corpus to draw emerging patterns. One of the main uses of the corpus is to identify what is central and typical in the language (Sinclair, 1991, as cited in Stubbs, 1996). Thus, linguistic patterns that reside in a language can best be judged when they are studied comparatively across the text corpora.

As noted above, the data were accumulated from five groups. The purpose behind the selection of different cohorts was to have a comparative standpoint whether linguistic features were spread across the board or concentrated in a particular cohort. Overall, the data were organized into lexical and syntactic tiers. Stubbs (1996) argues that the most powerful interpretation emerges if comparisons of texts across corpora are combined with the analysis of the organization of individual texts. Another approach would have been to view linguistic features in several different media genres (e.g., email, social networking, and text messaging), though compatible with this principle, seems to motivate relatively different linguistic choices (see, for example, Baron, 2008; Crystal, 2001, 2006; Thurlow & Poff, 2013) on formal and/ or informal end of the linguistic spectrum. Whereas the primary focus of this study was to collect naturalistic and observational data that were limited to Facebook communication on the wall.

Every language is governed by linguistic selections and restrictions. For example, both Urdu and English pursue linguistic conventions to transform and transmit a structure. We can, of course, say how are you? or *Ap ka kia hal hai*? But we may not say *you are how* or *Ap hai kia hal ka*. Hence, grammatical structure restricts the lexis that occurs in it, and conversely any lexical item can be specified in terms of the structures (Stubbs, 1996). The study presupposed that the linguistic flexibility that young people exercised in the mediated communication was based on the system of language more or less; however, it was the meaning that governed their apparently peculiar linguistic choices.

Semantic analysis of corpus has been of main concern in the Firthian school of thought. As Firth (1957, as cited in Stubbs, 1996) puts it, "You shall know a word by the company it keeps" (p. 11). Firth's notion of form and meaning in context was extended by Halliday (2002, 2003, 2004, 2005, 2007) in systemic linguistics. Systemic linguistics adopts a descriptive approach to language investigation that answers the questions: What is language? and How does language work? In the same way, the present study gives importance to the description of the characteristics of cyber-linguistic features. It may sound as though we hold deterministic attitude toward the relationship between form and meaning, but our intension is simply to claim the priority of meaning over form.

Minimal Linguistic Forms

Word Reduction

Linguistic reduction was the most common feature in the data. The participants mostly reduced linguistic forms from Urdu and English. However, it was English which acquired the impact of linguistic reduction the most. They minimized, seemingly either through conscious or subconscious efforts, words by compounding a number and lexeme or morpheme, for example, *some1* for someone, *gr8* for great, *b4* for before, and so on. Words that are composed of two syllables in which one syllable is substituted with a digit and the other remains constant may be called *lexo-numeric*. Conversely, the participants also customized words, may be labeled *digito-lexeme*, by substituting segment or segments of a word with a digit at the onset, for example, *2morow* for tomorrow, *4get* for forget, and so on. Among other categories of reduction, *digit word homophones* were prevalent, which were formed by replacing a full word with a digit, for example, 1 for "one," 2 for "to" or "too," 4 for "for," and so on. As many as 1,106 words were tabulated in these categories, which can be grouped under the basic term, namely, logogram. Thus, the logograms, in many ways, were the most frequently occurring features in cyber-communication of young Pakistani students.

The initial ambiguity that might have been as a result of minimal linguistic forms was seemingly compensated, if not altogether but mostly, through extralinguistic features. Apart from this, linguistic reduction can further be explained within the system of language. Furthermore, segmental reduction in English and Urdu is systematic, which indicates supposedly uniform patterns for the success of communication.

Reduction in the segments of English words can be characterized by phonological and morphological properties. As shown in Table 3, the derived forms conform to fixed phonological patterns, which are characterized by templates in which "C" stands for consonant sound and "V" represents vowel sound. Moreover, bold letters, for example, "**C**" stands for the sound that was picked up while committing the reduction in the original word. As is evident from the given template,

Table 3. Systemic Orthographic Reduction.

Base form	Template	Derived form
Am	V**C**	*m*
And	**V**C**C**/V**CC**	*n*/*nd*
Be	**C**V	*b*
Because	**C**V**C**V**C**	*bcz*
But	**C**V**C**	*bt*
Comment	**C**V**C**V**CC**	*cmnt*
Good	**C**V**C**	*gd*
Hospital	**C**V**CC**V**C**V**C**	*hsptl*
Love	**C**V**C**	*lv*
Now	**C**V**C**	*nw*
Please	**CC**V**C**	*plz*
tension	**C**V**CC**V**C**	*tnshn*
Thanks	**C**V**CCC**	*thnx*
That	**C**V**C**	*tht*
Was	**C**V**C**	*ws*/*wz*
What	**C**V**C**	*wt*

vowel sound or vowel-like sound is more susceptible to reduction. Vowel sounds seem to behave like the fillers in consonant sounds. We may speculate that English words can be read without vowels exerting a little effort. Conversely, there is no instance of reduction in consonant sounds; however, the consonant sounds which were reduced are primarily from consonant clusters. Hence, the consonant sound was used to cover the rest of the segment in a word. Moreover, diphthongs and weak vowel sounds, especially /ə/ to show a sound occurring at various segmental positions in a word, were occasionally truncated. As highlighted in the given template, it is the consonant sound in a word which the primary carrier of a meaning is. This can be compared with early Semitic languages such as Arabic and Hebrew in which the words' roots were isolated sets of consonants, for example, in Arabic the root word meaning "write" had the form *k–t–b*. It signifies that the part of linguistic principles have always been recycling since the first utterance was articulated as has been supported by Napoli and Lee-Schoenfeld (2010) who assert that our speech contains archaism—a little fossil from the past.

As noted above, the participants mostly reduced English forms; however, they simply substituted bi-syllabic Urdu words with the whole corresponding mono-syllabic English homophones or semi-homophones—may be considered a creative way of reducing morphemic properties of Urdu primarily in Romanized script. While substituting Urdu forms with its English counterparts, the overextended sound had no or less homophonous correspondence in some instances. It is evident that the substitution does not cover aspirated sound because this requires use of superscript [ʰ] that the participants simply avoided.

[b], pronounced as /bʰi/ is a semi-homophone of بھی which means "also"

Table 4. Substitution of Urdu Alphabets with Corresponding English Phoneme.

Urdu alphabets	English phoneme
ا، ع	/a/
ث، س، ص	/s/
د، ڈ	/d/
ر، ڑ	/r/
ذ، ز، ژ، ض،ظ	/z/
ت، ث، ط	/t/
ق، ک	/k/or/q/
غ، گ	/g/
ح،ہ، ھ	/h/

[c], pronounced as /si/ is a homophone of سی which means "also" or "like" or "of" or "for"

[i], pronounced as /ai/ is a homophone of ائی which means "coming"

[g], pronounced as /dʒi:/ is a homophone of جی which means "yes"

[k], pronounced as /keɪ/ is a homophone of کے which means "that"

[q], pronounced as /kjuː/ is a homophone کیون which means "why"

The logic behind these substitutions is that all the above-mentioned Urdu bi-syllabic words have the same or nearly the same sound in English alphabets. It seems as a matter of ease the participants replaced bi-syllabic forms with their mono-syllabic counterparts in their conversations. Similarly, they also substituted the following English words with their counterparts in Urdu.

[a], is pronounced as /ɑ://is a homophone of آ which means "come"	[may], is pronounced as /meɪ/ is a homophone of مین which means "in"
[gay], is pronounced as /geɪ/ is a homophone of گے which "went" or "past forms of will (would)"	[or], is pronounced as /ɔ:(r)/ is a homophone of اور which means "more")
[he/hi], is pronounced as /hi:/ is a homophone of ہی which is "intensifier"	[pass], is pronounced as /pæs/ is a homophone of پاس which means "near"
[her], is pronounced as /hɜ:(r)/ is a homophone of ہر which means "every"	[pay], is pronounced as /peɪ/ is a homophone of پے which means "on"
[key], is pronounced as /ki:/ is a homophone of کی which means "what"	[say], is pronounced as /seɪ/ is a homophone of سے which means "from"
[log], is pronounced as /lɒg/ is a homophone of لوگ which means "people"	[such], is pronounced as /sʌtʃ/ is a homophone of سچ which means "true"

Table 4 shows that another very interesting phenomenon, may be called reduction, is the substitution of Urdu graphemes with English phonemes as has been noted by Rafi (2013) and Ahmad (2011). There are 38 alphabets in Urdu but in Romanized Urdu 66% of them are reduced to 24%. Around 58% of Urdu alphabets find seemingly semi-homophonous corresponding letters in English. Mostly involved substitutions, between Urdu alphabets, for example, ق، ک، /ر، ژ، ع،/ا،/ د، ڈ، / and گ،غ/ and English phoneme, for example, /a/, /d/, /r/, /k or q/, and /g/, respectively. However, switches between two languages in writing Urdu letters, for example, ح،ہ، ھ/ and /ت، ث، ط/ ،ذ، ز، ژ، ض،ظ/ /س، ص ث،/ for the whole semi-homophonous English phoneme, for example, /s/, /z/, /t/, and /h/, respectively, were not infrequent. Surprisingly, the participants usually reduced English vowel sounds; however, when Romanizing Urdu, they preferred to minimize mostly if not always Urdu consonant sounds. More or less, these instances seem to be a reflection of linguistic accommodation, which can be used to support the verdict that CMC is a new discourse. Trudgill (2003) defines accommodation as "the process whereby participants in a conversation (usually face-to-face interaction) converge their accent, dialect, or other language characteristics according to the language of other participant(s)" (p. 3). Although convergence of Urdu graphemes with their English counterparts is apparently due to Roman transliteration of the Urdu language, this leads us to reveal that Urdu alphabets are reduced more or less to the size of English which are supposedly compensated with normal key stocks—to avoid complex application (shift, alt, and shift & alt) that these graphemes might have required. There is a fair chance that this trend may continue and consolidate with the present keyboard features (see, for example, Sperlich, 2005). Investigation of technological limitations and their impact on languages have access to Internet can be an interesting study to gauge how widespread the phenomenon is! However, investigation of this dimension is beyond the scope of the present study.

The second most noticeable lexical feature is the reduction of words to their initial letters. They are known as abbreviations. Abbreviations are commonly used and an accepted trend of writing the initials of words to make a new word that is not essentially pronounceable. However, there are instances of abbreviations, which obey phonetic property, for example, *AFAP* for as far as possible, *SOB* for son of bitch, *Yo* for your own, and so on. Abbreviations are usually spelled in capital letters; however, there are cases where the strings of words were abbreviated in lower case letters also. Like reduction in the base forms, abbreviations also involve loss of material. The principle of orthography is, however, of central importance in abbreviations, for example, *dp* for digital picture, *gf* for girlfriend, *np* no problem, and so on. As mentioned above, in limited cases, phonetic properties were also applied to derive abbreviations known as acronyms, for example, *asap* for as soon as possible, *lol* for lot of laughter, *afap* for as far as possible, and so on.

There are words that were derived from the first part of the base word. This process is labeled as clipping. Clippings appear as a mixed bag of forms reduced from base forms, which express familiarity with the denotation of the derivative. Thus, *pic* was used typically by the participants to refer to digital image and *bro* was part of their vocabulary to show probably an intimate relationship. There are clippings, for example, *add/addy* for address, *cos* for because, *fav* for favorite, *grats* for congratulations, *del* for delete, *dif* for different, *jus* for just, *moro* for tomorrow, *rehi* for hello again, *uni* for university, and *web* for website, which are characteristic features of cyber-communication.

Another unique way of segmental reduction is "g" omission in words ending with "g." Perhaps, the participants thought it an additional sound while texting words ending with "g." In native-English especially in some parts of the United Kingdom, for example, Norwich, Cardiff, and elsewhere, "g" dropping in colloquial speech is pervasive, as if there is simply an [n] on the end. This is indicative of some interesting facts such as migration of "g" dropping in the conversation of the participant that can be used as a correlate of foreign culture. This seems to explain that the so-called linguistic borders are melting down and spreading linguistic forms of the dominant language, which is English in our case, through new media communication.

Types of Structures

The participants used typically fragments and sentences (i.e., simple, compound, and complex) to carry out their conversation. Fragment can be referred to a segment of a sentence that may contain a single word or a string of words. The examples [4a-4k] indicate fragments consisting of a single word and a string of words. Three dots on either sides of the structure indicate a range of missing syntactic elements, for example, head word, verb (especially auxiliary verb), and complement. However, a sentence can be classified into a simple sentence (that contains a subject and a predicate, and it expresses a complete thought), a compound sentence (that contains two independent clauses joined by a coordinator), and a complex sentence (which consists of an independent clause joined by one or more dependent clauses). The snippets [4l-4n] show simple structures consisting of a subject and a predicate. The example [4o and 4p] highlights types of compound and complex structures present in the corpus. Table 5 shows that the use of fragmented and simple structures is common. The data show 45% use of fragments followed by 44% use of simple structure across the corpus; however, compound and complex structures constituted relatively a low percentage. Chi-square test of independence shows *p* value (.03), which is less than .05; we can reject the null hypothesis, and say that types of structures and languages are related. I have further explained the association between types of structures and languages in the following section.

Table 5. Frequency/Percentage of Types of Structures.

Language	Fragment	Simple	Compound	Complex
Urdu	231 (6%)	554 (16%)	64 (2%)	43 (1%)
English	870 (25%)	713 (20%)	66 (2%)	110 (3%)
Mixed Urdu and English	493 (14%)	283 (8%)	53 (1%)	68 (2%)
Percentage	45	44	5	6

Note. Chi-square test of independence (*p* = .03 < .05).

[4]
a. < . . . nice . . . >
b. < . . . Vry nice . . . > (Very nice.)
c. < . . . really nice . . . >
d. < . . . enjoyed . . . >
e. < . . . movie day . . . >
f. < . . . so sad . . . >
g. <*i wil* . . . :P> (I will . . .)
h. <will have more fun . . . >
i. < . . . definitely will do . . . >
j. < . . . tired and sick . . . >
k. < . . . breathless moments . . . hooooh>
l. < . . . *buss* copy paste *mera hai* . . . :-p> (. . . just have done copy and paste . . .)
m. <Baray log late aatey hain :p> (The prestigious people get late.)
n. <U r in my bradri too . . . hahaha> (You are my caste-fellow too . . . hahaha)
o <I think its first time that Pakistan and UK are celebrating Eid on same day>
p. <yeh dnt get me started on Morroco that and Tunisa is confusing!>

Syntactic Reduction

As discussed above, structural properties of English were found to be more susceptible to deletion. Nevertheless, the participants were relatively diligent about structural properties of Urdu. Figure 1 indicates syntactic constituents, which were regularly deleted. The structural properties of English, for example, pronoun and verb, were more frequently deleted than article, preposition, and inflection. Deletion of pronominal especially first person "I" was common. The participants frequently deleted "I" where it was preceded by an auxiliary "am." They occasionally truncated "I am" to "am" or "m" or simply omitted "I," for example, . . . *will do asap . . . m having my exams!! . . . hope u understand!! . . . love you mwaaaah*. Insertion of three dots signifies deletion of pronoun "I." Similarly, they omitted auxiliary verbs, such as *is, are, am, was*, and *were* in their conversations. Given the structural deletion, there was omission of capitalization in the beginning of an utterance or in the case of proper nouns and addition of toggle case. As noted above, only a few

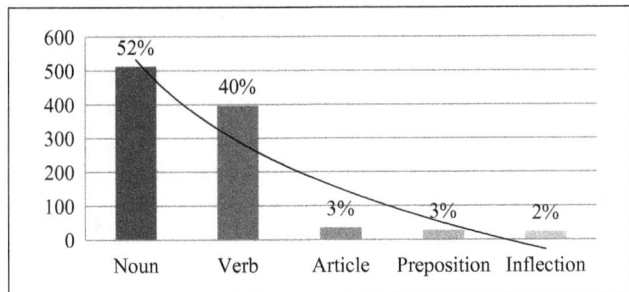

Figure 1. Frequency/percentage of structural properties deletion in the mediated communication of Urdu/English bilinguals.

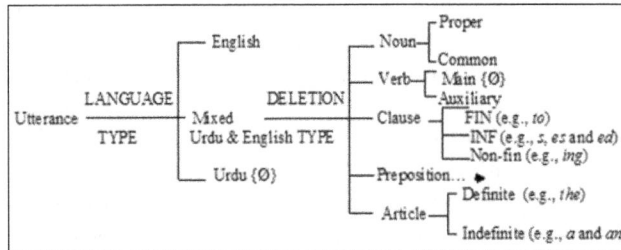

Figure 2. The network representation of structural deletion in the linguistic repertoire of Urdu/English bilinguals.

instances of cohesive ties were found across the corpus. As a result of this, communication was more a reflection of short structures then supported by linguistic nexuses. The omission of linguistic features shows that the participants might have used their pragmatic knowledge to presuppose that the receiver would know how to map out deleted expressions in their utterances.

The snippets in [5] show that utterances, even though shortened, are coherent and meaningful. Apart from obligatory words, the participants omitted optional words by assuming that obligatory words might be sufficient to express meanings. However, the English language resists omission of obligatory strings of words because they hold meaning and omission may damage intelligibility. In my observation, the deictic expressions, main verb, and attributive forms seem sufficient for the projection of deleted string of words. There is reason to believe that even though their communication was not structurally rich, concurrent exchanges revealed somehow that the participants might have inferred the deleted string of words for the success of communication. It is logical to argue that instead of structure, meaning was assumed to be a primary source of carrying a message. How minimal linguistic forms construe syntactic expression of mood and overall relevant to the theory of meaning making is examined in the next section.

[5]

a. <*yea* I *wid* U> (Yes, I . . . with you.)

b. <*examz* over . . . > (Exams . . . over . . .)

c. <*i* think paper little bit short> (I think . . . . short paper.)

d. <you so *qute*> (You . . . so cute.)

e. <*lyf goin gr8*> (Life . . . going great.)

f. <me cooking food for u> (. . . . . . cooking food for you.)

g. <me playing> (. . . . . . playing.)

h. <one of my favorite songs!> (. . . . . . one of my favorite songs!)

i. <awesome lyrics!!> (. . . . . . awesome lyrics!)

j. <sometimes nothing to say> (Sometime . . . . . . nothing to say.)

k. <more options?> (. . . . . . more options.)

l. <Sorry network problem> (. . . . . network problem.)

m. <now *feeLing bettEr* after> (Now . . . . . . feeling better.)

n. <not *geTTing sleEp* . . . :(> (. . . . . not feeling asleep.)

o <*BoRINg LIFeeeE* . . . !!!!! :(just *siTting alone* . . . :(:(:(:(:(:(:(:(:(> (. . . . . boring life . . . . just sitting alone.)

p. < . . . really missing u :-(:-(:-(> (. . . . . . really missing you.)

q. <this *iz make* by me> (This is make (past participle) by me.)

r. <she *like* it> (She like (inflectional "s") it.)

s. <Please not disturb> (Please . . . not disturb.)

t. <*u* not understand my *prob* . . . > (You . . . not understand my problem . . .)

Figure 2 reflects a classification of commonly deleted syntactic forms in the linguistic repertoire of Urdu/English bilinguals. The sign {Ø} in the figure nuances for the absence of a feature. As noted above, we did not find deletion in grammatical features of the Urdu language apart from superficial realization of Urdu graphemes. As shown in the figure, it is the English language that seems to give provision to deleting structural properties in digital discourse. It seems that the participants might have presupposed, of course to a certain extent, linguistic expressions which were deleted regularly. There is a fair reason to believe that the structural nativization of English is now read in the local context as supported by Michael (1993). In the next section, we will be concerned with the communicative intention of the types of structures that we have described in this section.

Syntactic Expression of Mood

In attempting to express themselves, the participants did not only exchange utterances containing the set of rules, they also performed structural moods. In many ways, they juxtaposed linguistic features for the success of communication. For instance, in [6a], imperative structure was used along with the insertion of exclamatory tone to enquire. Similarly in [6b and 6c], declarative structure was used to enquire. In [6c], use of a question mark at the boundary indicates that the participant had intention to enquire. In [6d], the structure is interrogative but the exclamation mark and comma seem to indicate the intention. Thus, the structure alone may not be sufficient to perceive the communicative function in digital discourse. The subsequent discursive practices they carried out uncover a success of their communication. They juxtaposed various linguistic and extralinguistic features to choose functional roles and to resolve a kind of ambiguity that may emerge as a result of apparently a flexible relationship between structure and function, for example, a statement can be read as a question or vice versa (see, for example, Sinclair & Coulthard, 1978).

[6]
a. *<Tell me abt it!>* (Tell me about it!)
b. *<Mje call pe btana Kal paper hai mera.>* (Let me know on phone . . . I have exam tomorrow.)
c. *<Kal match on?>* (. . . . . . game on tomorrow?)
d. *<Phr kon dayta hay apko, clases,,?!>* (Then, who does teach you?)

Conclusion

This study pursued a descriptive approach to analyze the linguistic repertoire of Urdu/English bilinguals to explain how CMC works. The findings reveal that the participants are prone to minimizing forms, apparently presupposed, in English and Urdu. The data indicate that they tend to delete optional words and segments more or less, which do not hide the base forms for the projection of missing elements. While mapping out minimal linguistic forms, we found unique configurations which were equally rich in their semantic manifestation.

Despite the fact that the participants committed structural irregularities in their communication, they adhered to grammatical rules and conversation principles to the extent that their communication was intelligible. The elements which were missed or omitted can be guessed at or perceived because, while communicating, they were not only concerned with just the structure of their message but with the meaning and context surrounding the message, as supported by Spears, Lea, and Postmes (2001). Perhaps meanings, more than receiver or mode, directed their linguistic choices as to how far they could manipulate the structure of English and Urdu. They marked moods with the types of structures we have discussed. Although we generally construe a mood both through the transformation of a structure along with insertion of punctuation marks, they did this simply by inserting punctuation marks and avoiding structural transformation. Apart from structure, the discourse roles the participants technically assumed tend to contribute to the theory of meaning in digital environment.

Unlike previous studies, which have revealed the ambivalent, if not weird, nature of cyber-linguistic features, this study helps define a broader classification of them, which supposedly project some unique configurations. The study reveals that even though Urdu and English are subject to morphological and syntactic reduction, their structural manipulation in the context of CMC does not obscure meaning. The study motivates future researchers to investigate linguistic reduction to reveal linguistic variation and change in the languages used for the Internet communication. This would further provide us with a foreground, so we can understand more about the linguistic behavior and social identity of e-communities.

Declaration of Conflicting Interests

The author(s) declared no potential conflicts of interest with respect to the research, authorship, and/or publication of this article.

Funding

The author(s) received no financial support for the research and/or authorship of this article.

References

Ahmad, R. (2011). Urdu in Devanagari: Shifting orthographic practices and Muslim identity in Delhi. *Language in Society*, *40*, 259-284.

Androutsopoulos, J. (2007). Language choice and code switching in German-based diasporic web forums. In B. Danet & S. C. Herring (Eds.), *The multilingual internet: Language, culture, and communication online* (pp. 340-361). Oxford, UK: Oxford University Press.

Axelsson, A. S., Abelin, A., & Schroeder, R. (2007). Anyone speak Swedish? Tolerance for language shifting in graphical multiuser virtual environments. In B. Danet & S. C. Herring (Eds.), *The multilingual internet: Language, culture, and communication online* (pp. 362-384). Oxford, UK: Oxford University Press.

Barasa, S. (2010). *Language, mobile phones and Internet: A study of SMS texting, email, IM and SNS Chats in computer mediated communication (CMC) in Kenya*. Janskerkhof, The Netherlands: LOT Publisher.

Barnes, S. (2003). *Computer mediated communication: Human-to-human communication across the Internet*. Boston, MA: Pearson Education.

Baron, N. S. (2008). *Always on: Language in an online and mobile world*. New York, NY: Oxford University Press.

Bodomo, A. B. (2009). *Computer-mediated communication for linguistics and literacy: Technology and natural language education*. Hershey, PA: IGI Global.

Boyd, D., & Ellison, N. (2007). Social network sites: Definition, history, and scholarship. *Journal of Computer-Mediated Communication*, *13*, 10-25.

Crystal, D. (2001). *Language and the internet* (1st ed.). Cambridge, UK: Cambridge University Press.

Crystal, D. (2006). *Language and the internet* (2nd ed.). Cambridge, UK: Cambridge University Press.

Danet, B., & Herring, S. C. (2007). Introduction: Welcome to the multilingual internet. In B. Danet & S. C. Herring (Eds.), *The multilingual internet: Language, culture, and communication online* (pp. 3-39). Oxford, UK: Oxford University Press.

Durham, M. (2007). Language choice on a Swiss mailing list. In B. Danet & S. C. Herring (Eds.), *The multilingual internet: Language, culture, and communication online* (pp. 319-339). Oxford, UK: Oxford University Press.

Firth, J. R. (1957). A synopsis of linguistic theory. In Stubbs, M. (1996). *Text and corpus analysis* (p. 35). Oxford, UK: Blackwell.

Grinter, R., & Eldridge, M. (2001, September 16-20). *Y do tngrs luv 2 txt msg?* Proceedings of the European Conference on Computer-Supported Cooperative Work, Bonn, Germany.

Halliday, M. (2002). *Linguistic studies of text and discourse.* London, England: Continuum Publisher.

Halliday, M. (2003). *The language of early childhood.* London, England: Continuum Publisher.

Halliday, M. (2004). *The language of science.* London, England: Continuum Publisher.

Halliday, M. (2005). *Studies in English language.* London, England: Continuum Publisher.

Halliday, M. (2007). *Language and society.* London, England: Continuum Publisher.

Herring, S. C. (1996). *Computer-mediated communication: Linguistic, social and cross-cultural perspectives.* Amsterdam, The Netherlands: John Benjamins.

Herring, S. C. (2011). Grammar and electronic communication. In C. Chapelle (Ed.), *Encyclopaedia of applied linguistics.* Hoboken, NJ: Wiley-Blackwell. Retrieved from http://ella.slis.indiana.edu/~herring/e-grammar.2011.pdf

Herring, S. C., Stein, D., & Virtanen, T. (2013). Introduction to the pragmatics of computer-mediated communication. In S. C. Herring, D. Stein, & T. Virtanen (Eds.), *Handbook of pragmatics of computer-mediated communication* (pp. 3-31). Berlin, Germany: Mouton de Gruyter.

Ilyas, S., & Khushi, Q. (2012). Facebook status update: A speech act analysis. *Academic Research International, 3*, 500-507.

Jørgensen, J. N. (2001). Multi-variety code-switching in conversation 903 of the Koge Project. In S. Leppänen (2007). Youth language in media contexts: Insights into the functions of English in Finland. *World Englishes, 26*, 149-169.

Jørgensen, J. N., Karrebæk, M. S., Madsen, L. M., & Møller, J. S. (2011). Polylanguaging in superdiversity. *Diversities, 13*, 23-37.

Kalman, Y. M., & Gergle, D. (2009, November 12-15). *Letter and punctuation mark repeats as cues in computer-mediated communication.* Paper presented at the National Communication Association's 95th Annual Convention, Chicago, IL.

Kalman, Y. M., & Gergle, D. (2010, September 12-14). *CMC cues enrich lean online communication: The case of letter and punctuation mark repetitions.* Paper presented at the 5th Mediterranean Conference on Information Systems, Tel-Aviv, Israel.

Ko, K. K. (1996). Structural characteristics of computer mediated language: A comparative analysis of interchange discourse. *The Electronic Journal of Communication, 6*, 13-22.

Leppänen, S. (2007). Youth language in media contexts: Insights into the functions of English in Finland. *World Englishes, 26*, 149-169.

Mann, C., & Stewart, F. (2000). *Internet communication and qualitative research.* London, England: SAGE.

Michael, H. (1993). *The metaphysics of virtual reality.* New York, NY: Oxford University Press.

Napoli, D. J., & Lee-Schoenfeld, V. (2010). *Language matters.* Oxford, UK: Oxford University Press.

Paolillo, J. C. (2007). How much multilingualism? Language diversity on the Internet. In B. Danet & S. C. Herring (Eds.), *The multilingual internet: Language, culture, and communication online* (pp. 319-339). Oxford, UK: Oxford University Press.

Paolillo, J. C. (2011). "Conversational" code switching on usenet and internet relay chat. *Language@Internet, 8.* Available from http://www.languageatinternet.org/

Peuronen, S.-R. (2008). *Bilingual practices in an online community: Code-switching and language mixing in community and identity construction at www.godspeed.fi* (Master's thesis in English). Retrieved from https://jyx.jyu.fi/dspace/bitstream/handle/123456789/18390/urn_nbn_fi_jyu-200805061436.pdf?sequence=1

Rafi, M. S. (2008). SMS text analysis: Language, gender, and current practices. *Online Journal of TESOL France.* Available from www.tesol-france.org

Rafi, M. S. (2010). The sociolinguistics of SMS: Ways to identify gender boundaries. In T. Rotimi (Ed.), *Handbook of research on discourse behavior and digital communication: Language structures and social interaction* (pp. 104-111). New York, NY: IGI Publishers.

Rafi, M. S. (2013). Urdu and English in an e-discourse variation in the theme of linguistic hegemony. *European Academic Research, 6*, 1260-1275.

Rafi, M. S. (2013). Urdu and English contact in an e-discourse: Changes and implications. *Gomal University Journal of Research, 29* (2), 78-86.

Raskin, R. (2006). Facebook faces its future. *Young Consumers: Insight and Ideas for Responsible Marketers, 7*, 56-58.

Seargeant, P., Tagg, C., & Ngampramuan, W. (2012). Language choice and addressivity strategies in Thai-English social network interactions. *Journal of Sociolinguistics, 16*, 510-531.

Segerstad, H. Y. (2002). *Use and adaptation of written language to the condition of computer mediated communication.* Göteborg, Sweden: Department of Linguistics, Goteborg University.

Sengupta, S., & Rusli, M. E. (2012, February 1). Personal data's value? Facebook is set to find out. *The New York Times.* Available from www.nytime.com

Sinclair, J. McH. (1991). Corpus, concordance, collocation. In M. Stubbs (1996). *Text and Corpus Analysis.* Oxford, UK: Blackwell.

Sinclair, J. McH., & Coulthard, M. R. (1978). *Towards an analysis of discourse.* Oxford, UK: Oxford University Press.

Spears, R., Lea, M., & Postmes, T. (2001). Social psychological theories of computer-mediated communication. In W. P. Robinson & H. Giles (Eds.), *New handbook of language and social psychology* (pp. 601-623). Chichester, UK: John Wiley.

Sperlich, W. B. (2005). Will cyberforums save endangered languages? A Niuean case study. *International Journal of the Sociology of Language, 172*, 51-77.

Stubbs, M. (1996). *Text and corpus analysis.* Oxford, UK: Blackwell Publishers.

Tagg, C. (2009). *A corpus linguistics study of SMS text messaging* (Doctoral dissertation). Retrieved from etheses.bham.ac.uk/253/1/Tagg09PhD.pdf

Thurlow, C., Lengel, L., & Tomic, A. (2004). *Computer mediated communication: Social interaction and the internet.* London, England: SAGE.

Thurlow, C., & Poff, M. (2013). Text messaging. In S. C. Herring, D. Stein, & T. Virtanen (Eds.), *Handbook of pragmatics of computer-mediated communication* (pp. 163-190). Berlin, Germany: Mouton de Gruyter.

Trudgill, P. (2003). *A glossary of sociolinguistics*. Oxford, UK: Oxford University Press.

Turkle, S. (2011). *Alone together: Why we expect more from technology and less from each other*. New York, NY: Basic Books.

Warschauer, M., El Said, G. R., & Zohry, A. (2007). Language choice online: Globalization and identity in Egypt. In B. Danet & S. C. Herring (Eds.), *The multilingual internet: Language, culture, and communication online* (pp. 303-318). Oxford, UK: Oxford University Press.

Weir, A. (2012). Left edge deletion in English and subject omission in diaries. *English Language & Linguistics*, *16*, 105-129.

Author Biography

Muhammad Shaban Rafi is a doctoral candidate in the Department of English Language and Literature, University of Management and Technology, Lahore, Pakistan. He teaches to postgraduate students. His research interests are language variation and change, sociology of cyber-communication, and language teaching.

7

Investigating the Effect of Communication Skills Training for Married Women on Couples' Intimacy and Quality of Life

Elnaz Farbod[1], Mohammad Ghamari[2] and Mojtaba Amiri Majd[2]

Abstract

The aim of this study was to investigate the effect of communication skills education for married women on couples' intimacy and quality of life. The subjects of this study were married female students enrolled in the spring semester of 2011 at the Applied Sciences and Technology University (Tourism Unit) of Kerman in Iran, along with their husbands. Of the students who expressed interest in taking part in this study, 30 subjects were selected and randomly assigned to either an experimental or a control group. Subjects in the experimental group participated in a 12-session training for the improvement of communication skills in the context of marriage and family therapy. Comparison of these two groups indicated that communication skills training for married women can increase their intimacy and quality of life in their relationships with their husbands ($p < .01$). This study further revealed that this sort of training programs could also increase the quality of life of the husbands in their relationships with their wives ($p < .01$).

Keywords

communication skills, marital intimacy, marital satisfaction, quality of life

Introduction

Family is the basic unit in every society and currently marriage as the foundation of this important unit has faced a lot of challenges in modern societies (Olson & Olson, 1999).

Marital satisfaction, marital quality, and dyadic adjustment are terms that have been used to express quality of relationship in married couples (Askari, Noah, Hassan, & Baba, 2012; Harper, Schaalje, & Sandberg, 2000).

Intimacy is one of the marital process variables that can be affected when the couples' relationship is dissatisfactory. It has been shown that intimacy is strongly connected to quality of life in married couples (Harper & Elliott, 1988; Harper et al., 2000; Tolstedt & Stokes, 1983; Waring, 1988). Based on a definition proposed by Waring (1988), intimacy is a "multifaceted interpersonal dimension which describes the quality of marriage relationship at a point in time" (Harper et al., 2000, p23). There are other definitions for intimacy which involve the notions of conflict resolution and sharing hurt feeling (Harper et al., 2000; Holley & L'Abate, 1979). Schaefer and Olson (1981) have defined intimacy in the context of sharing intimate experiences, which include five factors: emotional, social, sexual, intellectual, and recreational (Harper et al., 2000). Harper and Schaalje have proved that intimacy is a mediator of daily stress effects on couples' relationships (Harper et al., 2000).

Communication skills training has been extensively studied in the context of couples' relationships and interactions (Bellack & Hersen, 1979). Lack of proper communication skills has been shown to be a major issue leading to dissatisfaction in couples' lives (Bellack & Hersen, 1979; Boland & Follingstad, 1987; Burleson & Denton, 1997; Mitchell, Bullard, & Mudd, 1962). Communication skills have been among top three significant problems reported in distressed marriages along with difficulties in sexual expression and personality factors (Bellack & Hersen, 1979; Birchler, 1977). Therefore, communication skills training seems to be an essential part of treatment programs for couples complaining of dissatisfaction in their relationships (Jacobson & Margolin, 1979).

Efficient communication is considered in forms of Speaker Skills in which the term "I" is used to address a specific message and Listener Skills in which a person listens to his or her partner in an active dynamic form and try to make

[1]Kerman University of Applied Sciences and Technology, Iran
[2]Department of Counseling, Abhar Branch, Islamic Azad University, Abhar, Iran

Corresponding Author:
Mojtaba Amiri Majd, Department of Counseling, Abhar Branch, Islamic Azad University, Abhar, Iran.
Email: amirimajd@abhariau.ac.ir

the communication more clear through positive feedbacks (Hahlweg, Revenstorf, & Schindler, 1984).

Blanchard et al. (2009) have mentioned the correlation between Marriage and Relationship Education (MRE) and couples' communication skills. Based on their meta-analysis study, they have suggested the invention of better methodologies to evaluate the role of MRE in improvement of communication skills. Schilling et al. (2003) conducted a longitudinal study to investigate the effects of Prevention and Relationship Enhancement Program (PREP) on development of distress in marital relationships. They found that PREP can significantly increase the positive communication skills in men and women. An interesting finding in this study was the effect of increased positive communication skills in women which could lead to increased risk of male and female distress onset whereas increased positive communication skills in men predicted decreased risk of marital distress (Schilling et al., 2003).

Behavioral marital therapy (BMT) has been shown to be able to help couples improve their communication skills (Hahlweg et al., 1984). They have mentioned that lack of communication skills is one of the main complaints in couples seeking BMT and they have shown that BMT has a positive effect in increasing communication skills in couples. Boland and Follingstad (1987) have comprehensively reviewed all the literature related to relationship between communication skills and marital satisfaction, which has been addressed as a dominant and important correlation. The same sort of integrative review has been done by Burleson and Denton (1997).

Iranian couples live in a transitional society moving from being traditional to modern. In such a complex transitional state, couples' relationships, their quality of life and intimacy are affected. Problems related to communication skills in Iranian couples have been addressed in the literature but limited knowledge is available toward communication skills in relation with quality of life and intimacy of couples and also interventional methods to overcome these problems (Askari et al., 2012; Kalantarkousheh & Hassan, 2010; Rakhshani, Niknami, & Moghaddam, 2005).

Method

The research design in this study was pre-test and post-test research design with experimental and control groups. Subjects were 30 married female students enrolled in the spring semester of 2011 in Applied Sciences and Technology University (Tourism Unit) of Kerman in Iran and their husbands. They were randomly assigned to an experimental group (15 subjects) and a control group (15 subjects). We used pre-test and post-test to determine the effects of communication skills training on intimacy and quality of life of the women and their husbands.

Bagarozzi Intimacy Needs Survey Questionnaire (Bagarozzi, 2001) includes 41 questions and is a useful tool to measure total intimacy as a score. It uses eight factors for

scoring intimacy: Emotional Intimacy, Psychological Intimacy, Intellectual Intimacy, Sexual Intimacy, Physical Intimacy, Spiritual Intimacy, Aesthetic Intimacy, and Social-Recreational Intimacy (Bagarozzi, 2001).

The psychometric properties of a quality of life scale (World Health Organization [WHO], 1996, 2004), the WHO Quality of Life-BREF (World Health Organization Quality of Life Assessment-Brief, WHOQOL-BREF), has 26 items which measure physical health, psychological health, social relationships, and environmental factors. The WHOQOL-BREF is a shorter version of the original instrument, which can be used in larger studies and clinical trials (WHO, 1996, 2004).

We used Bagarozzi Intimacy Needs Survey Questionnaire and WHOQOL-BREF on all the married couples volunteered for this study among which all were female students enrolled in the spring semester of 2011 in Applied Sciences and Technology University (Tourism Unit) of Kerman in Iran. A total of 30 couples with the lowest scores were selected as subjects for this study.

Women in experimental group participated in a 12-session training on marital communication skills. The aim of this training was to increase positive communication skills and diminish negative communication behaviors. The control group received no training. We used analysis of covariance to test our hypotheses:

Hypothesis 1: Communication skills training for married women increases their intimacy.

Hypothesis 2: Communication skills training for married women increases their quality of life.

Hypothesis 3: Communication skills training for married women increases their husbands' intimacy.

Hypothesis 4: Communication skills training for married women increases their husbands' quality of life.

Results

After analysis of the data obtained through pre-test and postest, the effect of communication skills training had significantly ($p < .01$) increased women's intimacy, and Hypothesis 1 was approved. Hypothesis 2 was approved as well and communication skills training had significantly ($p < .01$) increased women's quality of life. Although communication skills training for women had not significantly increased the intimacy of their husbands ($p > .1$), it had a significant effect on increasing the husbands' quality of life ($p < .01$). The result is consistent with other studies done before (Askari et al., 2012), showing the importance of communication skills training program in treatment of distressed couples seeking different sorts of therapies.

Discussion

Sprecher (2002) has indicated that communication is an essential requirement for establishing an intimate relationship and giving rise to adaptability between couples. Effectiveness of communication skills training in fostering and maintaining positive intimate relationship between couples has been approved (Askari et al., 2012; Blanchard et al., 2009; Durana, 1997; Hahlweg et al., 1984; Halford, Markman, Kling, & Stanley, 2003; Miller, Nunnally, & Wackman, 1976; Schilling et al., 2003).

Communication skills training helps couples to send their messages more explicitly and hence come to a deeper understanding of one another. Systematic training of positive communication skills to couples and at the same time practice of communication skills by couples will add positive habits into their behavioral repertoire and these habits will be used in conflict situation and avoid engaging in stress-induced behaviors such as criticisms, blaming, and taunting.

Hrapczynski (2008) has shown that cognitive-behavioral therapy can reduce negative communication patterns and negative attitudes among couples, which will lead to increased satisfaction.

The result of this study showed the positive effect of communication skills education for married women on increasing their intimacy and quality of life. Although it has been observed that communication skills trainings have a gender-dependent pattern on affecting couples' behaviors (Schilling et al., 2003), our results showed that communication skills education to women can just increase the intimacy of their husbands and not their quality of life. Schilling et al. (2003) had somehow a similar observation and found that increased positive communication skills in women not only cannot decrease the risk of male and female distress onset but also increases it. The dynamics behind these observations need to be studied in detail.

Nowadays in every society, family is the main source of security for individuals. However, families encounter different challenges and problems in modern societies. One of the most common problems families face is diminished marital intimacy and satisfaction, which leads to decreased quality of life of families. Interventional programs to prevent these problems are very helpful and communication skills training programs for couples are good examples for interventional measures.

Although the impact of communication skills training on intimacy, satisfaction, and quality of life of couples has been studied, the contents of these trainings and mechanism to be used for inducing positive practice of communication skills via educational and training programs need to be studied.

Declaration of Conflicting Interests

The author(s) declared no potential conflicts of interest with respect to the research, authorship, and/or publication of this article.

Funding

The author(s) received no financial support for the research and/or authorship of this article.

References

Askari, M., Noah, S. B. M., Hassan, S. A. B., & Baba, M. B. (2012). Comparison the effects of communication and conflict resolution skills training on marital satisfaction. *International Journal of Psychological Studies, 4*, 182-195.

Bagarozzi, D. (2001). *Enhancing intimacy in marriage: A clinician's* Guide. Routledge New York, NY.

Bellack, A. S., & Hersen, M. (1979). *Research and practice in social skills training.* Springer, New York, NY.

Birchler, G. (1977). *A multimethod analysis of distressed and non-distressed marital interaction: A social learning approach.* Paper presented at the annual meeting of the Western Psychological Association, Seattle, WA, April 1977.

Blanchard, V. L., Hawkins, A. J., Baldwin, S. A., & Fawcett, E. B. (2009). Investigating the effects of marriage and relationship education on couples' communication skills: A meta-analytic study. *Journal of Family Psychology, 23*, 203-214.

Boland, J. P., & Follingstad, D. R. (1987). The relationship between communication and marital satisfaction: A review. *Journal of Sex & Marital Therapy, 13*, 286-313.

Burleson, B. R., & Denton, W. H. (1997). The relationship between communication skill and marital satisfaction: Some moderating effects. *Journal of Marriage and the Family, 59*, 884-902.

Durana, C. (1997). Enhancing marital intimacy through psychoeducation: The PAIRS program. *The Family Journal, 5*, 204-215.

Hahlweg, K., Revenstorf, D., & Schindler, L. (1984). Effects of behavioral marital therapy on couples' communication and problem-solving skills. *Journal of Consulting and Clinical Psychology, 52*, 553-566.

Halford, W. K., Markman, H. J., Kling, G. H., & Stanley, S. M. (2003). Best practice in couple relationship education. *Journal of Marital & Family Therapy, 29*, 385-406.

Harper, J. M., & Elliott, M. L. (1988). Can there be too much of a good thing? The relationship between desired level of intimacy and marital adjustment. *American Journal of Family Therapy, 16*, 351-360.

Harper, J. M., Schaalje, B. G., & Sandberg, J. G. (2000). Daily hassles, intimacy, and marital quality in later life marriages. *American Journal of Family Therapy, 28*, 1-18.

Holley, J., & L'Abate, L. (1979). Intimacy is sharing hurt feelings: A comparison of three conflict resolution models. *Journal of Marital & Family Therapy, 5*(2), 35-41.

Hrapczynski, K. M. (2008). The impact of *couple therapy for abusive behavior on partners' negative attributions about each other, relationship satisfaction, communication behavior, and psychological abuse.* Ann Arbor, MI: ProQuest.

Jacobson, N. S., & Margolin, G. (1979). *Marital therapy: Strategies based on social learning and behavior exchange principles.* Brunner/Mazel Publishers, Inc., New York, NY.

Kalantarkousheh, S. M., & Hassan, S. A. (2010). Function of life meaning and marital communication among Iranian spouses in Universiti Putra Malaysia. *Procedia-Social and Behavioral Sciences, 5*, 1646-1649.

Miller, S., Nunnally, E. W., & Wackman, D. B. (1976). A communication training program for couples. *Social Casework, 57*, 9-18.

Mitchell, H. E., Bullard, J. W., & Mudd, E. H. (1962). Areas of marital conflict in successfully and unsuccessfully functioning families. *Journal of Health and Human Behavior, 3*, 88-93.

Olson, D. H., & Olson, A. K. (1999). PREPARE/ENRICH program: Version 2000. In R. Berger & M. T. Hannah (Eds.), *Preventive approaches in couples therapy* (pp. 196-216). Philadelphia, PA: Brunner/Mazel.

Rakhshani, F., Niknami, S., & Moghaddam, A. A. (2005). Couple communication in family planning decision-making in Zahedan, Islamic Republic of Iran. *Eastern Mediterranean Health Journal, 11*, 586-593.

Schaefer, M. T., & Olson, D. H. (1981). Assessing intimacy: The Pair Inventory. *Journal of Marital & Family Therapy, 7*, 47-60.

Schilling, E. A., Baucom, D. H., Burnett, C. K., Allen, E. S., & Ragland, L. (2003). Altering the course of marriage: The effect of PREP communication skills acquisition on couples' risk of becoming maritally distressed. *Journal of Family Psychology, 17*, 41-53.

Sprecher, S. (2002). Sexual satisfaction in premarital relationships: Associations with satisfaction, love, commitment, and stability. *Journal of Sex Research, 39*, 190-196.

Tolstedt, B. E., & Stokes, J. P. (1983). Relation of verbal, affective, and physical intimacy to marital satisfaction. *Journal of Counseling Psychology, 30*, 573-580.

Waring, E. M. (1988). *Enhancing marital intimacy through facilitating cognitive self-disclosure*. Routledge, New York, NY.

World Health Organization. (1996). *WHOQOL-BREF: Introduction, administration, scoring and generic version of the assessment*. Geneva, Switzerland: Author.

World Health Organization. (2004). *The World Health Organization Quality of Life (WHOQOL)-BREF*, http://www.who.int/substance_abuse/research_tools/en/english_whoqol.pdf.

Author Biographies

Elnaz Farbod is a counseling psychologist who currently lives in New York City, USA. She has received her BSc and MA in counseling psychology and her research interests include child psychology, psychological services for cancer patients and couple and family therapy.

Dr. Mojtaba Amiri Majd is an assistant professor of special education at the Islamic Azad University, (Abhar Branch), Iran. His research interests are behavioral and learning problems in children and family threapy issues.

Dr. Mohammad Ghamari is an assistant professor of guidance and counseling at the Islamic Azad University (Abhar Branch), Iran. His research interests are adolescence, career counseling, and family threapy issues.

8

Mobile Public Memory: The (Digital/Physical) (Artifacts/Souvenirs) of the (Archiver/Tourist)

Shane Tilton[1]

Abstract

This article will look at the impact that mobile technologies have had on the ability of people to document their everyday lives. What is important to note about this documentation is that it has become a public display of events and experiences via mediated content. This documentation becomes more interesting to observe when it is put in context of how people can now record their travels. This "public memory" of private travel is assisted via mobile technologies with applications designed to record locations through Global Positioning System data and mediated content. After the documentation, the consumption of this content is conducted through social media services and other public outlets as opposed to the traditional means of showing where people have traveled via postcards, slides, and souvenirs. Through a mixed-method study, this process of documentation is analyzed via the thematic dichotomies that emerged throughout the course of surveys and interviews. This article will explore this contrast between the digital and physical through an analysis of the "traditional tourist" versus the "archiver of experiences." Finally, the impact of this documentation will be framed in the context of mobile communication.

Keywords

memory, artifacts, travel, mobile communication, mixed-method research

As people adjust to mobile technologies as part of the normal flow of society, those same people find new and innovative methods of applying these technologies beyond mere one-to-one communication or even one-to-many communication (Ling & Campbell, 2011). Some people are applying the channels of communication to become more of a self-documenting mode. This self-documenting mode is comprised of writing posts and comments and recording mediated content for the purpose of recording the experiences and mundane actions of everyday in a way that can be recalled or searched via a public social feed. This feed of information represents a kind of "public memory" for the individual who posted the content.

For the purpose of this analysis, the term *public memory* is simply shorthand for posting content onto a mediated platform where one of the purposes of posting is to journal or note information that has a personal significance. For example, one could take pictures of their meals not because they are showing the dinner to their friends but rather keeping a recorded journal of their favorite meals to remember them later. In this way, there is not necessarily an imagined audience that the individual is focusing their content toward, but rather it is designed to be an online scrapbook.

The purpose of this work is to look at how mobile media is changing the nature of the collection of memory via digital artifacts created by the hybrid of services that represent the social and hyper-local interactions. These interactions are best exemplified through the use of the SoLoMo (social, local, mobile) category of applications for mobile devices, especially services such as FourSquare and Instagram (Nelson, 2013). These services will be discussed within the context of their impact on public memory, specifically how they bridge the gap between the digital and the physical, the gap between an artifact and a souvenir, and its ability to combine the role of the tourist and the archiver. The remaining parts of this article will define these terms and describe the impact that the mobile media space has had on these concepts.

Public Memory and Rhetoric

While this article can only provide a basic understanding of memory as a theoretical concept, there are two elements that can be focused on to help explore this connection between the transmission of personal public memory via social networks and the binary nature of those creations as described in the introduction of this work. Both of these thematic elements can be explored by looking at the rhetorical

[1]Ohio Northern University, Ada, USA

Corresponding Author:
Shane Tilton, Assistant Professor of Multimedia Journalism, Ohio Northern University, 525 South Main Street, Ada, OH 45810, USA.
Email: s-tilton@onu.edu

elements of memory and the psychological elements of shared memories/shared social experiences.

Most of the nominative constructions of the term *public memory* tend to be embedded in the rhetorical aspects of communication. This idea of "public memory" goes to the practice of public remembrance to cultural events. Condit (1985) framed this concept within Aristotle's inartistic elements of rhetoric. Primarily, Condit made the argument that public memory performed an epideictic element of rhetoric as public memory was normally invoked through the ceremonial speeches surrounding a particular event (in Condit's example, the Boston Massacre). Communication actions using epideictic discourses allow for a definition of a public event by a speaker via the sharing understanding of the context of the event toward the larger whole shaped by the speaker. The speaker shaped shared symbols to have a community focus on one particular theme or image. The nature of these actions is normally for the purpose of display or entertaining the community. Those commemorative discourses are normally for a shared audience with a mass understanding.

Another aspect of public memory is the sense of placement. According to Dickinson, Blair, and Ott (2010), the construction of the public memory is tied to the rhetorical art of memory in contemporary public culture. Public memory exists "at the nexus of rhetoric, memory, and place" as this concept represents a deep complexity of events as they were occurring. The connection between rhetoric and public memory was enhanced by the works of Maurice Halbwachs (1992) and Michel Foucault, Bouchard, and Simon (1977) as their work focused on the forms of memory, which tended to be broken up and defined by the cultural norms and the "language of the memory" in the mind of society. Place within this context notes the

> terminologies of spaces, spatiality, sites, territory and deterritorializing, border cultures, the urban and exurban, social locations, geographies, zones, the archive . . . to reckon with the complexities of what anyone might mean when s/he deploys one or more of these terms. (Dickinson et al., 2010, pp. 22-23)

Public memory as a concept has a broad definition because the theoretical discourse around the term *public memory* has some semantical weight behind it. The definition fragments as newer communication and humanities theorists explore the central concept of memory. In communication, memory is one of the foundations of rhetoric. This foundation adds to this discussion of the chosen definition of "public memory" for this article because it exemplifies the need to observe the cultural aspects of memory. The technological aspects of memory related to the personal encoding and decoding of information can best be theorized in the scientific/psychological realm.

Public Memory and Technological Influence

The reason that the technological support of memory is noted for this article is because the nature of memory distorts

without some method of recording. One of the major influences to this sense of memory is social influence. People will normally conform to the "erroneous recollections of the group, producing both long-lasting and temporary errors, even when their initial memory was strong and accurate" (Edelson, Sharot, Dolan, & Dudai, 2011, p. 108).

This representation of communal public memory can be influenced through the normal discourse of communication actions or through social interactions. The record of an event can be "protected" from these social influences with a well-constructed archival system designed to limit this layer of social static and misinterpretation (Jacobsen, Punzalan, & Hedstrom, 2013).

This archival element to the technological influence on the public memory has often been framed in the scientific/psychological realm as a version of collective memory. This collective memory represents "an inherently mediated phenomenon . . . as much a result of conscious manipulation as unconscious absorption" (Kansteiner, 2002, p. 180). This aspect of collective memory means that the dissemination of events comes from channels of communication and messages embedded in those channels. Under the traditional understanding of these channels of communication, the messages related to collective memory encoded in these channels are often placed into these "linear, one-way channels of description" (Hall, 2013). New media channels create a more complete collective memory as the artifacts of the events are placed in a shared space with the purpose of "making claims about the subjective experience, as if claiming that a given representation is a authentic recollection (neither perception nor imagination)" (Haye, 2012, p. 27). This representation is crafted through the co-creation of the meaning of memories placed within the shared space under the guise that most of the artifacts added to the shared space come with their own set of bias and positional aspects. There is a sense of social static and misinterpretation can exist with this space, but there are embedded check and balances that can normalize those elements of distortion.

This version of public memory described in this article is not a representation of the communal public memory as collective memory, but rather the publication of private or semi-private events (e.g., traveling) into a semi-public or public sphere (e.g., Instagram) as a way of remembering the experience and categorizing the elements of the experience via the norms of the mode of communication (e.g., hashtags and check-ins). Hashtags have the potential to act as the "guideposts" toward the shared spaces described earlier as they function as a public link to a common set of pictures, posts, and links that is hosted within the social networks of choice. The common vernacular of hashtags has a shared lexicon with significance and meaning. Those tags place individual artifacts connected to the tag within a framework designed to either support a community or maintain a dialogue around a common theme or event (Alper, 2013). The second aspect of the hashtags that is rarely discussed is their ability to act as personal organizational tools. A person can search through his or her own feeds via typing his or her username and the selected hashtags to find post with a significant importance (e.g., the first trip abroad). These elements of selective filtering influenced the creation of the methods used for this research.

Method

The data from this article come from mixed-methods approach using a survey and follow-up interview. A snowball sample technique was used that led to a sample size of 98 people (Goodman, 1961). A snowball sample was used due to the point of analysis. The problematic in question is within the range of those who travel and use SoLoMo to document their travel. As this would be considered a specific population, the use of the snowball sample seemed appropriate.

The surveys included 12 prompts on a 9-point Likert-type scale. Those prompts focused on the use of social networks and the tools used to access those social networks. In addition, there were 10 prompts on a 7-point Likert-type scale that focused on audience awareness, travel practices, and cultural awareness. The rationale for conducting the survey was to make sure that the interview questions would fit the "real world" as understood by the people and at the same time would eliminate possible personal bias. The two different types of Likert-type scales were required for this research as the 9-point Likert-type scale prompts related to temporal elements (*once a year, once a month, 2 or 3 times a month, once a week, 2 or 3 times a week, once a day, 2 or 3 times a day, every hour*, and *multiple times an hour*) to get a sense of the survey taker relationship between themselves and the tools of posting (technology and services), while the 7-point Likert-type scale deals with the perceptions of survey takers toward the tools of posting and thus had a "more traditional" set of ranges within the scale (*strongly disagree, disagree, disagree slightly, neither agree or disagree, agree slightly, agree*, and *strongly agree*).

A structured interview was conducted after the survey was completed. The interviews ranged from 15 to 20 min. The interviewee was asked a few ethnographic questions that closely resembled a "guided ethnography" (Tilton, 2012), with the exception that the researcher would not be there to observe them as they used their mobile devices. Interview questions primarily dealt with how they use their mobile devices when they traveled and whether they altered their documentation practices because of the SoLoMo applications.

To identify participants, the following message was sent via Facebook feed and several group pages that dealt with academic research:

Facebook Hivemind: I'm looking to interview about 30 people about their use of FourSquare and Instagram, especially how they use it when they travel. If you are interested in being interviewed, please message me. If you are not interested in being interviewed, please share this.

In addition, an abbreviated version of this call was sent via Twitter. From this original call, 49 responses were received. From those requests, the respondents were asked if they knew anybody else that would fit both of the criteria. From those requests, there were 49 more responses. Between the two groups, the total sample size was 98.

Data Analysis

The sample size included 64 women and 34 men. Four of the research participants were college students, and 2 of the older participants were retired. All of the other participants were employed, with 16 of them being employed in social media/marketing-related positions. The participants all live in the United States, and the pool includes individuals from cities in the Northeast (Washington, DC; Boston, MA; New York, NY; Arlington, VA), the Southeast (Raleigh, NC; Chapel Hill, NC; Charlotte, NC; Atlanta, GA; Kennesaw, GA; St. Augustine, FL), the Midwest (Chicago, IL; Cincinnati, OH; Indianapolis, IN), and the Pacific Northwest (Seattle, WA; Portland, OR; Central Washington state).

The interviewees were frequent users of FourSquare, Instagram, or both. Frequent users were defined as people who had more than 100 posts on one of the sites and had checked in at least once in the previous 3 days. Of the survey population of 98 people, there were 26 people who fit this characteristic. Six of the participants were interviewed face-to-face. Seventeen people were interviewed using Skype and three people were interviewed using a phone with no video. The interviewer did not notice an appreciable difference between the face-to-face and Skype/phone interviews. A grounded theory approach was used throughout the data collection as a way of maintaining balance within the instrument between a loose interpretation of the data and a narrow deconstruction of the interviewee's statements (Charmaz, 2006; Glaser & Strauss, 1967).

The interviews were transcribed and placed within an open coding procedure. There were extensive notes taken to encode the interviews in a way that helped overlay the analysis of the survey "on top of" the transcription of the interviews. Throughout the back and forth of interviewing and coding, the data were organized into categories and eventually grouped categories based on thematic similarities. The coding included multiple themes that emerged from the data.

Those themes were gathered by transcribing the 26 interviews, for a total of 10 hr of interviews. The transcripts of those interviews were then coded to examine for common themes. The researcher was the lone coder for the process. The common themes were examined for patterns consistent with the trend found in the analysis of the surveys. Finally, those common themes were framed within the context of the current research in the field. The artifacts from this coding and selection process will be shown in the impact sections of the article. For the purpose of this article, the focus will be on three primary thematic dissidences that arose in the data.

Table 1. Percentage of Mobile Phone Use of Sample Population (*N* = 98).

	Less than once per day	Once a day	Two or three times a day	Every hour	Multiple times an hour	NA
Mobile phone use	0	4.08	13.27	33.67	46.94	0

Table 2. Documentation and Travel Practices of Sample Population (*N* = 98).

	Interviewee average (*n* = 26)	Interviewee SD.	Non-interviewee average (*n* = 72)	Non-interviewee SD.
Documentation	4.454	1.905	4.454	2.016
"I purchase a souvenir from the place I travel for myself."	4.000	1.837	4.060	1.951
"I purchase a souvenir from the place I travel for my family."	3.787	1.932	3.545	1.985
"I purchase a souvenir from the place I travel for my friends."	3.437	1.664	3.250	1.813
"I travel with my family."	4.363	1.635	4.000	2.031
"I travel with my friends."	3.667	1.707	3.606	1.599

The next two sections deal with the analysis of the surveys and interviews as a means of framing the thematics at the end of this work.

Survey Analysis

As this was a mixed-method study, the analysis took on two separate elements, an analysis of the survey and the coding analysis for thematic elements of the interview. The initial rationale for this process was to reinforce the qualitative research and analysis with the analysis of quantitative instrument. The survey instruments allowed for a clearer picture of the people completing the interview process and better identified the limitations that may be present in an interview that the researcher would not be aware of.

One of the most interesting data points that came to light from the survey was with the prompt "I use my mobile phone." All 98 of the participants responded by marking they used their phone once a day or more. Of the 98, 79 participants marked that they used their mobile phones once every hour, as noted in Table 1. This seems to be in line with other similar research dealing with current usage of mobile phones (Baron & Segerstad, 2010).

The next point of analysis is the documentation practices of the sample population during their travels. The sample population was divided between those who fit the characteristics of the interviewee pool (*n* = 26) and those who did not (*n* = 72). From these data, there was little difference in the documentation and souvenir practices of those who were interviewed versus those who were not. In terms of the analog aspects of travel, both groups tended to buy souvenirs about half of the times they travel. As noted in Table 2, more often those purchases would be for themselves over family

and friends. In addition, this group was more likely to travel alone than with friends or family.

The final point of analysis from the survey would be the posting practices of the sample population. The only point of divergence between the interviewee and non-interviewee that stands out is their awareness of their audiences when the post content about their travels. Those in the interviewee group were less likely to be concerned about their audience when posted travel content on their social media feeds compared with the non-interviewee group. This point of divergence is noted in Table 3.

This point of divergence helped shape some of the interview instrument to find out what the interviewees focused on when posted content from their trips.

Interview Analysis

With a smaller interview pool, it was necessary to craft the questions to not only work with the larger problematic but also the element raised during the survey regarding audience awareness. This major point of separation between those who were being interviewees and non-interviewees leads to a deductive approach of interviewing and analysis (Schadewitz & Jachna, 2007). The questions raised during the interview focused on the interviewee's understanding of memory and their audience awareness.

From the analysis of both methods, a series of thematic dichotomies emerged that positioned the collection of digital artifacts while traveling into uses and gratifications present when using any online services, especially those services that can act as "catalysts for personal narrative and recall" (Viegas, boyd, Nguyen, Potter, & Donath, 2004).

Table 3. Posting Practices of Sample Population (N = 98).

	Interviewee average (n = 26)	Interviewee SD.	Non-interviewee average (n = 72)	Non-interviewee SD.
Posts while traveling	5.334	1.756	4.698	1.836
Uses social media to remember trips	5.125	1.665	5.030	1.519
Collects digital artifacts (pictures, badges, check-ins, etc.)	3.177	2.193	3.505	1.869
Audience awareness	3.507*	1.574	5.400*	1.813
Audience empathy	4.003	1.838	4.712	2.031

*p < .05.

Digital Versus Physical

When a person captures the experience of travel and translates it into a mediated form, both the digital and the physical are the same. A person can see the Grand Canyon in front of him or her while taking the picture of it on a digital camera. The farther you move away from the physical location, the more the digital representation of that space becomes the "authentic" symbol of your experience at that space. The Instagram picture allows the person to experience the feelings of being in the location through a combination of observing the picture with the personally embedded textual and location data within the picture. The comments attached to the picture become the "vox populi" of the overall impression of the experience. This combination of all of the mediated content represents the digital contextualization of the experience.

In contrast, the physical represents a token, a takeaway that allows the person to tactually touch elements from a location where the person visited. This ability to touch the object means that the object can be placed in space defined by the person to be an area of significance based on other objects of travel. Objects allow the person to immerse himself or herself in the memories of the travel via the sensory relationship between himself or herself and the object. The smell of suntan lotion on a piece of sea glass allows the person to remember a trip to the beach via the smell.

The convergence of the physical and the digital is moving closer together via the social media platforms and SoLoMo apps. A person can look over his or her own feed on Instagram or FourSquare to find a great place to watch a sunset. They have the mediated experience of watching the sunset on Instagram, which is a social media service design to share a small (300 × 300 pixels in size) picture that can be framed by the "filters" that the users can use to overlay on top of the picture. This simple picture of the sunset also includes the location where the picture was taken (either in the form of Global Positioning System [GPS] data or location information "borrowed" from Facebook or FourSquare) and other textual provided by the Instagram community in the form of comments and tags.

It is in this way that the difference between the digital and analog is more than a "bits versus atoms" discussion; it is also a "dynamic versus static" discussion. It is more than the simple ability to store the elements and place them in some form of order and structure. Dynamic allows for different modes of selection and filtering. Static maintains a standard form of selection. A person can look over his or her boxes to view his or her collection of postcards. The digital elements can be organized in multiple albums. The ability to tag these elements allows for a more complex and comprehensive approach to memory selection. For example, a person can organize his or her elements based on the people associated with the elements ("my wife" or "Grandma Sofia") or the timing of the elements ("2013 Trips") or the physical location the elements were taken from. One element can fit multiple organization platforms (Hine, 2003).

Artifacts Versus Souvenir

Artifacts are the small constructed elements that have some cultural purpose (the postcards can be mailed to others to tell about the trip or communicate some other aspect of travel) or a social purpose (the Disney shot glasses can be used at a party to serve drinks) that contains the visual cues of the location that the artifact comes from. Sociologically, artifacts become the central point of analyzing the practices of organized communities.

The ways that artifacts are used can tell a lot about a society. Digital or physical artifacts have a sense of placement that reflects the discourses about how the culture or subcultures connect and interact with others. The slideshow is a communal sharing of pictures that have some meaning to the person or group sharing. In a similar fashion, the album or feed on social media has some significance to the person posting the photos without necessarily worrying about what type of audience is viewing the work.

Souvenirs are mass produced works that are created to fill the needs of the tourist as a trinket from location that has some significance to the tourist as a representation of the location (i.e., a piece of the Berlin Wall). In addition, souvenirs could be individually created and have a larger representation to societies in the form of collections or "placement." For example, you can look at the collective nature of the "pressed penny" as a self-created souvenir that not only represents the location (the person inserts a penny and some other change to have the penny transform into a souvenir via the pressing of a symbol of the location on the penny, thus flattening the penny) but also the penny becomes an addition to the penny collector collection.

Both the artifact and the souvenir in the area of public memory represent a type of signifier in the ritualization of travel. If ritual can be defined as a pattern of actions and the process of those actions via a type of performance, then the ability to collect digital artifacts or souvenirs has the same weight as taking physical pieces from a location (Rothenbuhler & Coman, 2005).

In the digital realm, the argument is that the poster of the digital work is the performer and there is an audience that can be actual (posting pictures for friends or family to see them) or imagined. It is in this regard to talk about services such as FourSquare as part of the ritual. FourSquare is a location-based social network service where the users of the site can go to locations in the real world and "check-in" on the site to indicate they are at the location. As a "ritual leader," FourSquare is responsible for mediating the artifacts of the experience (pictures taken at the location to prove aspects of the experience) and the souvenirs (user can collect "badges" based on their travel and check-ins). FourSquare also amplifies the ritual nature of travel via the expression of the experience through mediated channel to the imagined other ("the crowd").

As it relates to public memory, the important element to note in this analysis of ritual is that the performer and the audience can be one and the same. The performer in the present is collecting artifacts and/or souvenirs for the purpose of the future self to observe the aspects of the performance. A person performs the aspects of the ritual of taking pictures while travel. The person digs his or her phone out of his or her pocket or bag. She or he goes to their camera app, takes the pictures, and adds the metadata or other information relevant to their current location and/or experience.

Archiver Versus Tourist

The last conflict to consider is the role of the person collecting the elements related to travel. In the traditional mode of discussing their role, the person would be described as a tourist, with all of the semantic baggage associated with that term. The tourist is defined as a person who travels from his or her home to other place, while bringing all of his or her cultural bias and frameworks to the place of travel. All new experiences the tourist has while traveling are foreign based on the binary of familiar/non-familiar.

This foreignness forces the tourist to reconceptualize basic tenets of the mundane. Even the extremely basic function of getting up from sleep has the aspects of foreignness (the bed is not their own, the location of clothing and basic bathroom products would be different). It is also this foreignness that frames the tourist interactions with people and the inanimate.

The collections of memories are based on their relationship to the "known" ("it tastes like chicken") and the "unknown" ("why are they wearing those funny hats?"). The collections of memories to the tourist are in the form of mementos. Therefore, the idea of public memory for the tourists would be the scrapbook of their own experience via the mediated platforms.

Archivers are trying to understand the location of travel through points of analysis and cataloging their experiences within the context of the tools of collection. Analysis is conducted via the reflective examination of the collection of elements within a mediated distribution platform. The picture of the Grand Canyon is understood and placed within the context of similar experiences, either personally experienced by the archiver or through the aggregation of similar elements within a common platform. The tools of collection mediate the experience and transform the experience into an element of analysis.

Impact on Mobile Media Space

As the interplay between mobile communication and the real world is becoming more fluid, the ability to reflect on everyday experiences becomes a mode of coping and adjusting to the speed of life. The framework of this work was to explore public memory through the view of its impact to document travel for the individual. It is a natural connection as the mobile device is a person's lifeline as they explore places that she or he is not familiar with. It is their map, connection to the outside world, recommendation engine for trying new restaurants and, as explored in the course of this work, the mode of documenting those new experiences. However, public memory is more than a reflective mode of understanding the world. It is a way of filtering and organizing all information that an individual is exposed to on a daily basis. As people are exposed to newer stimuli and feeds of information, there needs to perhaps be agents that can help a person actively document his or her own experience. In addition, those agents could act as the extension of memory that can be recalled via mobile devices. Those technologies, platforms, and services could be thought of collectively as the public memory of the individual with the purpose of being aware of one's own environment.

Impact on Mobile Media Usage

One of the common themes from the interviews and the literature that impacts this concept of "mobile public memory" is the increase of usage of mobile devices over the past decade, specifically the consumption of mobile media. Mobile phones have changed how people in the developing world cope with their dynamic lives through the engagement of the larger community. Of the 26 interviewees, 20 discussed how their mobile phone was an extension of themselves. Amber, a 29-year-old sales professional from New

Orleans, stated, "If I left the house and realized that I left my (Samsung) Note there, I would be running back for it. I feel naked without it. My entire life is on my phone." Barry, an African American 36-year-old freelance artist from Columbus, discussed during his interview:

> I only use my laptop when working on client's projects. When I'm out, I have my DSLR (his digital camera) and my phone. Sometimes, I'll just use my phone to take pictures, even if I have my DSLR with me.

When pushed why he used his phone instead of his professional camera, he said "it's easier to post from my phone and it's quicker."

Part of the discussion related to mobile phone usage is the "social disconnection" supposedly created by the use of mobile device. The social disconnection within mobile device relates to the anecdotal evidence that the more a person consumes mobile media, the less likely that person is able to maintain relationship away from the device. Hampton, Sessions, and Her (2011) disputed this claim in the research stating,

> [i]n support of our hypotheses, findings reveal that neither Internet nor mobile phone use is associated with having fewer core discussion confidants, or having less diverse ties with whom to discuss important matters. As predicted, mobile phone ownership and specific Internet activities—the use of certain "social media"—were found to be associated with having a larger number of confidants. (p. 148)

This assertion was supported through the interview conducted during this research. In all, 19 out of the 26 interviewees mentioned through the course of this research that they felt the same number of friends or more when compared with a decade ago.

These social connections were important to note in this research because despite the observational point that the interviewees felt they had more social connections, the postings on SoLoMo apps and Instagram were more for the interviewee rather than for the imagined audience. In all, 16 of the 26 interviewees very rarely use hashtags to attract audience. Of the 26 interviewees, 19 would use a hashtag or GPS metadata on their postings as a way to remember an event or a location they were at. Charlie, a 40-year-old educator, encapsulated this point when he said, "I remember my grandparents showing me their vacation slides and they were organized in this carousels. I sort of see using the map on Instagram or putting a hashtag in as organizing my photos in those carousels."

Impact on Documenting Experiences

The second thematic of documenting experiences versus collecting souvenirs was also reinforced from the interviews and the previous literature. Both documenting experiences and collecting souvenirs fit into what Neubaum, Rösner, Rosenthal-von der Pütten, and Krämer (2014) would describe as the "psychosocial functions of social media usage" and framed such functionality within the uses and gratifications of these services. The context of the psychosocial was broken down into information gathering and sharing and observing emotions. Neubaum et al. were studying how emotions in a disaster situation were transmitted and expressed via social media connections and public interactions online. The psychosocial was also described in the documenting experiences (the private interpretation of the SoLoMo postings) in the observation of emotions and mentioned in the collection of souvenirs (the public display of the SoLoMo apps) when describing the sharing emotions. David, a 27-year-old male, described these psychosocial elements by saying, "When I travel, I will normally look at the pictures and comments about a location from FourSquare, Yelp and Instagram. I find this seems to be more of a realistic look at a place." This would correspond to information gathering. He would later talk about sharing the picture not with his followers on Instagram, but rather with his family.

> The last time I was in Boston, I went to a Red Sox (baseball) game. I took a few pictures and post them on Instagram... I wanted to show my son the pictures I took and I couldn't find them on my phone. I went back through my Instagram feed and found them. He likes Jonny Gomes and I wanted to show him the pictures I took of (Gomes). He kind of giggled when (my son) saw the pictures.

Impact on Posting Practices

The final thematic that was explored was the posting practices of the individuals on Instagram. Most of the literature that studied posting practices on Instagram were looking how the individuals captured everyday experiences with the service (Memarovic et al., 2014) or how the foreign and strange were "normalized" via the service (e.g., war photography; Alper, 2013). In the case of Memarovic et al.'s research, the service was mutated to a networked public display that could post snapshots taken on the display of an individual at a location onto any service or website the individual wished. The other option was to leave the picture on the public display as a part of the example pictures on the display. Of 26, 18 chose to remove the pictures from public viewing and place those snapshots onto private services. The public display, and therefore a more public audience for the photograph, was rejected by a majority of the individuals. Those individuals preferred a more private display of the photographs. The interviewees reinforced this private desire, specifically when discussing hashtags. Ellen, a 22-year-old female sales representative from Austin, Texas, commented on this by saying,

Most of the time when I take pictures when I travel, I normally don't post "#postforlikes" or "#foodpron." I like having the pictures on Instagram. But, I don't need a large group of followers validating my pics or likes. Most of the time, I post the pictures for me.

Limitations

The major limitation to this work was the snowball sample that led to the survey takers and the interviewees. Both the units for inclusion in the sample were not based on random selection. Therefore, it is impossible to determine the possible sampling error and make statistical inferences from the sample to the population. This research is not fully representative of the population being studied. The rationale for conducting the survey and the interviews with a snowball sample allowed for a more select group that would provide better artifacts within the context of the thematics of the research. The sample group was more familiar with tools of distribution being described in the research and could therefore articulate the elements of study needed to conduct this research; specifically the sample could discuss the connection between audience, memory, and the tools of distribution in a sensible manner.

Conclusion

One of the common themes that has made up the current conceptualization of the idea of memory is its connection to mediated content. A by-product of this thematic understanding of memory is how digital content has impacted this discourse. The collection of slides and postcards from trips long ago have become the feeds of pictures from social network services as we travel. The accessibility and ubiquity of this content has changed the way that memories are captured. There are multiple public systems and social networks designed to catalog and distribute experiences in the form of mediated content. Layers of metadata are embedded within the content posted on these networks and services. As the academy and scholars study the influence and impact of mobile and social media on society, it is important to note how these services and networks change people's memories on previous experiences.

Declaration of Conflicting Interests

The author(s) declared no potential conflicts of interest with respect to the research, authorship, and/or publication of this article.

Funding

The author(s) received no financial support for the research and/or authorship of this article.

References

Alper, M. (2013). War on Instagram: Framing conflict photojournalism with mobile photography apps. *New Media & Society*.

Retrieved from http://nms.sagepub.com/content/early/2013/09/16/1461444813504265.full.pdf+html

Baron, N., & Segerstad, Y. (2010). Cross-cultural patterns in mobile-phone use: Public space and reachability in Sweden, the USA and Japan. *New Media & Society, 12*, 13-34.

Charmaz, K. (2006). *Constructing grounded theory: A practical guide through qualitative analysis*. London, England: SAGE.

Condit, C. M. (1985). The functions of epideictic: The Boston Massacre orations as exemplar. *Communication Quarterly, 33*, 284-298.

Dickinson, G., Blair, C., & Ott, B. L. (2010). *Places of public memory: The rhetoric of museums and memorials*. Tuscaloosa: University of Alabama Press.

Edelson, M., Sharot, T., Dolan, R. J., & Dudai, Y. (2011). Following the crowd: Brain substrates of long-term memory conformity. *Science, 333*, 108-111.

Foucault, M., Bouchard, D. F., & Simon, S. (1977). *Language, counter-memory, practice: Selected essays and interviews*. Ithaca, NY: Cornell University Press.

Glaser, B. G., & Strauss, A. L. (1967). *The discovery of grounded theory: Strategies for qualitative research*. Chicago, IL: Aldine.

Goodman, L. A. (1961). Snowball sampling. *The Annals of Mathematical Statistics, 32*(1), 148-170.

Halbwachs, M. (1992). *On collective memory*. Chicago, IL: University of Chicago Press.

Hall, J. (2013). Internet media & collective memory: Protesting a historical signboard in Okinawa. *Waseda University Journal of the Graduate School of Asia-Pacific Studies, 26*, 45-61.

Hampton, K. N., Sessions, L. F., & Her, E. J. (2011). Core networks, social isolation, and new media: How Internet and mobile phone use is related to network size and diversity. *Information, Communication & Society, 14*, 130-155.

Haye, A. (2012). Continuing commentary: Beyond recollection: Toward a dialogical psychology of collective memory. *Culture & Psychology, 18*, 23-33.

Hine, C. (2003). *Virtual ethnography* (3rd ed.). London, England: SAGE.

Jacobsen, T., Punzalan, R. L., & Hedstrom, M. L. (2013). Invoking "collective memory": Mapping the emergence of a concept in archival science. *Archival Science, 13*, 217-251.

Kansteiner, W. (2002). Finding meaning in memory: A methodological critique of collective memory studies. *History and Theory, 41*, 179-197.

Ling, R. S., & Campbell, S. W. (2011). *Mobile communication: Bringing us together and tearing us apart*. New Brunswick, NJ: Transaction Publishers.

Memarovic, N., Fels, S., Anacleto, J., Calderon, R., Gobbo, F., & Carroll, J. M. (2014). Rethinking third places: Contemporary design with technology. *The Journal of Community Informatics*. Retrieved from http://uc.inf.usi.ch/sites/all/files/rethinking-JoCI-01-web.pdf

Nelson, D. (2013). *Social media enactment study for the District of North Vancouver*. Retrieved from http://hdl.handle.net/1828/4517

Neubaum, G., Rösner, L., Rosenthal-von der Pütten, A. M., & Krämer, N. C. (2014). Psychosocial functions of social media usage in a disaster situation: A multi-methodological approach. *Computers in Human Behavior, 34*, 28-38.

Rothenbuhler, E. W., & Coman, M. (2005). *Media anthropology*. London, England: SAGE.

Schadewitz, N., & Jachna, T. (2007). *Comparing inductive and deductive methodologies for design patterns identification and articulation*. Retrieved from http://crossculturalcollaboration.pbworks.com/f/IASDR_PAPER_schadewitz_final.pdf

Tilton, S. (2012). *First year students in a foreign fabric: A triangulation study on Facebook as a method of coping/adjustment* (Doctoral dissertation). Ohio University, Athens.

Viegas, F., boyd, d., Nguyen, D., Potter, J., & Donath, J. (2004). Digital artifacts for remembering and storytelling: Posthistory and social network fragments. In *Proceedings of the Annual Hawaii International Conference on System Sciences.* Retrieved from http://alumni.media.mit.edu/~fviegas/papers/posthistory_snf.pdf

Author Biography

Shane Tilton is an assistant professor of mulitmedia journalism at Ohio Northern University. He was the former chair of the Communication and the Future interest division (National Communication Association) and Two-Year/Small School interest division (Broadcast Education Association). In 2013, he earned the Kenneth Harwood Outstanding Dissertation Award from the Broadcast Education Association for the best doctoral dissertation in field of broadcasting and electronic media.

9

Viewing Race in the Comfort Zone: Acceptance and Rejection of Black-Centered TV Programming via Nielsen Ratings 1964-1994

Brenda L. Hughes[1]

Abstract

Carter suggests the concept of a "comfort zone" to explain the inability of dramatic African American programs to be successful on television. He argues that a workable formula has been developed for successful African American series, "portray black people in a way that would be acceptable to the millions of potential purchasers (whites) of advertised products. That is, non-threatening and willing to 'stay in their place.'". Using a data set constructed from television ratings and shares, this study examines "black-centeredness" within the context of program success and failure. The comfort zone concept argues Black-centered television series are only successful in a comedic genre because White audiences, who have the majority of the ratings power, will only watch Black-centered series with which they are comfortable. The findings suggest that, in general, race, that is Black-centeredness, did not negatively influence program ratings or shares.

Keywords

communication, culture, and technology, communication technologies, mass communication, social sciences, radio/TV/film, journalism, media and society, media consumption, sociology of race and ethnicity, sociology

Prior to the proliferation of 24-hr cable television, DVDs, TiVo, and streaming media content, prime-time television programming served as the primary vehicle for an evening's entertainment in American homes. Since its beginnings in the 1940s, television has taken viewers on a fanciful journey, as they have been invited into the homes of upper-, middle-, and lower-class families, into neighborhood bars and diners, into corporate boardrooms and courts, into junkyards and taxi garages, and into police stations. Audiences visited the mansions of the Ewing family in *Dallas* (1978-1991), the Chicago ghetto-apartment of the Evans family in *Good Times* (1974-1979), and the New York brownstone of Dr. and Mrs. Heathcliff Huxtable of *The Cosby Show* (1984-1992). Good always triumphed over evil in the television world, week after week, as viewers watched characters such as Mannix, Kojak, and MacGyver nab the bad guys. Captain Stubbing, Julie, Gopher, and Doc took audiences aboard the *Love Boat* (1977-1986) to tropical island paradises, and everyone wanted to buy tickets to Fantasy Island where all dreams come true.

On occasion, the socially constructed world of network television has turned from the scripts of fluff and fancy, to mirror real life through the dramatization of relevant social issues. The scope and breadth of those depictions have encompassed many aspects of the American experience, some more realistically than others. However, as several observers of the world of television have argued, at least one aspect of American life has not found its proper place in the scripted world of network television: the world of Black culture, the world of African American life. It has very rarely received sensitive dramatic portrayal in a regularly appearing network television series. In its 1978 report, *Window Dressing on the* Set, the U.S. Commission on Civil Rights reminded readers of the absence and its gravity: "It should be taken for granted that the fantasy land of television does not represent reality, occupational or otherwise. So long as television is going to portray fantasy, however, all groups should benefit similarly from fantasy-acquired status" (p. 17).

That network television offered very few dramatic series featuring Black actors and Black-centered (BC) experiences at that time and, subsequently, has been well-established. Simple proportionate counts tell that story. Whatever the willingness of network executives to support the development of BC shows, drama or comedy, those shows that did gain access to prime-time scheduling faced the immediate hurdle of television ratings and shares. In his well-known

[1]Florida A&M University, Tallahassee, USA

Corresponding Author:
Brenda L. Hughes, Florida A&M University, 1740 S. Martin Luther King Jr. Blvd., 403 Perry Paige, Tallahassee, FL 32307, USA.
Email: Brenda.hughes@famu.edu

article, *TV's Black Comfort Zone for Whites*, published in 1988, Richard Carter introduced the concept of "comfort zone" to account for the low audience ratings and consequent lack of success of BC dramatic series. He argued that BC television series are only successful in a comedic genre because White audiences, who have the majority of the ratings power, will only watch BC series with which they are comfortable (Carter, 1988). Whether due to bigotry or ignorance-sustained prejudice, White audiences are uncomfortable with serious Black issues in real life, and they are unwilling to learn more about those issues even in the vicarious distance of representations that appear as television entertainment (Carter, 1988). Taking Carter's (1988) supposition of a comfort zone within the context of media choice theories, I argue that viewing is an active process through which audience members decide to accept or reject media content as they perceive its compatibility with their personal preferences for not merely an acceptable world but a *comfortable* world.

Prior Studies

No prior study known to this author has undertaken the empirical investigation reported in this article. However, various studies have examined conditions and aspects relevant to the present inquiry. For example, several reports have confirmed the obvious—that television network executives, afraid of offending segments of their audience, avoided BC programming (MacDonald, 1992). The exceptions—shows such as *Beulah* and *Amos 'n' Andy*, both of the 1950s—were relatively timid situational comedies, which, while featuring Black actors, could as well have been comedies with Black actors in White faces (to use Carter's, 2007, more recent depiction). By the late 1960s, with the country moving toward assimilation, African American actors reappeared in series such as *I Spy* and *Julia*, which presented "whitewashed" African American characters. These characters were nearly devoid of references and depiction of issues relevant to African American culture. The next trend in programming for BC series was segregated situation comedies and a revamping of minstrelsy. Series such as *Good Times, The Jeffersons,* and *Sanford and Son* focused on Black humor and placed African American characters in a predominantly Black world. In 1984, *The Cosby Show* broke the mold of these stereotypical representations of African Americans. *The Cosby Show* presented a view of African American life rarely seen: an upper-middle class, dual-career family with near perfect children, focusing on mainstream values, which appealed to a mass audience. It could be argued that a few other comedy series of this period including *Frank's Place, A Different World,* and *Roc,* followed the innovativeness of *The Cosby Show* by presenting new and non-traditional representations of blackness. Yet, most BC series of this era featured stereotypical characters and plots (e.g., *227* and *Amen*). The mid-1990s brought a move back toward segregation and

"ghettoization" via the concentration of BC series on up-start commercial networks such as United Paramount Network (UPN).

Success in shows is measured primarily by two indices, rating and shares, because both factors tell commercial sponsors how much of an audience their product advertisements are reaching. The two indices differ only by the denominator of a ratio: Rating is the number of households with a television tuned to Show S, in ratio to the total number of households having at least one television set, whereas share is the same numerator in ratio to the total number of households watching television during the time slot of Show S. Thus, the former measure defines its universe as including people who are engaged in an activity other than TV viewing, while the universe of the latter measure is restricted to people who are watching television during the given time slot. The fact that the two measures are highly correlated testifies to the strength of the medium in capturing attention, a fact that advertisers were quick to discern. Either of the measure has been the most influential proximate factor in the success or failure of a series (Webster, Phalen, & Lichty, 2000; Wimmer & Dominick, 1991). Various other factors are significant conditions to that success, however, and the following summary, drawn from results of previous studies, emphasizes three conditions that will be examined in the data analyses below.

The following review of prior studies emphasizes conditions that were extant during the three decades of television shows examined in this article. It was a time of momentous changes in U.S. society, much of it reported in television newscasts and some of it reflected in network decisions about program content. It was also a time of growing interest in the study of television as a window on larger society, and some of that interest resulted in isolation of a number of specific effects relevant to variations in program success—in particular, effects of a show's genre (comedy vs. drama), of the network on which it appeared (ABC, CBS, NBC, and FOX), and of the period during which it debuted.

Genre Effects

The situation comedy is the foundation of prime-time programming. During the years of the present study, the average rating was consistently higher for comedic shows than for other program types (Mintz, 1985). The other primary genre in episodic television programming is the 1-hr drama, of which there have been several sub-genres including police procedurals, detective stories, action-adventure dramas, fantasy, science fiction, family drama, the Western, medical drama, and prime-time soap opera. The sub-genres tend to be cyclical in popularity, initially debuting as a result of innovation, then evolving via imitation, and ultimately declining as a result of saturation. Situation comedies were more popular than dramatic series among network executives and advertisers, as well as (and partly because) more popular among viewing audiences. Situation comedies provided a form of escapism from the tensions of daily life, whereas dramatic shows made more demands on the viewer's imagination, created suspense, and could put viewers in a frame of mind that would counteract sponsors' hopes for their advertising investments. Advertisers agreed that it was easier to sell products when viewers were in a relaxed frame of mind, and comedy prepared viewers for consumption (Kuhns, 1970). For network executives, economics figured in yet another way: Production costs were lower for situation comedies than for dramas (Eastman, Head, & Klein, 1985). Thus, when examining the possibility of a racial factor in program success, it will be important to control difference of genre.

Network Effects

The broadcast histories and programming preferences of networks probably influenced the types of BC programs aired as well as their ratings, shares, and rates of survival. Gitlin (1983) and Atkin (1992) discussed differing programming trends of each network. NBC, the oldest commercial network that began broadcasting in 1926, was considered the most innovative of the major three and was known for introducing forms such as the "desk and sofa talk show" (e.g., *Today Show, Tonight Show*) as well as the free-form comedy variety format (e.g., *Saturday Night Live* and *Laugh-In*), the 1-hr soap opera, and the miniseries (Brown, 1982). CBS began broadcasting 1 year later and was considered the most conservative commercial network, in part, because during its earliest years of existence, CBS was able to secure prominent broadcast signals for its owned stations and for many of its affiliates, which reached further than NBC signals. CBS had more rural affiliates and greater penetration of rural markets than the competition, and it was ratings leader into the mid-1970s (Brown, 1982). ABC began in 1943 when NBC was forced by the Federal Communications Commission to sell its Blue Network. The youngest of the Big 3 commercial networks with fewer affiliates in small markets and more in big cities, ABC developed programming for young urban audiences (e.g., *Happy Days, Laverne and Shirley, Charlie's Angels*, and *The Love Boat*). In 1986, the FOX network debuted and initially focused on young urban audiences. This strategy was partly targeted toward a niche audience often ignored by the older networks—young African Americans—with programs such as *Martin, Living Single*, and *New York Undercover*. By the mid-1990s, however, as FOX gained greater presence in the general market, its program content moved away from minority-centered shows and toward White-centered programs that appealed to younger audiences (e.g., *Melrose Place, X-Files*, and *Beverly Hills 90210*).

In an analysis of series from 1950 to 1991, Atkin (1992) offered two economically based arguments to explain network trends in minority-lead series programming.[1] He suggested that during periods of greater internal competition among the networks, minority-lead programming should increase, as the networks become more willing to try innovative programming to attract audiences. According to Atkin, network competition peaked between 1976 and 1979. In addition, he suggested that periods of external competition (mainly from cable television and videocassette recorders) should correlate positively with minority-lead series. Atkin found partial support for both arguments, noting that increases in minority-lead programming peaked during the late 1970s and again after 1985. Furthermore, he found statistically significant differences among the three major networks in the number of minority-lead series broadcast: NBC had 45% of all minority-lead series; ABC, 31%; CBS, 24%. This is consistent with MacDonald's (1992) report that in the early years of television, NBC, in an effort to improve the network's image with African Americans, published guidelines for the equitable portrayal of minorities on television. The network noted that "henceforth all programs treating aspects of race, creed, color and national origin would do so with dignity and objectivity" (MacDonald, 1992, p. 4).

Those reports support the claim that CBS was the most conservative of the commercial networks and NBC the most progressive. Consequently, the network on which a BC program aired might have affected its success. The more innovative networks might have been willing not only to experiment with unconventional programming but also to be more patient while it tried to build its audience.

Period Effects

Another factor that could affect program ratings and shares is historical period—that is, the year a program began. In general terms, period effects are often manifest in the production of a cultural commodity; for, as Peterson (1982) put it, "the nature and content of symbolic products are shaped by the social, legal and economic milieus in which they are produced" (p. 145). An example specific to the present investigation is the effect of the Kerner Commission's citation of the media as having played a role in the "creation of a schism between black and white America" (Stroman, Merritt, & Matabane, 1989, p. 44).[2] Whereas even the most progressive executives of the television industry had been reluctant to place African Americans in prominent roles, for fear of offending large segments of the viewing audience, thus reducing ad revenues, the stance of the Kerner Commission opened a safer space by recommending that "television should develop programming which integrates Negroes into all aspects of televised presentations . . . In addition to news-related programming we think that Negroes should appear more frequently in dramatic and comedy series" (Stroman et al., 1989, p. 45). In response, networks developed BC programming that had a better chance of being palatable to White audiences by representing of Black life in comedic genres or in supporting roles. At the time of the Commission's report, network television offered only one Black-featured drama: *I Spy* (1965-1968), with Bill Cosby as the first African

American in a leading role in a dramatic series. In the years following the Commission's recommendations, three BC programs debuted: two situation comedies, *Julia* (1968-1971) and *The Bill Cosby Show* (1969-1971), and a Western, *The Outcasts* (1968-1969).

There has also been evidence of a technology-related period effect. While numerous observers had been noting a downward drift in average ratings and shares, Lewine, Eastman, and Adams (1989) found that the threshold level of ratings or shares at which a program would be canceled had declined about 10% during the 1980s. Whereas prior to the 1980s, a rating "below 20 (or an audience share of less than 30) almost always resulted in cancellation on any network," the threshold then dropped to a rating of 17 (or a share of 27). Network decision makers were being more patient with "slow builders"—"programs that acquire a loyal audience only after months of patient nurturing" (e.g., *Cheers*). This shift coincided with the introduction and increased use of cable television as well as satellite systems, which cut into the major network's share of the viewing audience.[3]

In sum, the period during which a series was telecast—and more especially, the year of its debut—probably had an impact on its success. Programs that aired during the first two decades of the interval of this study had to survive within an environment of relative high average ratings and shares, but they could do so within a more restricted (by comparison with later years) number of market competitors. That situation changed during the 1980s, for a variety of reasons. Thus, the present analyses must be alert to period effects. This especially includes careful attention to series that were broadcast in the years immediately following the Kerner Commission's report.

Theoretical Guidance and Hypotheses

Media choice theories seek to explain why audience members select specific media content.[4] For purposes of the present study, selection of content is treated within the context of theoretical perspectives that seek to explain media choice as an active rather than passive process. Given the finite options from which to choose (e.g., radio, magazines, televisions, newspapers, books) and the finite options of content available within each of those options, audience members not only consciously accept or reject program content but are also attentive to the content selected.[5] According to the "selective exposure" thesis, audience members are more likely to choose content that is consistent with their prior opinions and beliefs and to avoid content they assume to be, or discover is, incompatible with those beliefs and opinions (Hartman, 2009; Webster et al., 2000; Wright, 1986). Experimental research in political campaigns, for example, has shown that preferences in program content tend to determine who listens or watches which messages (Wright, 1986). During the period of the present study, Zillman and Bryant (1985) found that selection of entertainment programming could provide comfort and/or

reduce discomfort via pacifying information. Individuals anxious about crime, for instance, were more likely than others to select program content that made ideals of justice readily apparent and victorious and avoided content that featured unanswered victimization (Severin & Tankard, 1992). Other factors that affect media choice include demographic and cognitive-emotive characteristics of potential and actual audience members, as well as factors of opportunity and availability (Hartman, 2009; McQuail, 1997; Webster et al., 2000; Wright, 1986). The present study focuses on characteristics of available program content and on responses by potential and actual audience members as measured by their collective voice in ratings and shares data. Consequences of choice of content to view are for present purposes of the survival of a given show. Consequences of choice extend far beyond that, of course. In a longitudinal panel study of the differential effects of television exposure, for example, Martins and Harrison (2012) found that, in general, the level of self-esteem was adversely affected among Black children and among White girls. White boys, however, displayed improved self-esteem. This and other consequences are undeniably important. They are beyond the scope of the present investigation.

This study looks behind the "comfort zone" thesis to examine detailed evidence about audience reactions to hundreds of shows that were telecast between 1963 and 1994 on the major commercial networks. To what extent do representations of blackness in fictionalized episodic television programs affect success as measured by ratings and shares of audience? Mean ratings and means shares, calculated for the lifetime of each of the television shows, are compared within a 2 × 2 grid defined by BC versus not and drama versus comedy. Other variables that affected program success will be introduced as controls, to gauge sensitivity of success to the two primary variables, race and genre. Accordingly, and in view of the fact that the large majority of the television audience was not Black (much less BC), my main hypotheses are as follows:

Hypothesis 1: BC programs received significantly lower average ratings and shares than programs that were not BC.

Hypothesis 2: Dramas received significantly lower average ratings and shares than comedies.

Hypothesis 3: BC dramatic programs received significantly lower average ratings and shares than other programs (dramas and comedies not BC as well as BC comedies).

Hypothesis 4: Each of the expected differences was robust enough to withstand controls for the effects of network and period.

Method

The data analysis is based on a working sample drawn from a full sample of 410 prime-time television programs—87 BC and 323 non-BC—telecasted on the four major commercial networks between 1963 and 1994 (Brooks & Marsh, 1995). BC programs (hereafter, BC vs. non-BC) are operationally defined as (a) the cast *predominately* consisted of regularly appearing African American characters (e.g., *Good Times, The Cosby Show*) or (b) the cast included at least one *major* regularly appearing African American character and the program sometimes featured plots that focused on African American issues (e.g., *I'll Fly Away*). To check the validity of my own identifications, I consulted three sources in addition to Brooks and Marsh (1995), MacDonald (1983), Woll and Miller (1987), and Jackson (1982).

The 1963-1964 television season is the starting point because it marks the beginning of network attempts to broadcast BC dramatic series. The 1994-1995 television season is the end point because of the predominance of BC programming on Warner Brothers (WB) and UPN, two new networks that initially featured content designed to appeal to African American audiences.[6] All of the 410 prime-time programs offered fictionalized content, whether comedic or dramatic. News, sports, movies of the week, variety programs, "talk shows," and "specials" are excluded. The comfort zone thesis is most relevant to fictionalized and regularly appearing television characters.

The approximate universe of programs among the four networks between 1963 and 1995 was 1,233, almost evenly split between genres (618 comedies and 615 dramas).[7] Given the large disproportion between BC and non-BC shows, sample selection was split: All BC prime-time television programs, comedy or drama, telecast between 1963 and 1994 were selected; a random sample of all other prime-time comedies and dramas of the same time period on the four major networks was drawn, resulting in a selection of 323 non-BC programs, using Brooks and Marsh (1995) as the sampling frame. This produced the full sample of 410 shows.

The sampled observations consist of the ratings and shares for each program from birth to death, extracted from the Nielsen Television Index (NTI). However, 18 of the programs were right-censored on ratings and shares data (i.e., they remained on air at end of observation) and have been excluded. In addition, ratings and shares data were missing for 36 programs. The Library of Congress holds the most comprehensive, publicly available collection of the NTIs.[8] For purposes of this study, I was able to review the NTI journals from September 1963 through December 1977. However, the Library of Congress collection was missing NTIs from January 1978 through August 1985. In addition, data are missing for some programs (e.g., *Shaft* and *Tenafly*) that were part of rotating series.[9] Excluding cases because of right censorship and missing data resulted in a working sample of 351 cases. While sample-selection bias cannot be ruled out conclusively, the comparisons displayed in Table 1 suggest that the bias is minimal. The ratio of BC to non-BC shows are nearly identical between the full and the working samples (1 to 4.8 vs. 1 to 4.5), and the distributions by network are very similar. Other comparisons tell the same story: The exclusions very likely do not account for analytic results reported below.

Table 1. Distribution of Samples (Full, Working, and by BC Status) Across the Four Networks.

	CBS	NBC	ABC	FOX	Total
Full sample					
n by network	123	109	153	25	410
As % of Σ	30.0	26.6	37.3	6.1	100.0
n BC by network	26	25	28	7	86
As % of Σ	30.2	29.1	32.6	8.1	21.0
As % of network Σ	21.1	22.9	18.3	28.0	—
Working sample					
n by network	102	91	134	24	351
As % of Σ	29.1	26.0	38.0	6.9	100.0
n BC by network	22	24	25	7	78
As % of Σ	28.2	30.8	32.1	9.0	22.2
As % of network Σ	21.6	26.4	18.7	29.2	—

Note. BC = Black-centered.

Dependent Variables

Audience acceptance or rejection is measured by program ratings and shares, as previously described. Ratings and shares are reported on a biweekly or weekly basis in the NTI.[10] Both measures focus on the number of television households viewing a specific program. This number is then relativized to a different base: all households with a television in the case of ratings and only those households viewing *any* program during the relevant time slot in the case of shares. Advertisers have often been especially interested in ratings, as it indicates the penetration of an advertiser's message into the total *potential* audience (i.e., the general population of consumers). Shares are better suited for comparing program performance relative to the competition (Webster et al., 2000).

An average rating value and an average share value were calculated for the lifetime of each of the sampled programs, using the NTI data as described above. Although these means are highly correlated ($r = .97$), each distribution has been used in analyses, in case one of the measures was more sensitive to viewer choices on the BC versus non-BC dimension.

Independent Variables

In addition to the main variable of interest (BC, coded 1, vs. non-BC), four categorical variables comprise the list of conditions available to analysis: genre (drama, coded 1, vs. comedy); network (with CBS as reference category when all four networks are considered); period, which was coded as three partially overlapping dummy variables to explore different possible effects (more below); and season of program debut (with fall as reference category, covering debuts from mid-August to January vs. winter, January through April, and summer, May to mid-August). Season of debut is potentially important as a control, because shows that debut in winter typically had lower chances of survival, as did those debuting during summer, despite the fact that their competition often consisted of "summer reruns."

Table 2. Mean Ratings and Mean Shares of TV Shows, by Race and Genre.

	M	SD	n
Mean ratings			
BC dramas	11.28	3.105	20
BC comedies	11.89	4.145	58
Non-BC dramas	11.32	3.763	139
Non-BC comedies	12.65	4.242	134
Mean shares			
BC dramas	19.57	6.024	20
BC comedies	20.61	7.378	58
Non-BC dramas	19.76	6.678	139
Non-BC comedies	21.63	6.887	134

In the present study, the main hypothesized period effect has to do with the report of the Kerner Commission. Period 1 (1968-1975, coded 1) is designed to tap possible responses in the immediate aftermath of that report. Setting 1975 as the end of that period is arbitrary: If pertinent responses decayed rapidly, 1975 is perhaps too late, but if pertinent responses were slow to occur, 1975 is perhaps too early. However, lengthening the interval increases the probability that other factors, unrelated to the report, will appear as responses to it, while decreasing the interval could work against the fact that program development, including the recruitment of commercial sponsors to pay for the development costs, takes time. (In fact, however, moving the upper terminus did not result in a qualitative change of outcome, and the quantitative difference was very small.)

Findings

Basic Analyses

As point of departure, let us revisit the distributions by network of the full sample and of the working sample, both in total and by BC status (Table 1). In addition to the lack of evidence of selection bias, two conclusions stand out. First, the BC shows were distributed across networks in the same proportions as the total of BC and non-BC shows. Second, within network, the proportion of shows that were BC was greatest on FOX; otherwise, the networks were not different. While the FOX difference is in keeping with prior comments about the network's early strategy, it is well short of statistical significance. Also in keeping with prior reports, dramas were scarcer among BC than among non-BC shows (12.6% vs. 30.2%, $\chi^2 = 15.6$, $p < .001$). Table 2 reports that comparison as well as a comparison of mean ratings and mean shares within the 2 × 2 of race and genre. The main conclusion from the latter comparison is indifference. The main effect of genre could be considered significant (at $p < .10$, or $p < .05$ in a one-tailed test), with comedies slightly more successful. But the difference by race (i.e., BC vs. non-BC) was

essentially zero (ratings: 11.8 vs. 12.0, $p = .54$; shares: 20.4 vs. 20.7, $p = .77$), and the interaction effect (BC × Drama) was also far short of inferential significance (ratings: $p = .53$, shares: $p = .68$).

In sum, the analyses so far have demonstrated support for the second hypothesis (the genre effect) but have led to rejection of the first and third hypotheses. Before resting with that conclusion, however, some possible confounding conditions must be examined. For instance, present data confirm that there was a downward trend in ratings and shares (Figure 1). Most of the decline occurred after 1975. During the first several years of observation, ratings and shares fluctuated around a flat (trendless) line, the only notable divergence being an increased volatility during the early 1970s. But from 1975 onward, both ratings and shares drifted downward.[11] Was some aspect of the fact of BC programming implicated in that downward drift? One might ask, for instance, whether a proportionate increase in BC shows, perhaps in network responses to the Kerner Commission report, had an overall negative effect on audience choice of television as an evening entertainment. There is no evidence for it in the present analyses, although this data set is not ideally suited to answer that question. What can be said is that the decline occurred in both BC and non-BC categories, however, without significant difference in rate of decline. Furthermore, if, in line with the comfort zone thesis, White audiences were deflected from television viewing because of programming responses to the Kerner Commission report, that deflection was a long-delayed effect (beginning in 1975, 7 years after the report).

Suppression of a BC effect on ratings and/or shares success is unlikely but should not be ruled out of hand. The question is whether results of the difference-of-means tests have been confounded by other conditions—in particular, network, period, and debut season, as well as genre. Given the relatively small number of BC cases, a multivariate ordinary-least-squares analysis will yield pertinent tests while conserving degrees of freedom. Table 3 reports these tests for the shares measure. Results for the ratings measure are qualitatively identical (and quantitatively different in insignificant degree); thus, they are not shown to conserve page space.[12]

Again, there is no evidence in support of any but the second research hypothesis, now including the fourth one. Consistent with results of the bivariate analyses reported above, the genre effect is maintained net of other conditions. Network also made a difference in shares (and ratings), with ABC shows somewhat performing better, and FOX shows performing much worse, than shows on either CBS or NBC. Season of debut was also a significant condition of success, inasmuch as shows debuting in winter (i.e., January through April) performed slightly less well than shows debuting during the fall season. This is not an unexpected result, of course, given the structure of commercial network program development.

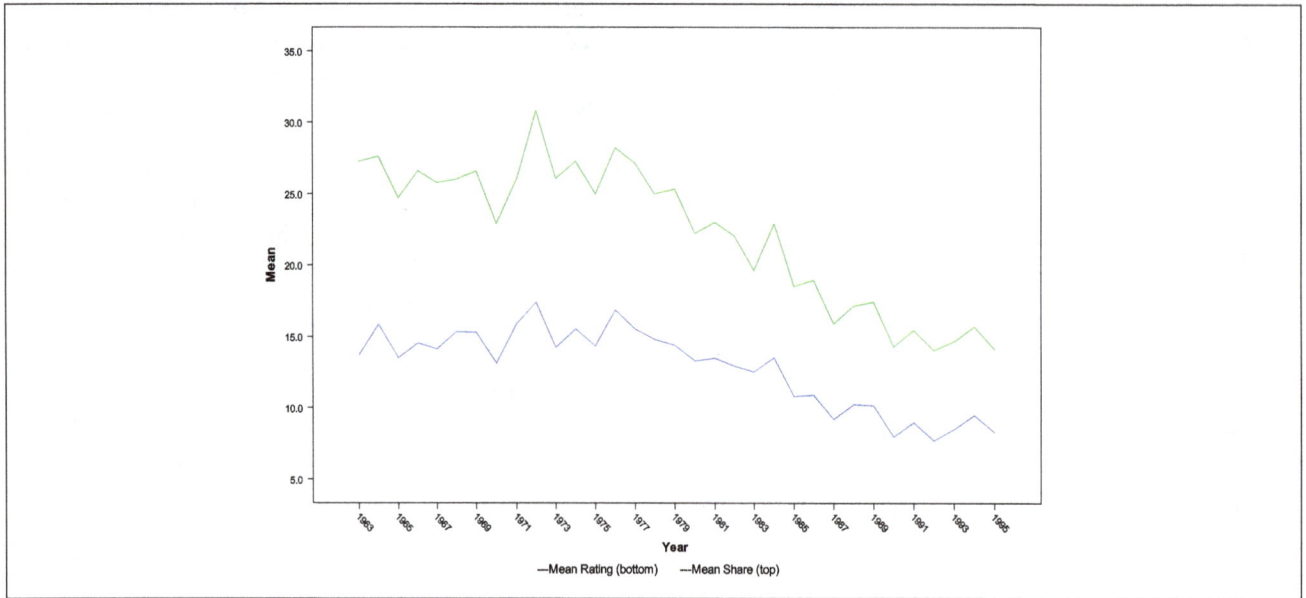

Figure 1. Mean rating and share by year.

The regression coefficient for Period 1 is positive and significant, indicating that program inauguration during the years following the Kerner Commission report fared better. However, the weight of this condition was the same for non-BC as for BC programs. Moreover, there is reason to conclude that the coefficient is in fact not describing a period effect so much as it is capturing a chunk of the information shown in Figure 1—that is, the downward trend in shares (as in ratings). When trend is factored out, the coefficient for Period 1 collapses (to .84, with a standard error of .80). The same result holds for each of the other period terms (hence, no point in pursuing them).[13] If a period effect was in fact associated with the Kerner Commission report, it was too subtle to be manifest within the general trend in mean shares and mean ratings.

The chief conclusion from the multivariate analysis is that the empirical observations do not support the main prediction regarding BC programs, as based on Carter's comfort zone thesis. As can be seen in Table 3, the coefficient for BC (vs. non-BC) never reached parity with its standard error. If there was an effect of the status of a program as BC versus not, it was not manifest in these measures of program success.

Discussion

The finding on genre effect fits with prior studies and with the sort of gratification theory perspective discussed by Wright (1986). Television viewers, perhaps wanting to escape the daily realities of life, may seek temporary escape through the comic antics of a favorite television character in a situation comedy, which is less tension-filled or anxiety-ridden than one finds in the typical dramatic program that may include a

Table 3. Predicting Mean Shares of TV Shows as a Function of Race, Genre, Network, Historical Period, and Season of Debut: OLS Estimations ($n = 391$).

	1	2	3	4	5	6
Constant	20.68***	21.32***	21.54***	21.02***	20.26***	21.58***
	(0.42)	(0.49)	(0.56)	(0.71)	(0.69)	(0.75)
	49.66	43.21	38.16	29.56	29.37	28.90
Black	-0.26		-0.71	-0.32	-0.67	-0.39
	(0.89)		(0.91)	(0.81)	(0.77)	(0.80)
	-0.30		-0.78	-0.39	-0.87	-0.49
Drama		-1.55*	-1.68*	-1.77**	-1.50*	-1.70**
		(0.74)	(0.75)	(0.67)	(0.64)	(0.67)
		-2.12	-2.23	-2.64	-2.35	-2.55
ABC				2.22**	1.66*	2.13**
				(0.80)	(0.77)	(0.80)
				2.77	2.16	2.68
NBC				1.41	1.63	1.38
				(0.88)	(0.84)	(0.85)
				1.60	1.95	1.58
FOX				-10.79***	-10.04***	-10.84***
				(1.38)	(1.32)	(1.38)
				-7.82	-7.60	-7.86
Period 1					5.22***	
					(0.87)	
					5.97	
Winter						-1.68*
						(0.72)
						-2.34
Summer						-0.62
						(1.31)
						-0.47
Adjusted R^2	.00	.01	.01	.22	.29	.23

Note. Cell entries are regression coefficient (top), standard error (middle), t ratio (bottom). OLS = ordinary least squares.
*$p < .05$. **$p < .01$. ***$p < .001$.

murder mystery, police chase, or courtroom drama. Furthermore, situation comedies are shorter in duration, and one does not have to be an "active" viewer to follow the plot of a situation comedy. Hour-long dramatic programs require more intellectual engagement to follow the plot.

The fact that programs airing on the FOX network tended to have lower mean ratings and shares than programs on the other networks is likely driven, in part, by the date of FOX's debut, which came after the general plateau in ratings and shares and which was followed by reduced scheduling in prime-time hours during its early years. Then, too, of course, a new network would fare less well in the competition for new programs, especially those likely to be winners, and would be less able to poach proven winners from the established networks. So the network effect was no more unexpected than the effect of season of program debut. Programs beginning during Fall months benefited from the long-established fanfare the latest "new TV season," which for television audiences often meant a build-up of anticipation after a long summer of reruns. Such programs had time to build an audience base, whereas replacement programs coming after the holiday season and New Year were less likely to draw viewers away from their established viewing patterns, which are generally more unconventional,

Finally, the fact that the multivariate model even in strongest version did not account for a large portion of the variation in ratings or shares is not an unexpected outcome. While this could well be partly because of the proverbial "missing variable problem" (e.g., variables that should have been included but were unobtainable, such as a direct measure of the motivation for viewing a television program), it is probably mainly an indication of a large random component in the process that determines the longevity and popularity of specific television programs. Simply put, as Bielby and Bielby (1994) observed years ago, "All hits are flukes." A mixture of structural factors within the television industry affects the types of television programs that are produced and ultimately their success or failure. Bielby and Bielby found that the success and ultimately the longevity of television programs occur after the fact, so to speak. The industry attempted to minimize risks of failure by using proven formulas and personnel with successful reputations. Programs that were given the "green light"—those that made it into a network's schedule—were likelier to reside within the comfort zone as programming executives had come to know it. For BC programs, this meant stereotypical representations in situation comedies and marginalized representations in dramatic programs. Supported by advertisers, America's commercial television sought to reach as many potential consumers as possible. This interdependent relationship between advertisers, who desire mass audiences, and commercial television, which depends on advertisers for support, resulted in the television industry creating programs that appealed to the largest audiences possible. Therefore, the programming strategies adopted by the industry involved finding common themes and content that appealed to a heterogeneous society. Television content generally reinforced the core norms and values of American society. To challenge these norms and values might offend segments of the mass audience and result in a loss of the large audience needed by advertisers (Wilson & Gutierrez, 1995).

That argument is in many ways compelling. So why was it not manifest in the results of the analysis of mean shares and ratings in terms of the contrast between BC shows, and especially BC dramas, and shows that were not BC?

One possibility is that the argument itself is not subtle enough. It paints too much in the stark contrast of Black and White, not allowing room in the middle, so to speak. By the measures considered here, BC dramas succeeded on average as well as non-BC dramas. Likewise, BC comedies succeeded on average as well as non-BC comedies. There may well have been a substitution effect accounting for part of that lack of difference: Black households, although not a large portion of the total, could increase the odds of a BC show, perhaps especially a BC drama, achieving equivalent success, by viewing that show in much higher proportion than the proportion of non-Black households viewing non-BC fare during that same time slot. But in addition, it could also have been true that a substantial portion of non-BC households also viewed the BC dramas that did succeed. At least judging by the present data on ratings and shares, there must have been a substantial "cross-over" viewership. It was not large enough to give *Shaft* a run of more than a handful of episodes, but then the program that alternated with it, *Hawkins*, staring Jimmy Stewart (an image about as different from Richard Roundtree's *Shaft* image as most viewers at the time could have pictured), also failed to survive more than a handful of outings.

This is not to deny the continued presence of stereotypical characterizations. They were rampant in BC comedies (as, differently, in many non-BC comedies). Indeed, Roundtree's *Shaft* was in its own way stereotypical, with a characterization that for much of the White audience was seen as an "in your face" flaunting, which accounted for part of its appeal to Black audiences. What was discomforting to one segment of viewing audience could be, and was, comforting to another, and the edge of that contrast worked in both directions.

Conclusion

The present study found no support for Carter's comfort zone thesis with regard to BC television programming. One could argue that the definition of "Black-centered," used to support reliability of analysis, was too loose, and that a tighter definition would have yielded evidence in support of Carter's thesis. Indeed, the definition of BC programs included not only those in which *all* regularly appearing casts members were African American (e.g., *The Cosby Show* and *Amen*) but also those with plots that focused on issues relevant to African American issues, though with mixed casts (e.g., *I'll Fly Away* and *In the Heat of the Night*).

But a narrower definition of "Black-centered" would have reduced the power of the statistical tests. Not surprisingly, therefore, a replication of the main analysis using the narrower definition did not lead to a different conclusion.

A dimension of program success that has not been directly addressed is duration. All television shows fail; some fail sooner than others. While duration is no doubt a function of mean ratings and mean shares, that relationship is far from complete. This study found no race-based distinction related to ratings or shares, but it was not designed to detect differences in duration as such. Future research should explore this dimension of program success.

Declaration of Conflicting Interests

The author(s) declared no potential conflicts of interest with respect to the research, authorship, and/or publication of this article.

Funding

The author(s) received no financial support for the research and/or authorship of this article.

Notes

1. "Minority-lead" indicated a program in which only the lead or co-lead character, whether adult or child, was Black, Hispanic, or Asian.
2. The Kerner Commission—formally, the National Advisory Commission on Civil Disorders—held the media partially responsible for creating a social environment that resulted in the urban summer riots of the mid-1960s. The limited visibility of representations of Blackness in entertainment portrayals reinforced the inequalities of American society. Even in the make-believe world of television, African Americans were on the periphery when not excluded altogether. The commission's report was released February 29, 1968.
3. Note, too, that FOX was becoming a strong competitor to what had been the three major networks.
4. For a recent extensive overview, theoretical and empirical, see Hartman (2009).
5. The context of this study, bear in mind, is the pre-Internet world. Options have greatly expanded, leading some to argue that greater media richness has posed new opportunities and new problems for media choice, though the "richness" claim has been disputed. See, for example, Robert and Dennis (2005) and Trevino, Lengel, and Daft (2012).
6. Both the Warner Brothers (WB) network and United Paramount Network (UPN) began broadcasting in January 1995.
7. That near-equivalence is telling, as most comedies were 30-min long, while many dramas were of hour length: In general, the a priori chance of success was much lower for a dramatic series.
8. The Nielsen Television Index (NTI) is a publication of Nielsen Media Research that issues biweekly or weekly national network television ratings and shares.
9. Rather than appearing weekly, *Shaft* alternated with *Hawkins* and *The New CBS Tuesday Night Movies. Tenafly* was one of four rotating series on the *NBC Tuesday/Wednesday Mystery Movie* (Bogle, 2001; Brooks & Marsh, 1995). The NTI did not identify rating and share with a specific program in the rotating sequences.
10. The NTI was published every 2 weeks until August 1987; subsequent issues were published on a weekly basis.
11. The missing data problem for the 1978-1985 interval might be confounded in the downward drift, but the fact that the decline continued at roughly the same pace after 1985 suggests that the observed pattern is not an artifact.
12. The corresponding table is available from the author. The correlation between measures was the same within the BC category of cases as within the non-BC category.
13. Factoring out the trend effect also reduced the weight of the FOX effect (to −7.25). The genre effect was unchanged, and although the point estimate for BC was higher (.79), it remained far below significance level.

References

Atkin, D. (1992). An analysis of television series with minority-lead characters. *Critical Studies in Mass Communication, 9,* 337-349.

Bielby, D. D., & Bielby, W. T. (1994). "All hits are flukes": Institutionalized decision making and the rhetoric of network prime-time program development. *American Journal of Sociology, 99,* 1287-1313.

Bogle, D. (2001). *Prime time blues: African Americans on network television.* New York, NY: Farrar, Straus and Giroux.

Brooks, T., & Marsh, E. (1995). *The complete directory to prime time network and cable TV shows.* New York, NY: Ballantine Books.

Brown, L. (1982). *Les Brown's encyclopedia of television.* New York, NY: Zoetrope.

Carter, R. G. (1988). TV's black comfort zone for whites. *Television Quarterly, 23*(4), 24-34.

Carter, R. G. (2007). How roots and Black.White. Broke racial TV ground. *Television Quarterly, 37*(1), 55-60.

Eastman, S. T., Head, S. W., & Klein, L. (1985). *Broadcast/cable programming: Strategies and practices.* Belmont, CA: Wadsworth.

Gitlin, T. (1983). *Inside prime time.* New York, NY: Pantheon Books.

Hartman, T. (2009). *Media choice.* London, England: Routledge.

Jackson, H. (1982). *From "Amos 'n' Andy" to "I Spy": Chronology of blacks in prime time network television programming, 1950-1964.* Ann Arbor, MI: University Microfilms.

Kuhns, W. (1970). *Why we watch them: Interpreting TV shows.* New York, NY: Benziger.

Lewine, R. F., Eastman, S. T., & Adams, W. J. (1989). Prime-time network television programming. In S. T. Eastman, S. W. Head, & L. Klein (Eds.), *Broadcast/cable programming: Strategies and practices* (pp.134-172*).* Belmont, CA: Wadsworth.

MacDonald, J. F. (1983). Blacks and White TV: Afro-Americans in Television Since 1948. Chicago: Nelson-Hall Publishers.

MacDonald, J. F. (1992). Blacks and white TV: Afro-Americans in television since 1948. Chicago: Nelson-Hall Publishers.

Martins, N., & Harrison, K. (2012). Racial and gender differences in the relationship between children's television use and self-esteem. *Communication Research, 39,* 338-357.

McQuail, D. (1997). *Audience analysis.* Thousand Oaks, CA: SAGE.

Mintz, L. E. (1985). *"Situation comedy" in TV genres: A handbook and reference guide* (B. Rose, Ed.). West Port, CT: Greenwood Press.

Peterson, R. A. (1982). Five constraints on the production of culture: Law, technology, market, organizational structure and occupation careers. *Journal of Popular Culture, 16,* 66-74.

Robert, L. P., & Dennis, A. (2005). Paradox of richness: A cognitive model of media choice. *Professional Communication, 48*(1), 10-21.

Severin, W. J., & Tankard, J. W., Jr. (1992). *Communication theories: Origins, methods, and uses in the mass media.* White Plains, NY: Longman.

Stroman, C., Merritt, B., & Matabane, P. (1989). Twenty years after Kerner: The portrayal of African Americans on prime-time television. *The Howard Journal of Communications, 2,* 44-56.

Trevino, L. K., Lengel, R. H., & Daft, R. F. (2012). Media symbolism, media richness, and media choice in organizations. *Communication Research, 14,* 553-574.

U.S. Commission on Civil Rights. (1978). *Window dressing on the set: An update.* Washington, DC: U.S. Government Printing Office.

Webster, J. G., Phalen, P. F., & Lichty, L. W. (2000). *Ratings analysis: The theory and practice of audience research.* Mahwah, NJ: Lawrence Erlbaum.

Wilson, C., & Gutierrez, F. (1995). *Race, multiculturalism, and the media: From mass to class communication.* Thousand Oaks, CA: SAGE.

Wimmer, R. D., & Dominick, J. R. (1991). *Mass media research: An introduction.* Belmont, CA: Wadsworth.

Woll, A. L., & Miller, M. (1987). *Ethnic and racial images in American film and television: Historical essays and bibliography.* New York, NY: Garland.

Wright, C. R. (1986). *Mass communication.* New York, NY: McGraw-Hill.

Zillmann, D., & Bryant, J. (1985). Affect, mood, and emotion as determinants of selective exposure. In D. Zillmann & J. Bryant (Eds.), *Selective Exposure to Communication* (pp.157-190). Hillsdale, New Jersey: Lawrence Earlbaum Associates.

Author Biography

Brenda L. Hughes, PhD, is an assistant professor of sociology at Florida A&M University. She is the program coordinator for the sociology program. Her research interests include mass media, stratification, and popular culture.

10

Authentic Language Input Through Audiovisual Technology and Second Language Acquisition

Taher Bahrani[1], Sim Shu Tam[2], and Mohm Don Zuraidah[2]

Abstract

Second language acquisition cannot take place without having exposure to language input. With regard to this, the present research aimed at providing empirical evidence about the low and the upper-intermediate language learners' preferred type of audiovisual programs and language proficiency development outside the classroom. To this end, 60 language learners (30 low level and 30 upper-intermediate level) were asked to have exposure to their preferred types of audiovisual program(s) outside the classroom and keep a diary of the amount and the type of exposure. The obtained data indicated that the low-level participants preferred cartoons and the upper-intermediate participants preferred news more. To find out which language proficiency level could improve its language proficiency significantly, a post-test was administered. The results indicated that only the upper-intermediate language learners gained significant improvement. Based on the findings, the quality of the language input should be given priority over the amount of exposure.

Keywords

audiovisual programs, authentic language input, exposure, second language acquisition

Introduction

In today's audiovisually driven world, technology has become the track upon which the train of education is heading toward its destination quickly. The growth of its application and its rapid development in transforming the process of learning is unbelievable (Mayya, 2007). Computer-assisted language learning (CALL, hereafter), computer and audiovisual equipped classes, the Internet, e-mail, chat, and mobile-assisted language learning (MALL, hereafter) are just a few examples of the application of technology in language teaching/learning. As a result, language learners can have access to various types of authentic language inputs through various technologies such as computers, TV, and CDs/DVDs, apps among others for language learning.

However, there are many internal as well as external factors that influence second language acquisition (SLA). Among them, language input that learners receive in SLA is one of the external factors that plays a fundamental role. However, while the important role of language input in SLA has been advocated by various language learning theories, there has been a controversy in the field of language acquisition between those theories that attribute a small or no role to language input and those attributing it a more important role (Ellis, 2008).

Language input has also been considered to be a major source of data for language learners to construct their competence or mental representation of the language (Patten &

Benati, 2010). Indeed, language acquisition process is dependent upon the availability of appropriate input. Considering the fact that some sort of language input is necessary to acquire the language in-and-outside the classroom, various audiovisual programs have the potential to be utilized as sources of authentic language input for SLA.

Taylor (1994) defined authentic language material as any material in English which has not been specifically produced for the purpose of language teaching. Similarly, Nunan (1999) defined authentic language materials as spoken or written language material that has been produced in the course of real communication and not specifically produced for the very purpose of language teaching.

In the last few years, various audiovisual technologies have dominated the world by massive developments in providing language learners/teachers with sources of authentic language input for SLA. Indeed, audiovisual technologies have provided many possibilities for teachers to construct activities for language learners. Accordingly, language learners can have access to various authentic language inputs

[1]Department of English, Mahshahr Branch, Islamic Azad University, Mahshahr, Iran
[2]University Malaya, Kuala Lumpur, Malaysia

Corresponding Author:
Taher Bahrani, Faculty of Languages and Linguistics, University Malaya, Kuala Lumpur 43300, Malaysia.
Email: taherbahrani@yahoo.com

through different technologies such as computers, TV, and CDs/DVDs for language learning, particularly outside the classroom settings. On similar lines, the integration of different audiovisual programs such as news, films, comedy, cartoons, and songs as sources of authentic language input into language learning has attracted the attention of many researchers (Bahrani & Tam, 2011; Brinton & Gaskill, 1978; Kaiser, 2011; Mackenzie, 1997; among others).

In view of the above, the present research aims at finding out the relationship between greater exposure to a particular type of audiovisual program and language proficiency development in informal settings.

Review of the Related Literature

The effect of exposure to audiovisual news broadcasts as a source of authentic language input has been the focus of research since the 1980s. Brinton and Gaskill (1978) conducted a study concerned with the effectiveness of listening to audiovisual news programs on enhancing English as a foreign/second language (EFL/ESL) learners' listening skills. The study was carried out in Germany in an EFL context and in the United States in an ESL context. During the study, videotaped news broadcasts from TV were incorporated in advanced EFL/ESL classes once a week for approximately 6 months. According to Brinton and Gaskill (1978), exposure to audiovisual news language input has the potential to improve the listening skill because different newscasts bring reality into the classroom and enable the students to focus on substantive issues. In addition, because of the recycling feature of vocabulary in different audiovisual news, EFL/ESL language learners become more familiar with many contextualized vocabulary items during a long period of exposure.

In the 1990s, the possibility of using audiovisual news reports as language input for lower-proficiency-level language learners has been scrutinized by Mackenzie (1997). Without providing empirical evidence, Mackenzie (1997) rejected the idea that because the newscasters speak very fast, the content is very multifaceted, and the vocabulary is very difficult, audiovisual news cannot be integrated into low levels of language learning classrooms. In contrast, with the careful selection of audiovisual news items and applying some simple techniques such as selecting the content of the news reports based on the language learners' interest and background knowledge, news reports can be used even at elementary or intermediate levels (Mackenzie, 1997).

More recently, Bahrani and Tam (2011) conducted an experimental research to gauge the effectiveness of exposure to audiovisual news broadcasts materials and non-news materials on improving the speaking proficiency of intermediate language learners. The research was carried out with 60 intermediate language learners who were assigned to two groups. During the experiment, the participants in the first group were exposed to authentic materials from audiovisual news, whereas the second group participants were exposed

to non-news materials. The findings indicated that exposure to audiovisual news promotes intermediate language learners' speaking proficiency more than exposure to non-news materials. According to Bahrani and Tam (2011), the intermediate participants showed their enthusiasm in the creative use of various vocabularies, sentences, and structures in talking about the topics during the interviews (speaking test) in the post-test.

In short, the review of the literature on the use of audiovisual news as a source of authentic language input for SLA reveals that most of the studies were descriptive and examined the pedagogical value, the possibility of using news at all levels of language learning, and the selection criteria without empirical evidence.

Movies have been also regarded as an important source of language learning for language instructors because it is an authentic source of material (Kaiser, 2011). In fact, movies provide language learners with opportunities of exposure to the real language uttered in authentic settings (Stempleski, 1992). The spoken language of movies often includes various types of speeches such as those of various educational levels, the speeches of children and non-native speakers, slang and jargon, rural and urban speeches, and a range of regional dialects that language learners will encounter in the target language country (Kaiser, 2011). However, the review of the quantitative studies regarding the incorporation of movies as a source of authentic language input for language learning is limited.

In a research conducted by Huang and Eskey (2000), the effectiveness of exposure to closed captioned TV on intermediate language learners' listening comprehension was addressed. The findings of the study showed that captions improved listening comprehension skill. On similar lines, Gebhardt (2004) and Heffernan (2005) anecdotally considered movies to be utilized as motivating materials, which have the potential to enhance language learning.

Supported by empirical evidence, Yuksel (2009) highlighted the effectiveness of viewing movie clips with/without captions on EFL learners' vocabulary enhancement. The research was carried out with 120 language learners in a preparatory class. The participants were randomly divided into two groups. Before the treatment, the participants in both groups took a sample 20-item vocabulary knowledge scale pre-test. During the study, participants in Group 1 were exposed to some movie clips with captions, whereas the second group participants were exposed to the same movie clips without captions. One month after the treatment, participants in both groups were given another vocabulary knowledge scale test with 20 words as a post-test. The results obtained from the pre–post-tests of both groups revealed that both groups demonstrated significant gains. In fact, viewing the movie clips reinforced the expansion of the vocabulary knowledge of the language learners regardless of the presence or the absence of captions (Yuksel, 2009). According to Yuksel (2009), the development in the vocabulary knowledge stems from the importance of encountering the vocabularies in the real context. In fact, incidental vocabulary learning can be facilitated through contextual cues.

Other studies have mostly focused on the effectiveness of exposure to movies with or without captions/subtitles on developing listening skills (Huang & Eskey, 2000; Markham, 1999; Markham & Peter, 2003). Huang and Eskey's (2000) research considered the effectiveness of exposure to closed captioned TV on intermediate language learners' listening comprehension. The findings of the study showed that captions improved listening comprehension skills. Similarly, Markham (1999) carried out a research on the effectiveness of captioned videotapes on listening vocabulary recognition skills. Markham concluded that the captions significantly helped language learners develop their listening word recognition skill.

In a nutshell, the review of the related literature on the use of movies as a source of authentic language input is mostly limited to either some qualitative studies that have considered movies to be pedagogically valuable authentic motivating material for language learning or a few quantitative studies that have focused on the effects of exposure to movies on enhancing listening skill.

Cartoons as authentic language materials have also been considered as excellent teaching tools because they not only add humor to a topic but also illustrate the idea in a memorable way. In this regard, many studies underscore the point that cartoons make learning an enjoyable and, importantly, memorable experience, because the activities in the teaching and learning process using cartoons are interesting and interactive for the students.

In an anecdotal study conducted by Clark (2000), it was highlighted that cartoons can engage the attention of the learners and present information in a non-threatening atmosphere. Besides, cartoons have the potential to reinforce thinking processes and discussion skills (Clark, 2000). Another study was carried out by Doring (2002) focusing on the effect of exposure to cartoons on language learning. The results of the study revealed that the language learners who had exposure to cartoons could produce oral answers that were very proactive and interesting in different discussions held in the classes, which creates non-threatening learning environment for them. Moreover, the discussions were rich and the students had high confidence. It seems that the high confidence that the language learners acquire is due to exposure to cartoons that create low affective filter atmosphere for learning.

Rule and Auge (2005) conducted a study providing evidence of the students' preferences to use cartoons in language learning. Accordingly, cartoons are preferred because they create a high degree of motivation to recognize and produce humor for the students, enhance the memory, and make connection between the new materials and the prior knowledge through analogy. However, Rule and Auge (2005) did not go further to provide empirical evidence on the effect of exposure to cartoons on language development.

In a recently conducted research, Bahrani and Tam (2012) compared the effectiveness of exposure to audiovisual news, movies, and cartoons as three different sources of language input on language proficiency development of the low-level language learners. To do so, Bahrani and Tam (2012) assigned 60 low-level language learners into three groups. The three groups ran by one of the researchers and met in three different classes once a week for a period of 4 months. Each group had exposure to different audiovisual materials. Group 1 had exposure to news, Group 2 had exposure to movies, and Group 3 had exposure to cartoons. The results of the study showed that those participants who had exposure to cartoons could enhance their language proficiency to a significant extent. In contrast, the participants who had exposure to either news or movies could not gain significant language proficiency development.

A review of the related literature in the area of the incorporation of various songs as another type of authentic material in language teaching provides limited empirical evidence. Schoepp (2001) anecdotally proposed songs in foreign language classrooms to lower anxiety and increase motivation, provide physiological benefits, guide lesson planning and practical classroom, and enhance cultural awareness and sensitivity.

Documentary films as authentic and communicative teaching materials that reflect the real situation have also attracted the attention of some researchers. Recently, Soong (2012) has supported the use of documentary films over other types of audiovisual materials by highlighting the pedagogical values of utilizing documentary films in language learning, particularly oral interpretation classes. The author underscores, unlike other types of audiovisual materials such as movies that use a great number of slangs or even abusive languages in the dialogues, that the language of documentaries is carefully scripted and delivered in Standard English, which makes them more suitable as teaching materials for EFL students. However, documentary films should be carefully selected and even tailor-made for the oral interpretation class, so as not to discourage students (Soong, 2012). Soong (2012) emphasizes the point that language learners can learn how to pronounce new words correctly while watching documentary films. Besides, the key vocabulary items tend to recur throughout the soundtrack, providing EFL students with valuable repeated encounters with the lexical content (Soong, 2012).

To determine the appropriate duration of documentary films in terms of time and content to be utilized in 2 hr oral interpretation class especially with low and pre-intermediate-level students, Soong surveyed 129 students who had been taught oral interpretation for half a year using documentary films. The survey included four questions related to their viewpoints on the application of documentary films as teaching materials. Another source of data came from five teachers who also used documentary films in their teaching for at least a year. The results of the questionnaires and the interview revealed that documentary films are more interesting than textbooks, long documentary films (more than 20 min) should not be used in the classrooms because they may frustrate the students, and documentary films meet the demands of authenticity and communication.

However, Soong's study lacked empirical evidence on the effect of utilizing documentary films in oral interpretation classes on the students' scores. Indeed, the conclusions were only based on the students' and the teachers' opinions.

In short, wide arrays of audiovisual programs are available as authentic sources of language input for SLA in EFL and ESL contexts. However, the related studies are mostly anecdotal. The quantitative studies have also been mostly investigated in the formal language learning setting. Indeed, informal language learning setting that has great potential for SLA has not been investigated. Importantly, language learners' exposure to their preferred type of audiovisual program and language proficiency development has not been investigated.

Research Questions

The present research was set to answer the following questions:

Research Question 1: What type of audiovisual program is watched most by the low-level/upper-intermediate-level language learners in informal settings?
Research Question 2: What is the correlation between the low-level language learners' amount of exposure to the most viewed audiovisual program and their language proficiency development (if any improvement is gained)?
Research Question 3: For the upper-intermediate-level language learners, what is the correlation between the amount of exposure to the most viewed type of audiovisual program and their language proficiency development (if any improvement is gained)?

Method

Participants

Initially, 134 language learners aged 21 to 26 years majoring in Teaching English as a Second Language (TESL), including both males and females, from Malaysia went through the research voluntarily. Then, a smaller population of 60 participants was selected based on a sample International English Language Testing System (IELTS) pre-test. The 60 selected participants were divided into low level ($n = 30$) and upper-intermediate level ($n = 30$) based on the scores which they obtained in the pre-test.

Instruments

The two instruments that were utilized to accumulate the necessary data for the present research were a set of two

Table 1. Descriptive Statistics Related to the Administration of the Two Tests to the Same Group.

	n	Minimum	Maximum	M	SD	Variance
First test	20	4.00	7.00	5.433	0.673	0.424
Second test	20	4.00	7.00	5.254	0.745	0.502

parallel IELTS language proficiency tests and a self-report sheet. Prior to the main data collection, the sample IELTS language proficiency tests were verified to be parallel to ensure the internal validity of the findings. Both tests were also verified in terms of reliability.

To obtain quantitative data on the type and the amount of the low and the upper-intermediate participants' most preferred type of audiovisual program in informal settings, a self-report sheet consisting of news, films, cartoons, songs, documentary films, series, game shows, talk shows, speeches, and sport programs as 10 different types of audiovisual materials was prepared and given to the 60 selected participants after the pre-test. The validity of the self-report sheet was also verified through a pilot study.

Procedure

The present research used quantitative method and a pre–post-tests design. The first step to take before the study was carried out was to verify that the two sample IELTS language proficiency tests were parallel to assure the internal validity of the data obtained from pre–post-tests. Parallel tests are two tests of the same ability that have almost the same means and variances when administered to the same group after a short interval (Bachman, 1990).

Accordingly, both sample IELTS language proficiency tests were administered to 20 trial language learners majoring in TESL with a short interval (1 week). Then, the means and the variances of both tests were calculated separately (Table 1).

The statistical analysis of the data obtained from the administration of both tests to the 20 participants showed that the means and the variances of both tests were almost the same, which indicated that the two-sample IELTS tests were parallel. Moreover, the reliability coefficient of the two tests was calculated as .943 using Cronbach's alpha formula.

The second instrument to be verified was the self-report sheet. To this end, all the 20 trial language learners were asked to have exposure to various audiovisual programs outside the classroom setting for 1 week and report their preferred type and amount of exposure to various programs by filling out the self-report sheet. In relation to the format and entries, the researchers asked the participants to report any possible difficulty or problem they might face while filling out the self-report sheet. The input extracted from the analysis of the participants' self-report sheets indicated that it could be used for collecting the necessary data to answer Research Question 1.

Table 2. Low-Level Language Learners' Total Amount of Exposure to Each Program During the Period of the Study.

Type of program	News	Films	Documentary films	Songs	Talk shows	Game show	Sport programs	Series	Speeches	Cartoons
Amount (min)	6,450	9,485	2,350	3,462	260	346	650	1,750	238	18,038

Table 3. Upper-Intermediate-Level Participant' Total Amount of Exposure to Each Program During the Period of the Study.

Type of program	News	Films	Documentary films	Songs	Talk shows	Game shows	Sport programs	Series	Speeches	Cartoons
Amount (min)	16,685	8,548	1,380	3,580	220	120	550	870	1,180	4,805

After verifying the instruments, the actual data collection started with the selection of the participants. To do so, one of the parallel IELTS language proficiency tests was given to a population of 134 language learners including both males and females majoring in TESL to select 60 participants initially. Then, 30 participants who obtained 4 or 4.5 of 9 in the overall band score were selected as low level and 30 participants who obtained 6 or 6.5 of 9 in the overall band score were selected as upper-intermediate level. The selection criteria were based on the IELTS band score categories.

Following the selection of the participants, the participants of both proficiency levels were asked to have exposure to their preferred audiovisual programs outside the classroom (for 8 weeks) and keep a diary of the type and the amount of the program(s) they watch. At the end of the study, the second IELTS test was administered to all the participants. The data obtained from the pre–post-tests and the self-report sheets provided answers to Research Questions 2 and 3.

Results

To provide answers to the research questions, the data obtained from the pre–post-tests and the self-report sheets were analyzed and tabulated using SPSS 19 statistical analysis software.

Research Question 1

In relation to the first research question concerning the low-level language learners' most preferred type of audiovisual program, a one-way repeated-measures ANOVA was conducted to assess whether there were significant differences between the mean amounts of exposure to each type of audiovisual program in the low proficiency level. The results indicated a significant main effect of program, $F(2.840, 68.160) = 199.974$, $p < .001$, partial $\eta^2 = .893$, which was a great effect size. This indicated that the mean amounts of exposure to various types of audiovisual programs were not equal (Table 2). Accordingly, low-level language learners' preferred cartoons more than other types of programs.

With regard to the intermediate-level participants, a one-way repeated-measures ANOVA was also conducted to assess whether there were significant differences between the mean amounts of exposure to each type of audiovisual program in intermediate proficiency level. The results indicated a significant main effect of program, $F(3.142, 43.201) = 185.205$, $p < .001$, partial $\eta^2 = .749$, which was a great effect size. The results indicated that the mean amounts of exposure to various types of audiovisual programs were not equal (Table 3). As a result, the intermediate-level language learners preferred various news programs more than other types of audiovisual programs.

Research Questions 2 and 3

To answer the second and the third research questions and determine the relationship between the low-/upper-intermediate level language learners' amount of exposure to cartoons/news and their language proficiency development, the data obtained from the administration of the pre–post-tests to the low-/upper-intermediate-level language learners were first analyzed by means of a paired-sample t test to find out whether the participants in each proficiency level could improve their language proficiency. Then, the correlation between the low-/upper-intermediate-level language proficiency improvement and the amount of exposure to cartoons/ news was calculated separately using the means of correlation coefficient.

In relation to the raw scores obtained from the pre–post-tests by the participants in the low-level group, there was a minor increase in the mean score in the post-test. A paired-sample t test was conducted to find out the significance of the improvement. According to the results of the paired-sample t test (Table 4), the value of the t-observed was not statistically significant ($p > .05$).

Considering the second research question, the correlation coefficient between the amount of exposure to cartoons and the low-level participants' language proficiency improvement was low and not significant ($r = .122$, $p = .906$).

In relation to the scores obtained from the pre–post-tests by the participants in the upper-intermediate level, there was

Table 4. Descriptive Statistics Related to the Low-Level Participants IELTS Pre-test and Post-test Results.

Low level	n	M	SD	t test
Pre-test	30	4.09	0.53	−0.25
Post-test	30	4.28	0.46	p > .05

Note. IELTS = International English Language Testing System.

Table 5. Descriptive Statistics Related to the Upper-Intermediate-Level Participants IELTS Pre-test and Post-Test Results.

Intermediate level	n	M	SD	t test
Pre-test	25	5.53	0.50	−3.20
Post-test	25	6.25	0.46	p < .05

Note. IELTS = International English Language Testing System.

a bigger increase in the mean score in the post-test compared with that of the low level. To find out whether the increase in the mean score was significant, a statistical analysis of a paired-sample t test was conducted. The analysis of the paired-sample t test (Table 5) showed that the t-observed was statistically significant ($p < .05$), which was indicative of the fact that the increase in the mean score was significant enough to lead to significant language proficiency improvement.

The amount of exposure to news broadcast significantly correlated with the upper-intermediate-level significant language proficiency improvement ($r = .429$, $p = .002$).

Discussion

While comparing the amount of exposure of the low- and the intermediate-level participants to their preferred type of program, it was found that the low-level language learners' total amount of exposure to cartoons as their preferred program was 18,038 min and upper-intermediate-level language learners' total amount of exposure to news as their preferred program was 16,685 min. However, regardless of the greater amount of exposure, the low-level language learners' language proficiency improvement was not significant. It can be hypothesized that merely greater amounts of exposure to various sources of authentic language input may not contribute a lot to SLA.

The reason behind the difference in the language proficiency development of both proficiency levels may be due to the quality of the language input embedded in the type of audiovisual programs rather than the quantity. Although both cartoons and news programs are pedagogically valuable authentic language materials that have the potential to be used as sources of authentic language input, the type of language input embedded in cartoons may be, to the most extent, modified or simplified to ease comprehension. In contrast, authentic audiovisual news consists of more unmodified type of language input than the modified one.

While comprehending modified or simplified language input is easier and requires less cognitive processing because of the type of data that is presented in a way to facilitate comprehension, it may not contribute a lot to SLA. In contrast, although the language input embedded in the type of unmodified input such as news requires much more input processing for comprehension, it may contribute much more to SLA than modified input.

The fact that the low-level language learners of the present research showed greater interest in viewing cartoons more than other types of programs can be supported by what Rule and Auge (2005) put forth regarding the high degree of motivation created through viewing cartoons as a type of authentic material in a non-threatening atmosphere. On similar lines, the low-level participants of the present research might have watched cartoons more because they might have had very few problems comprehending the language of cartoons. However, the problem with this type of simplified or modified input may be that it may not include those linguistic aspects that the low-level language learners need to acquire to enhance their language proficiency and go to a higher level of proficiency in short term. As Gass (1997) put forth, simplified input is created on the assumption to facilitate comprehension rather than causing acquisition.

In contrast to cartoons, audiovisual news broadcasts include more unmodified input, a type of authentic language input that is not simplified or modified for the sake of comprehension. While having exposure to unmodified input rather than the modified one, language learners may experience more difficulty comprehending some parts of the input. However, language learners may benefit more from unmodified input because it includes much more linguistic aspects that they had not acquired. Comprehending the new linguistic aspects embedded in unmodified input requires much more input processing effort that can enhance language proficiency. This is supported by White's (1987) incomprehensible input hypothesis.

In her hypothesis, White (1987) underscored the point that when language learners come across language input that is incomprehensible because, for example, their interlanguage rules cannot analyze a particular second language structure, they have to modify those interlanguage rules to understand the structure. In this way, the incomprehensible input enhances the process of SLA. The fact that the intermediate-level language learners improved their language proficiency through exposure to news broadcasts as an authentic source of language input sheds more light on the studies conducted by Bahrani and Sim (2011), Brinton and Gaskill (1978), and Mackenzie (1997). Accordingly, the present study showed that greater exposure to news might be the reason for the intermediate participants' significant language proficiency development. The reason behind this might be that there are some characteristics observed in developing any type of audiovisual news as a type of authentic material, which makes it a rich source of unmodified linguistic input to be utilized for language learning.

One of the characteristics of audiovisual news is vocabulary recycling that is considered as redundancy of input that can boost language learners' comprehension. Another essential characteristic of audiovisual news is speech fluency, which is considered as the use of appropriate pausing, rhythm, intonation, stress, and rate of speaking. Fluency of speech is a marked linguistic feature that can be observed in utterances designed and developed for audiovisual news stories to be read by newscasters. One more important feature of audiovisual news is the special discourse which is used throughout the issue. In this regard, essential factors such as the nature of the news; the cognitive, affective, and social status of both the news items and the audience; and the structure of the news should be of focus.

The point should be highlighted at the end that although the low-level language learners who had exposure to cartoons could not improve their language proficiency in general, further analysis of their language proficiency through pre–post-tests revealed that they could significantly enhance their listening skills. This indicates that cartoons as most preferred type of audiovisual language materials may prove effective in improving low-level language learners listening skills.

With regard to the upper-intermediate-level language learners who could significantly enhance their language proficiency during the period of the research, a closer look at their pre–post-test scores revealed that they could enhance their speaking skill more than other language skills.

Conclusion

It is a proven fact that in any form of language acquisition, input is essential for the success. Although a lot has been written on the role and importance of input in second language acquisition, limited studies have provided empirical data on the source and the type of language input. Considering this point and with the impressive developments in technology and the accessibility of different audiovisual programs that can provide authentic language input, the present article aimed at providing further empirical support on low and intermediate language learners' preferred type of audiovisual programs as authentic sources of language input and language proficiency enhancement.

Accordingly, it was found out that while the low-level language learners' preferred type of audiovisual program was cartoons, intermediate language learners preferred news more among other types of programs. The results of the study were indicative of the fact that the low-level language learners' amount of exposure to cartoons was more than intermediate-level language learners' amount of exposure to news. However, the higher quantity of exposure to cartoons

as a preferred source of authentic language input did not cause more improvement in language proficiency of low-level language learners. This was indicative of the fact that the quality of the type of exposure contributes more to the language proficiency enhancement than the quantity of the language input.

In a nutshell, the results of the study may be important to language teachers, practitioners, and institutions for investment in audiovisual technologies for language learning by exposing the language learners more than before to the most effective types of authentic audiovisual materials.

The point should be noted that this study has some limitations to be considered. First, it has addressed low- and upper-intermediate-level language learners. Consequently, the findings of the study are limited to these levels of language proficiency. The need to conduct the study with intermediate- or advance-level language learners is warranted. Second, the researcher did not further investigate how both group language learners dealt with the comprehension of the type of language input which they had exposure to in informal settings. Finally, it should be mentioned that the present research was conducted in informal settings where many variables which might have influenced the results could not be controlled. Hence, different results might be obtained if the study is conducted in formal settings with control and experimental groups.

Declaration of Conflicting Interests

The author(s) declared no potential conflicts of interest with respect to the research, authorship, and/or publication of this article.

Funding

The author(s) received no financial support for the research and/or authorship of this article.

References

Bachman, L. F. (1990). *Fundamental considerations in language testing*. Oxford, UK: Oxford University press.

Bahrani, T., & Tam, S. S. (2011). Technology and language learning: Exposure to TV and radio news and speaking proficiency. *Kritika Kultura, 17*, 144-160.

Bahrani, T., & Tam, S. S. (2012). Audiovisual news, cartoons, and films as authentic language input and language proficiency development. *The Turkish Online Journal of Educational Technology, 11*(4), 56-64.

Brinton, D., & Gaskill, W. (1978). Using news broadcasts in the ESL/EFL classroom. *TESOL Quarterly, 12*, 403-413.

Clark, C. (2000). Innovative strategy: Concept cartoons. *Instructional and Learning Strategies, 12*, 34-45.

Doring, A. (2002). The use of cartoons as a teaching and learning strategy with adult learners. *New Zealand Journal of Adult Learning, 30*, 56-62.

Ellis, R. (2008). *The study of Second language acquisition* (2nd ed.). Oxford, UK: Oxford University Press.

Gass, S. M. (1997). *Input, interaction, and the second language learner*. Mahwah, NJ: Lawrence Erlbaum.

Gebhardt, J. G. (2004). Using movie trailers in an ESL CALL class. *The Internet TESL Journal, 10*(1). Retrieved from http://iteslj.org/Techniques/Gebhardt-MovieTrailers.html

Heffernan, N. (2005). Watching movie trailers in the ESL class. *The Internet TEFL Journal, 9*(3). Retrieved from http://iTESLj.org/Lessons/Heffernan-MovieTrailers.html

Huang, H., & Eskey, D. (2000). The effects of closed-captioned television on the listening comprehension of intermediate English as a second language students. *Educational Technology Systems, 28,* 75-96.

Kaiser, M. (2011). New approaches to exploiting film in the foreign language classroom. *L2 Journal, 3*(2), 232-249.

Mackenzie, A. S. (1997). Using CNN news video in the EFL classroom. *The Internet TEFL Journal, 3*(2). Retrieved from http://www.aitwech.ac.jp/~iTESLj/

Markham, P. (1999). Captioned videotapes and second language listening word recognition. *Foreign Language Annals, 32,* 321-328.

Markham, P., & Peter, L. (2003). The influence of English language and Spanish language captions on foreign language listening/reading comprehension. *Journal of Educational Technology Systems, 31,* 331-341.

Mayya, S. (2007). Integrating new technology to commerce curriculum: How to overcome teachers' resistance? *The Turkish Online Journal of Educational Technology, 6*(1), 8-14.

Nunan, D. (1999). *Second language teaching and learning.* Boston, MA: Heinle & Heinle.

Patten, V. B., & Benati, A. G. (2010). *Key terms in second language acquisition.* London, England: Continuum international publishing group.

Rule, A. C., & Auge, J. (2005). Using humorous cartoons to teach mineral and rock concepts in sixth grade science class. *Journal of Geosciences Education, 53,* 548-558.

Schoepp, K. (2001). Reasons for using songs in EFL/ESL classrooms. *The Internet TEFL Journal, 7*(2). Retrieved from http://iTESLj.org/articles/Schoepp-Songs.html

Soong, D. (2012). Using documentary films in oral interpretation class what is the appropriate length? *International Journal of Applied Linguistics & English Literature, 1*(6), 131-141.

Stempleski, S. (1992). Teaching communication skills with authentic video. In S. Stempleski & P. Arcario (Eds.), *Video in second language teaching: Using, selecting, and producing video for the classroom* (pp. 7-24). Alexandria, Egypt: Teachers of English to Speakers of Other Languages.

Taylor, D. (1994). Inauthentic authenticity or authentic inauthenticity? *Teaching English as a Second or Foreign Language, 1,* 1-10.

White, L. (1987). Against comprehensible input: The input hypothesis and the development of L2 competence. *Applied Linguistics, 8,* 95-110.

Yuksel, D. (2009). Effects of watching captioned movie clip on vocabulary development of EFL learners. *The Turkish Online Journal of Educational Technology, 8*(2), 48-54.

Author Biographies

Taher Bahrani is an assisstant professor in TEFL. His field of interest is technology and language learning. He has published many ISI-indexed papers related to second language acqisition.

Sim Shu Tam is a senior lecture at the University of Malaya. Her filed of interest is CALL.She has published many ISI-indexed papers.

Mohm Don Zuraidah is a full professor in Applied linguistics. Her field of interest is discourse studies. She is one of the famous scholars in discourse studies.

11

Engaging With Patient Online Health Information Use: A Survey of Primary Health Care Nurses

Jean Gilmour[1], Sue Hanna[2], Helen Chan, Alison Strong[3], and Annette Huntington[1]

Abstract

Internet health information is used by patients for health care decision making. Research indicates this information is not necessarily disclosed in interactions with health professionals. This study investigated primary health care nurses' engagement with patient online health information use along with the respondents' disclosure of online sources to their personal health care provider. A questionnaire was posted to a random sample of 1,000 New Zealand nurses with 630 responses. Half the respondents assessed patients' online use ($n = 324$) and had encountered patients who had wrongly interpreted information. Health information quality evaluation activities with patients indicated the need for nursing information literacy skills. A majority of respondents (71%, $n = 443$) used online sources for personal health information needs; 36.3% ($n = 155$) of the respondents using online sources did not tell their personal health care provider about information obtained. This study identifies that there are gaps in supporting patient use but more nursing engagement with online sources when compared with earlier studies.

Keywords

nurses, health education, primary health care, information technology, information literacy

Introduction

Internet health information is an important education resource commonly sought by patients to support their health care decision making. An online survey of 12,262 respondents from 12 countries found that three fifths used online health information and 46% of this group used the information to self-diagnose health problems (McDaid & Park, 2011). Internet use in New Zealand is almost universally accessed by under 40-year-olds and is cited as the most valued medium for information (Gibson, Miller, Smith, Bell, & Crothers, 2013).

People independently seeking out online health information are motivated to develop their knowledge about health conditions and possibly adopt new lifestyle patterns (Fox, 2007). Supporting the development of confident decision-making skills is an important aspect of health education to strengthen beliefs about self-efficacy (Coleman & Newton, 2005). Self-efficacy is an aspect of the social learning theory developed by Bandura (1986) and is defined as "beliefs in one's capabilities to organize and execute the courses of action required to produce certain attainments" (Bandura, 1997, p. 3).

The Internet enables information access to support decision making but the medium has "noisy" characteristics. Noise in a communication sense is defined as interference

with message exchange and common understandings of the message (Bartol, Tein, Matthews, Sharma, & Scott-Ladd, 2011). DeVito (2012) identifies four noise types: physical, physiological, psychological, and semantic. These noise types can interfere with accessing and understanding online health information. Physical noise is external interference such as poor computer or mobile device screen visibility as a consequence of small font or popup advertisements. Physiological barriers include poor vision and cognitive issues. Psychological noise occurs when there is mental interference to the message being understood such as an emotional response to particular information or pre-existing biases. Semantic barriers are produced through complex and specific terminology and the lack of shared meaning between the sender and receiver.

External and intrinsic barriers to health information message reception can be mitigated with health literacy skills. Health literacy is underpinned by knowledge and skills in

[1]Massey University, Wellington, New Zealand
[2]Middlesex University, London, UK
[3]Hawke's Bay District Health Board, Hastings, New Zealand

Corresponding Author:
Jean Gilmour, School of Nursing, Massey University, Wellington, Private Box 756, Wellington 6021, New Zealand.
Email: J.A.Gilmour@massey.ac.nz

accessing, understanding, discerning, and applying information to make effective health management decisions and to seek appropriate health care (Keleher & Hagger, 2007). The need for online health literacy skills is demonstrated in a study testing online information literacy by van Deursen and van Dijk (2011). In this study only 35% of the participants were able to use the information they found for beneficial decision making. Education was the most important contributor to strategic Internet use by the study participants.

Nurses have a role to play in developing patient health literacy skills to support the efficacious use of online health information. Nurses' education activities require an understanding of resources available, including online information, which will support people's knowledgeable engagement in care decisions. There is the need to enquire about the information patients are independently sourcing to build skills and efficacy, as well as developing their evaluation skills in judging the quality of the information. Teaching about information evaluation provides the opportunity to model assessment strategies to ascertain the quality and applicability of online information. Along with self-efficacy, modeling is an important concept in social learning theory, the central tenant being that learning is stimulated through observing behaviors and that "coded information serves as a guide for action" in the future (Bandura, 1986, p. 47).

This study is located in the primary health care (PHC) setting where increasingly long-term conditions are being managed. The PHC sector is the key site for patient education as the first point of contact with the health system (King, 2001) along with providing the majority of ongoing care for people who have chronic conditions (Caughey, Vitry, Gilbert, & Roughead, 2008). In New Zealand, PHC nurses contribute to structured education for people with chronic illnesses with assessment, care planning, and regular monitoring activities (Ashworth & Thompson, 2011; Henty & Dickinson, 2007). Teaching patients how to productively use online health information is an opportunity to develop reciprocal knowledge-based relationships as reported in a study of oncology nurses' experiences of patients' online information use (Dickerson, Boehmke, Ogle, & Brown, 2005). On one hand nurses can alert patients to critical information assessment techniques, and on the other, patients can increase nurses' understanding of the information they have independently sourced and found as being meaningful and useful in their illness diagnosis and management.

Background

Patients use the Internet to clarify and extend the information provided by doctors and other health professional (Bowes, Stevenson, Ahluwalia, & Murray, 2012; Kivits, 2006; Knapp et al., 2011; Pletneva, Cruchet, Simonet, Kajiwara, & Boyer, 2011) and value the anonymity and accessibility attributes of Internet searches (Horgan & Sweeney, 2010).

Evidence over the last decade shows that many nurses have not yet made the transition to acknowledging patients' independent information-seeking activities and incorporating online health information in patient education. A survey of 1,170 Spanish nurses found 72.8% felt online material was relevant to very relevant to patients and 73% had discussed online information with at least some patients, but 54.4% would not recommend online information to patients (Lupiáñez-Villanueva, Hardey, Torrent, & Ficapal, 2011). The issues around nursing engagement with online information are longstanding as illustrated by an older Scottish survey of 130 general practice nurses where 73% of the nurses accessed the Internet but only 29% would refer patients to Internet information (Wilson, 1999).

New Zealand studies have also traced broader nursing engagement with a notable feature being the minority of respondents who assessed patients' use of online information (37% of postgraduate nursing respondents [Gilmour, Scott, & Huntington, 2008], 11% of undergraduate nursing respondents [Scott, Gilmour, & Felden, 2008], and 24.4% of nursing respondents working in medical wards [Gilmour, Huntington, Broadbent, Strong, & Hawkins, 2012]). These studies also consistently found that a very small minority of the nurses in the study settings worked to develop patients' information evaluation skills.

There is research interest on the effect of disclosure of independently accessed Internet information on relationships between patients and health professionals. Patients' reasons for discussing health information during medical consultations include wishing to make the best possible use of the time, along with seeking clarification and reassurance (Bowes et al., 2012). However, some patients, including those who are health professionals, choose not to discuss the information they found with their doctor: 47% of patients and 38% of health professional patients in a Health on the Net survey of 524 participants (Pletneva et al., 2011) and 31% of nurse respondents working in medical wards (Gilmour et al., 2012).

Imes, Bylund, Sabee, Routsong, and Sanford (2008) surveyed 714 Americans about constraining factors in discussing online health information with health care providers. Important factors were fear of disrupted relationships with the provider, intruding upon the health professionals' authority, worries about being perceived in a negative way for bringing up information, and concerns about the health care professional dismissing the information as invalid. McMullan (2006) argues that health professionals vary in their responses to health information users. Threatened health professionals use their positioning as expert to guide the consultation, in contrast patient-centered practitioners work collaboratively with clients. McMullan also suggested there is a group of health professionals who guide patients to useful sites. The validity of this latter approach is supported by survey findings where 80% of patients wanted health professionals to provide trustworthy online information (Pletneva et al., 2011).

Patients' concerns about locating trustworthy information are supported by the large body of published work focused on the quality of online health information. In a systematic review of research assessing the quality of online material, 70% of the studies suggested there were quality issues (Eysenbach, Powell, Kuss, & Sa, 2002). More recent empirical work shows that quality concerns are an enduring theme in the literature. Recent studies report information quality concerns with websites on health topics such as oncology information (Lawrentschuk et al., 2012), online infant sleep recommendations (Chung, Oden, Joyner, Sims, & Moon, 2012), and common pediatric issues (Scullard, Peacock, & Davies, 2010). A study of the compliance of websites with asthma education guidelines found that only 8.8% met all the guideline criteria with the implication that nurses needed to be knowledgeable about the most accurate sites so they could be recommended to their patients (Meadows-Oliver & Banasiak, 2010).

Primary health nurses who are working in the first point of health contact need to be alert for opportunities to strengthen clients' beliefs about self-efficacy and develop information evaluation skills. Research findings highlight a range of issues with evidence of (a) limited nursing engagement with online health information sources in patient education, (b) substantial patient use of online sources as a backup to health professional education, (c) patient perceived barriers to the communication of information sources with health professionals, and (d) quality concerns with some online material.

Study Aims

A premise of this research, informed by the issues raised in the literature review, is that patients may not volunteer to health professionals their information sources for a variety of reasons. There is also the possibility that the information sources, or patients' interpretations of information, may be flawed in nature. The aims of this research were to establish PHC nurses' assessment of patient online health information use and their support with patient evaluation of online education material, and explore PHC nurse's reasons for nondisclosure of information sources to their personal health care providers. Personal health care providers are defined for the purposes of this study as health professionals who provide personal health care to the respondents and their families.

Methods

The study used a cross sectional survey research design. A cross sectional study systematically collects quantifiable data from the population of interest at one point in time (Bryman, 2012). The data were collected from a random selection of PHC nurses using a mailed questionnaire. The

findings reported in this study on nurses' engagement with patient's online use were from one section of the questionnaire which also included sections on nurses' online access and heart failure education resources and activities. The survey questions elicited predominately quantifiable data through the use of fixed response items. There were two open-ended items generating textual data. The textual data were analyzed qualitatively using a content analysis approach. Qualitative content analysis involves the categorization of textual data through the identification of patterns and themes (Julien, 2008).

Study Sample

Sample inclusion criteria were currently working in PHC settings and selection of the Nursing Council of New Zealand categories "Primary health care" and "Practice nursing" at the time of annual practicing certificate renewal. Exclusion criteria were not currently working in the PHC sector and not agreeing to be contacted for research purposes. There were 4,673 practicing nurses in the relevant categories at the time of the survey and 2,780 of the group had agreed to be contacted for research purposes. The minimum survey sample size was calculated as 197 with an 80% probability of getting a statistically significant result with a population correlation effect size of .2, based on an alpha level of .05. A sample of 1,000 nurses was randomly selected from the Nursing Council of New Zealand data base.

Instrumentation

The study questions were based on a questionnaire used previously with nurses working in medical wards (Gilmour et al., 2012) and informed by a review of research literature. The questionnaire was further refined for the PHC sector but changes from the questionnaire used in the earlier study (Gilmour et al., 2012) were minimal. New questions were developed about the frequency that respondents' asked clients if they accessed the Internet for health information, and their awareness of patients' misconceptions because of incorrect Internet health information. The questionnaire was piloted with five PHC nurses to check for face validity along with questionnaire flow, length, and clarity.

The questionnaire had 12 items (Table 1). Four questions collected ordinal data; the topics were respondents' perceptions of clients' online information use, frequency of client discussion about online information, frequency of assessment of patient use, and confidence with computers. Six questions collected categorical data; topics included assessment of patient use, awareness of patients' misconceptions, and respondents' use for personal health care. Two open-ended questions generated qualitative data; respondents were asked how they worked with patients to develop information evaluation skills and why respondents did not discuss information found on the Internet with their personal health care

Table 1. Survey Questions and Response Categories.

How often do you believe that the clients who use your service access online information about their health?
(Response categories: often, sometimes, rarely, never, don't know)

How often do clients discuss online health information with you?
(Response categories: everyday, several times a week, several times a month, every few months, never)

Do you ask clients if they access the Internet for health information?
(Response categories: yes, no)

If yes:
How often do you ask clients if they access the Internet for health information?
(Response categories: always, most of the time, sometimes, occasionally, extremely rarely)

Do you assist your clients to evaluate the quality of Internet health information?
(Response categories: yes, no)

If yes:
Please explain how you help clients evaluate the quality of Internet health information.
(Open question with text box)

Are you aware of any of your clients having had misconceptions about their illness because of a **wrong interpretation** of correct Internet health information?
(Response categories: yes, no)

Are you aware of any of your clients having had misconceptions about their illness because of **incorrect** Internet health information?
(Response categories: yes, no)

Do you use the Internet to access health information for **your own or your family's health**?
(Response categories: yes, no)

If yes:
Do you tell your own or your family's health professional/s about the information you have found on the Internet?
(Response categories: yes, no)

Please explain any reasons why you **don't discuss** this information with your health professional.
(Open question with text box)

How do you rate your relationship to computers?
(Response categories: expert, confident, average, not confident, terrified of them)

provider where applicable. In addition, demographic characteristics were collected on age, gender, ethnicity, and nursing qualifications.

Data Collection

The 1,000 questionnaires were mailed out in April 2010. There was a follow up 2 weeks after the first mail out with a thank you and brief reminder, and a second mail out after 4 weeks to those who had not replied with a replacement questionnaire. Data collection was carried out from April until June 2010.

Ethical Considerations

The University Human Ethics Committee (Application 09/68) approved the study. An information sheet was posted with the questionnaire covering the survey aim and a statement that the respondents were anonymous to the researchers. An administrator separate to the research team managed the mail out and the follow up processes.

Analysis

The questions generated categorical, ordinal, and qualitative data. SPSS 20.0 (IBM SPSS Inc., Chicago, IL, USA) for Windows was used for the statistical analysis. Summary statistics are presented along with a categorical data test of proportions (Pearson chi-square), the Mann–Whitney test for ordinal data group differences, and a non-parametric data correlations test (Spearman's rank correlation co-efficient, $[r_s]$). The respondent total number varies according to the topic as answering questions was voluntary. Nurses who did not work with patients who accessed online health information ($n = 12$) are not included in the analysis of relevant topic areas.

The qualitative data were analyzed using an inductive content analysis approach whereby interpretation was informed by the respondents' textual data rather than a preexisting theoretical viewpoint (Julien, 2008). The analytic process began with a close reading of the texts followed by the development of codes, where extracts of text are named, and then the codes were grouped into categories. The qualitative data were initially coded and grouped into seven categories for each open-ended question by one researcher, the codes and categories were then confirmed by a second researcher. The textual data categories were also quantified with a count of the comments by categories (Bryman, 2012). Quotes are included in the findings to illustrate and validate the choice of categories.

Results

Response Rate

The final response rate was 65.5% (630 valid responses); 39 nurses informed us they were no longer eligible as they were not working in the PHC sector. The sample statistical power is calculated as being a 99% probability of getting a statistically significant result with a population correlation effect size of .2, based on an alpha level of .05.

Sample Description

Almost all respondents were female (99%, $n = 620$), the mean age was 49.45 years with ages ranging from 23 to 70 (Table 2). The major ethnic groups were New Zealand European (80.7%, $n = 501$) and Māori (9.2%, $n = 57$). The sample differs from the overall New Zealand registered nurse

Table 2. Sample Demographic Characteristics.

		n	%
Gender (n = 626)	Female	620	99.0
	Male	6	1.0
Age (n = 624)	M (SD)	624, M = 49.45(SD = 9.274)	
Ethnicity (n = 621)	New Zealand European	501	80.7
	Māori	57	9.2
	Pacific	11	1.8
	Other	52	8.3
Highest professional qualification (n = 621)	Certificate	162	26.1
	Diploma	94	15.1
	Degree	112	18.0
	PG Certificate	150	24.2
	PG Diploma	65	10.5
	Masters	38	6.1

Table 3. Online Health Information and Patient Use.

Question	Categories	n	%	Median
Belief about how often patients use online health information (n = 615)	4 = often	189	30.7	3.00
	3 = sometimes	299	48.6	
	2 = rarely	90	14.6	
	1 = never	6	1.0	
	0 = don't know	31	5.0	
Frequency patient discuss online health information with respondent (n = 610)	4 = everyday	19	3.1	2.00
	3 = several times a week	115	18.9	
	2 = several times a month	239	39.2	
	1 = every few months	174	28.5	
	0 = never	63	10.3	
Frequency assess patient online use (n = 319)	4 = always	2	0.6	2.00
	3 = most of the time	63	19.7	
	2 = sometimes	178	55.8	
	1 = occasionally	70	21.9	
	0 = extremely rarely	6	1.9	

workforce, 92% of the New Zealand overall nursing workforce is female, the average age of the workforce is 45.6 years, 68% are New Zealand European, and Māori are 7% of the workforce (Nursing Council of New Zealand, 2011). However, there is a close match with the PHC workforce where 97% are female, 76% New Zealand European/Pakeha, and 9.8% are Māori (Nursing Council of New Zealand, 2011).

Assessing Patient Use of Online Health Information

Most respondents (79.3%, n = 488) believed that patients used online health information at least sometimes and 61.2% (n = 373) encountered patients requesting to discuss this information with them several times a month or more (Table 3). Half the respondents (53%, n = 314) had

encountered patients wrongly interpreting Internet health information and 44.4% (n = 266) were aware of patients' misconceptions due to incorrect information. About half the group (52.8%, n = 324) asked their patients if they accessed online information with 20.3% (n = 65) of this group assessing patients use most of the time. Nurses who assessed patient's use scored significantly higher on confidence with computers ($U = 39746.0$, $N1 = 323$ [mean rank 327.95], $N2 = 289$ [mean rank 282.53], $p < .001$) compared with the non-assessing respondents answering these questions. The assessment of patients use was significantly associated with nursing qualifications, $\chi^2(1, N = 605) = 16.408$, $p < .001$. Two thirds (62.9%) of nurses with postgraduate qualifications assessed patients' online use as compared with 46.1% of nurses educated to degree level or less. There was also a significant positive correlation between the frequency of asking patients if they accessed the Internet for health infor-

Table 4. Assisting Patients to Evaluate the Quality of Internet Health Information.

Categories	Comments	Number of comments
Refer to reputable sites	Refer them to reputable sites (Govt and non-governmental organizations)	113
	Recommend reputable sites	
	Refer them to reputable sites before they make their final decision	
	Offer individual sites (not Google/Wikipedia)	
Evaluate and discuss with reference to own knowledge and/or check patient's sources of information	Review publisher, research quality	66
	Ask open questions, encourage discussion	
	Encourage patients to discuss information from all sources with us	
Encouraging/giving people tools to think critically about quality of information	Explain how to check the source	42
	Teach how to check validity of information	
	Educate them about guidelines for assessing information/sites	
	Caution author/source of information (whose interests)	
Warn/caution about trusting sites	Caution as to use of some sites	28
	Tell them not to believe all they are told by some websites	
	Caution about Wikipedia	
	Warn them to be careful of information on Internet	
Refer/encourage patient to talk to health professional	Need to talk to a health professional	13
	Refer to other specialities where necessary	
	Encourage them to talk to a health professional	
Discuss sites/sources with colleagues	Discuss with colleagues	5
Doesn't do it/not applicable	My clients are all over 80 years of age	3
	Difficult—how do you teach a client to access research-based medicine	

mation and level of qualification (r_s = .196, N = 315, $p \leq 0.001$). Assessing patients' use was significantly associated with awareness of patient misconceptions because of incorrect interpretations of online health information, $\chi^2(1, N$ = 591) = 12.292, $p < .001$; 59.9% of assessing nurses were aware of incorrect interpretations compared with 45.5% of non-assessing nurses.

The mean age difference between the group that assessed patient use (49.47 years) and the group that did not (49.38 years) was similar as was the mean nursing years of experience (25.52 years for group that assessed, 24.46 years for the group that did not).

Evaluating Information Quality

Under half of the total group (46.9%, n = 282) helped patients evaluate information quality. Comments about evaluation strategies were collated into seven categories, a small group of three respondents commented that it was too difficult to teach evaluation skills. The respondents who did assist with evaluation used various strategies (Table 4). Referral to reputable sites such as ministry and non-governmental organization sites was the most cited specific approach (n = 113) with one respondent excluding Google and Wikipedia as search possibilities.

The other categories were driven by patient provided material and two different positions were discernible. One category centered on teaching patients evaluation skills with 42 responses. Respondents provided education about how to check sources and use established guidelines to assess sites and information. The other categories took a protective stance. One major response category (n = 66) centered on information evaluation from the viewpoint of the respondents' knowledge. The information was reviewed in term of research quality and respondents' "encouraged discussion" with patients. The other major categories were focused on respondents' cautioning patients to be careful of online sources (n = 28) and referring the patient on to other health professionals and specialities (n = 13).

Nurses and Personal Online Information Use

The majority of respondents (71%, n = 443) used online health information for their personal needs and 36.3% (n = 155) of that group did not tell health professionals about online information they had found. The reasons for non-disclosure were varied (Table 5). The most frequently commented on category was not having a reason to discuss the information (n = 35), other related categories were confidence in the personal health care provider's knowledge base

Table 5. Respondents Reasons for Not Discussing Online Health Information With Their Health Professional.

Categories	Comments	Number of comments
No need to discuss	Occasion never arisen, would rather discuss info they give me Need hasn't arisen, no opportunity	35
Concern about disrupting relationship with health professional	Don't want to appear to know more than my health professional Depending on health professionals' initial response Some doctors are not happy to be questioned by nurses	18
Confident with health professionals knowledge	Their knowledge base is more accurate, unbiased, and up to date Not required . . . good feedback from my own GP(General Practitioner) I have trust in my Dr. and believe he is updated with recent relevant evidence-based research	16
Access to health professionals	Don't go to health professional very often Difficulty getting in to see GP. Feel able to make own judgment using Internet info	15
Complementary information only	It's just another way of gathering info. If there was a conflict I would be happy to gather evidence as required My research is for my own knowledge only	11
Lack of provider time	Time restraints on time available for appointment Response too complicated Doctor's are too busy	10
Feeling judged	As a health professional myself, I am embarrassed by this display of a lack of knowledge She expects me to be knowledgeable Fear of being judged	8

($n = 16$) and the complementary nature of the information ($n = 11$). Contrasting concerns were also expressed about disrupting the relationship with the health professional through volunteering information ($n = 18$). A respondent commented that "some doctors are not on happy to be questioned by nurses" suggesting that being a nurse in this instance complicates the relationship when more information is required from personal health care providers. Some respondents ($n = 8$) felt they could be judged in a negative sense as not being knowledgeable enough. There were also comments about lack of time to fully discuss the material as the "response are too complicated" and "doctor's are too busy." A small group had little or no opportunity for discussion because of lack of contact with a health provider.

Discussion

Developing health literacy requires collaborative relationships between health professionals and patients. Enquiring about information sources is important as patients will not necessary initiate a discussion. Nearly half the respondents in this study were aware of patients' misconceptions about information they had read, findings congruent with an earlier study in the medical ward context (Gilmour et al., 2012). Internet users can decide to discontinue medical treatment when it conflicts with information they have found (Weaver, Thompson, Weaver, & Hopkins, 2009); it is therefore prudent that health professionals are proactive in checking with patients their information sources. Patients with very limited

knowledge and little access to online information can be identified and fully supported using a variety of media such as hard copy resources, visual and plain language information catering for limited health literacy, and face-to-face individual and group sessions.

The research findings showed greater engagement with patients' online health information use in the PHC sector compared with an earlier study of medical ward nurses (Gilmour et al., 2012). More PHC nurses (61.2%) were involved in patient discussions about online information several times a month or more compared with medical ward nurses (30.7%), asked their patients if they accessed online information (52.8% compared with 24.4% of medical nurses), and helped patients evaluate its quality (46.9% compared with 24.9% of medical nurses). One possible explanatory factor is the differences in the proportion of nurses with advanced education in the two groups. More than 40% of the PHC sample had postgraduate qualifications compared with 24.2% of the medical ward sample (Gilmour et al., 2012). There was a significant association between postgraduate qualifications and assessment of patient online use in both studies.

Evidence suggests postgraduate study to master's level does influence practice approaches. A systematic literature review of the relationship of master's-level education with patient care identified themes of "increased confidence and self-esteem; enhanced communication; personal and professional growth; knowledge and application of theory to practise; and analytical thinking and decision making"

(Cotterill-Walker, 2012, p. 57). The knowledge-base and critical appraisal skills gained through in-depth scholarly work support efficient access and evaluation of relevant information sources. Opening up conversations about knowledge sources also demands careful communication so as not to alienate patients who may otherwise feel examined about their information sources.

The assessment and evaluation differences between PHC and medical wards can also be attributed to work organization and relationship with patients. In the PHC setting, there is the opportunity to work with patients intensively over a period of time. In contrast, in a medical ward environment, many nurses may care for a patient during short hospitals stays: a New Zealand study finding that on average medical ward stay, patients were cared by 10.7 nurses (Whitt, Harvey, McLeod, & Child, 2007). Minimal or no continuity of care will be a barrier to developing a comprehensive understanding of patients' information needs even though hospitalization is an ideal time to provide educational resources (Driscoll, Davidson, Clark, Huang, & Aho, 2009).

Just under half of the respondents ($n = 282$) assisted patients to evaluate the online information. This aspect of education is the key element in developing patient self-efficacy in using online information in a personally productive way. Nurses require information literacy skills to be able to develop patient evaluation skills. Information literacy includes the ability to recognize information needs, identify and find information sources, assess quality and applicability, and then "analyze, understand, and use the information to make good health decisions" (Medical Library Association, 2003). The evaluation activities conveyed by nurses in this study centered around three key activities: opening up a two-way dialogue through inviting discussion about information sources, referring patients to reputable sites, and empowering through teaching how to judge the validity of online material. The process of evaluation was underpinned by knowledge and critical appraisal skills to judge the authority of the sites along with the use of guidelines and evaluation tools. Useful evaluative tools discussed in nursing literature include (a) the *GATOR* approach, an acronym for genuine, accurate, trustworthy, origin, and readability (Weber, Derrico, Yoon, & Sherwill-Navarro, 2010) and (b) the 5Cs website evaluation tool which covers credibility, currency, content, construction, and clarity (Roberts, 2010).

The study finding that 36% of the nurses did not discuss their personal health knowledge sources with their doctors is congruent with other research findings (Gilmour et al., 2012; Pletneva et al., 2011). Some respondents were concerned about disrupting relationships and feeling judged. These barriers to disclosure suggest a communication style by some health professionals where there is little or no invitation to develop a two-way dialogue exploring patients perceptions of their condition. When nurses are concerned about sharing their use of online resources for personal health information with their family doctor, these feelings may well influence their decisions related to engaging in such a discussion with the patients.

The study limitations include the response rate of 65.5%. The information that accompanied the questionnaire stressed that the responses from non-users of online information was important to the study but response bias is likely. Nurses familiar with Internet health information will, in all probability, be more interested in completing the questionnaire and therefore be over represented. The study also relies on the respondents self-report of practices and behavior. Most importantly in terms of limitations, this study focuses on one aspect of health education only, nurses' engagement with patients independently seeking online material. Education strategies are wide ranging depending on patient preferences and the availability of online and hard copy educational resources.

Study Implications

Skill in the evaluation of online sources and their active incorporation into practice needs to be considered a basic competency for nurses in the PHC setting. Targeted professional development activities can be aimed at improving the incorporation of online resources into practice, developing nurses' knowledge of useful patient websites and applying evidence-based approaches to determine the reliability of the sites. It is inevitable that people's use of the Internet to support decision making in all aspects of life including health will rapidly increase and nurses' engagement with this significant change in the practice landscape can no longer be optional. Patients' self-directed information-seeking activities provide new opportunities for the development of health and information literacy skills.

Conclusion

The Internet is a powerful information and communication modality in today's societies. Online information is used by the public for self-diagnosis of health issues and to supplement information provided by health professionals. The appropriate use of Internet informatics may empower and assist health care consumers to achieve better health. There is also evidence of quality concerns which coupled with patient reluctance to discuss information with health professionals raises questions about potential harm through information omissions and lack of expert peer review. This study contributes to the literature on nurses and their engagement with online resources for the purposes of patient education in the PHC setting where increasingly long-term conditions are managed. More nurses in this study engaged in proactive assessment of patient use of online material and supported the development of evaluation skills as compared with reports by nurses in earlier studies. Active engagement in the medium by nurses and patients has the potential to contribute to the development of health literacy and increase patients' positive and knowledgeable engagement in health care decisions.

Declaration of Conflicting Interests

The author(s) declared no potential conflicts of interest with respect to the research, authorship, and/or publication of this article.

Funding

The author(s) disclosed receipt of the following financial support for the research and/or authorship of this article: The study was funded through a Strategy to Advance Research Grant provided by the New Zealand Tertiary Education Commission.

References

Ashworth, N., & Thompson, S. (2011). Long-term condition management: Health professionals' perspectives. *Journal of Primary Health Care, 3*, 16-22.

Bandura, A. (1986). *Social foundations of thought and action: A social cognitive theory.* Englewood Cliffs, NJ: Prentice-Hall.

Bandura, A. (1997). *Self-efficacy: The exercise of control.* New York, NY: W.H. Freeman.

Bartol, K., Tein, M., Matthews, G., Sharma, B., & Scott-Ladd, B. (2011). *Management: A Pacific Rim focus* (6th ed.). North Ryde, Australia: McGraw Hill Education.

Bowes, P., Stevenson, F., Ahluwalia, S., & Murray, E. (2012). "I need her to be a doctor": Patients' experiences of presenting health information from the internet in GP consultations. *British Journal of General Practice, 62*, e732-e738. doi:10.3399/bjgp12X658250

Bryman, A. (2012). *Social research methods* (4th ed.). Oxford, UK: Oxford University Press.

Caughey, G., Vitry, A., Gilbert, A., & Roughead, E. (2008). Prevalence of comorbidity of chronic diseases in Australia. *BioMed Central Public Health, 8*, Article 221. doi:10.1186/1471-2458-8-221

Chung, M., Oden, R. P., Joyner, B. L., Sims, A., & Moon, R. Y. (2012). Safe infant sleep recommendations on the Internet: Let's Google it. *Journal of Pediatrics, 161*, 1080-1084.

Coleman, M. T., & Newton, K. S. (2005). Supporting self-management in patients with chronic illness. *American Family Physician, 72*, 1503-1510.

Cotterill-Walker, S. M. (2012). Where is the evidence that master's level nursing education makes a difference to patient care? A literature review. *Nurse Education Today, 32*, 57-64.

DeVito, J. A. (2012). *Human communication: The basic course* (12th ed.). Boston, MA: Allyn & Bacon.

Dickerson, S., Boehmke, M., Ogle, C., & Brown, J. (2005). Out of necessity: Oncology nurses' experiences integrating the Internet into practice. *Oncology Nursing Forum, 32*, 355-362.

Driscoll, A., Davidson, P., Clark, R., Huang, N., & Aho, Z. (2009). Tailoring consumer resources to enhance self-care in chronic heart failure. *Australian Critical Care, 22*, 133-140.

Eysenbach, G., Powell, J., Kuss, O., & Sa, E. R. (2002). Empirical studies assessing the quality of health information for consumers on the world wide web: A systematic review. *Journal of the American Medical Association, 287*, 2691-2700.

Fox, S. (2007). *E-patients with a disability or chronic illness.* Washington, DC: The Pew Internet and American Life Project. Retrieved from http://www.pewinternet.org/2007/10/08/e-patients-with-a-disability-or-chronic-disease

Gibson, A., Miller, M., Smith, P., Bell, A., & Crothers, C. (2013). *The internet in New Zealand 2013.* Auckland, New Zealand: Institute of Culture, Discourse & Communication, AUT University.

Gilmour, J. A., Huntington, A., Broadbent, R., Strong, A., & Hawkins, M. (2012). Nurses' use of online health information in medical wards. *Journal of Advanced Nursing, 68*, 349-358.

Gilmour, J. A., Scott, S. D., & Huntington, N. (2008). Nurses and Internet health information: A questionnaire survey. *Journal of Advanced Nursing, 61*, 19-28.

Henty, C., & Dickinson, A. (2007). Practice nurses' experiences of the Care Plus programme: A qualitative descriptive study. *New Zealand Family Physician, 34*, 335-338.

Horgan, A., & Sweeney, J. (2010). Young students' use of the Internet for mental health information and support. *Journal of Psychiatric and Mental Health Nursing, 17*, 117-123.

Imes, R. S., Bylund, C. L., Sabee, C. M., Routsong, T. R., & Sanford, A. A. (2008). Patients' reasons for refraining from discussing internet health information with their healthcare providers. *Health Communication, 23*, 538-547.

Julien, H. (2008). Content analysis. In L. M. Given (Ed.), *The SAGE encyclopedia of qualitative research methods* (pp. 121-123). Thousand Oaks, CA: SAGE. doi:10.4135/9781412963909.n65

Keleher, H., & Hagger, V. (2007). Health literacy in primary health care. *Australian Journal of Primary Health, 13*(2), 24-30.

King, A. (2001). *The primary health care strategy.* Wellington, New Zealand. Ministry of Health.

Kivits, J. (2006). Informed patients and the Internet: A mediated context for consultations with health professionals. *Journal of Health Psychology, 11*, 269-282.

Knapp, C., Madden, V., Marcu, M., Wang, H., Curtis, C., Sloyer, P., & Shenkman, E. (2011). Information seeking behaviors of parents whose children have life-threatening illnesses. *Pediatric Blood & Cancer, 56*, 805-811.

Lawrentschuk, N., Sasges, D., Tasevski, R., Abouassaly, R., Scott, A. M., & Davis, I. D. (2012). Oncology health information quality on the Internet: A multilingual evaluation. *Annals of Surgical Oncology, 19*, 706-713.

Lupiáñez-Villanueva, F., Hardey, M., Torrent, J., & Ficapal, P. (2011). The integration of information and communication technology into nursing. *International Journal of Medical Informatics, 80*, 133-140.

McDaid, D., & Park, A. (2011). *Online health: Untangling the web: Bupa Health Pulse 2010.* London, England: London School of Economics and Political Science for The British Provident Association Limited. Retrieved from www.bupa.com/media/44806/online_20health_20-_20untangling_20the_20web.pdf

McMullan, M. (2006). Patients using the Internet to obtain health information: How this affects the patient-health professional relationship. *Patient Education and Counseling, 63*, 24-28.

Meadows-Oliver, M., & Banasiak, N. C. (2010). Accuracy of asthma information on the world wide web. *Journal for Specialists in Pediatric Nursing, 15*, 211-216.

Medical Library Association. (2003). *Health information literacy.* Retrieved from http://www.mlanet.org/resources/healthlit/define.html

Nursing Council of New Zealand. (2011). *The New Zealand nursing workforce: A profile of nurse practitioners, registered nurses, nurse assistants and enrolled nurses 2011*. Wellington: Nursing Council of New Zealand.

Pletneva, N., Cruchet, S., Simonet, M. A., Kajiwara, M., & Boyer, C. (2011). *Results of the 10th HON survey on health and medical Internet use*. Paper presented at Medical Informatics Europe. Retrieved from http://www.hon.ch/Project/publications.html

Roberts, L. (2010). Health information and the internet: The 5 Cs website evaluation tool. *British Journal of Nursing, 19*, 322-325.

Scott, S. D., Gilmour, J. A., & Felden, J. (2008). Student nurses and internet health information. *Nurse Education Today, 28*, 994-1002.

Scullard, P., Peacock, C., & Davies, P. (2010). Googling children's health: Reliability of medical advice on the internet. *Archives of Disease in Childhood, 95*, 580-582.

van Deursen, A. J., & van Dijk, J. A. (2011). Internet skills performance tests: Are people ready for eHealth? *Journal of Medical Internet Research, 13*(2), e35. doi:10.2196/jmir.1581

Weaver, J. B., Thompson, N. J., Weaver, S. S., & Hopkins, G. L. (2009). Healthcare non-adherence decisions and internet health information. *Computers in Human Behavior, 25*, 1373-1380.

Weber, B. A., Derrico, D. J., Yoon, S. L., & Sherwill-Navarro, P. (2010). Educating patients to evaluate web-based health care information: The GATOR approach to healthy surfing. *Journal of Clinical Nursing, 19*, 1371-1377.

Whitt, N., Harvey, R., McLeod, G., & Child, S. (2007). How many health professionals does a patient see during an average hospital stay. *New Zealand Medical Journal, 120*(1253), U2517.

Wilson, S. (1999). Impact of the internet on primary care staff in Glasgow. *Journal of Medical Internet Research, 1*, e7. doi:102196/jmir.1.2.e7

Author Biographies

Jean Gilmour is a senior lecturer in the School of Nursing at Massey University. Her research interests include the experience of chronic illness, health education, and nursing workforce issues.

Sue Hanna is a senior lecturer in the Department of Mental Health, Social Work and Integrative Medicine at Middlesex University London. She is trained social worker with a PhD in social work. Her research interests are varied, but are currently focused round interdisciplinary working and the recreational reading patterns of children in care

Helen Chan's areas of expertise include primary health care nursing. Her research interests are in the use of complementary/ alternative medicine in primary care.

Alison Strong is a Clinical Nurse Specialist in Heart Failure in Hawke's Bay. Alison completed a Master of Nursing in 2007 and her research project examined heart failure patient information on the Internet.

Annette Huntington is a Professor of Nursing with extensive experience in nursing regulation, education and research. Her research has been mainly in the areas of nursing workforce and related practice issues.

12

Trafficking in Human Beings in the European Union: Gender, Sexual Exploitation, and Digital Communication Technologies

Donna M. Hughes[1]

Abstract

In this article, the intersection of gender, trafficking for sexual exploitation, and use of digital communication technologies are analyzed based on data from the European Union (EU). Over the past two decades, an increase in trafficking in human beings in the EU has been accompanied by an increase in the development and availability of digital communication technologies. The first statistical analysis of trafficking in human beings (2008-2010) carried out by the European Commission found 23,632 victims of human trafficking in the reporting member states. Eighty percent of victims were women and girls; 20% were men and boys. The majority of the victims (62%) were trafficked for sexual exploitation. Digital communication technologies are widely used for trafficking for sexual exploitation, and more rarely for trafficking for forced labor. This article concludes that the combination of gender, trafficking for sexual exploitation, and use of digital communication technologies has created a nexus of victimization for women and girls. Based on this analysis and other sources of information, the European region is the world's leading region for trafficking for sexual exploitation.

Keywords

trafficking in human beings, sexual exploitation, forced labor, digital communication technologies, Internet, European Union

Introduction

Over the past 20 years, the use of digital communication technologies, particularly the Internet, has greatly expanded criminals' capacity to traffic human beings for different types of exploitation (Council of Europe, 2003; Hughes, 1996, 1999a; Latonero, 2011; Sykiotou, 2007). As new technologies have been developed, criminals have quickly adopted them to assist with their criminal enterprises. Seekers of justice have countered with new laws and policies to lawfully combat the misuse of new technologies. Law enforcement agencies have become skilled in the use of digital technologies and forensics to combat these serious crimes. The development of new types of communication media and devices keeps the race going between traffickers and law enforcement.

Although communication technologies are frequently used by perpetrators of trafficking, the underlying crime of trafficking in human beings remains the same: a trafficker tricks, coerces, or exploits the vulnerability or the age of a victim to compel the victim to work, provide services, engage in commercial sex acts, beg, or commit criminal acts. Trafficking in human beings became a European Union (EU) crime in 2004 with the implementation of the Council Framework Decision (2004) on combating trafficking in human beings. Trafficking in human beings is a serious violation of the law and human rights that is frequently compared with slavery. Therefore, the EU is committed to eradicating it (European Commission, 2012).

This article will analyze the trafficking in human beings in the EU and the use of communication technologies. Recent statistics on the number and type of victims will be used as a basis for the analysis. Finally, the nexus of gender, trafficking for sexual exploitation, and use of digital communication technologies will be discussed.

Trafficking in Human Beings in the EU

Trafficking in human beings is recognized as a serious criminal enterprise in the EU. It has been called one of the "most prevalent" types of organized crime activities in the EU by Steve Harvey, then Acting Head of Europol's Operational

[1]University of Rhode Island, Kingston, USA

Corresponding Author:
Donna M. Hughes, Professor & Eleanor M. and Oscar M. Carlson Endowed Chair, Gender and Women's Studies Program, University of Rhode Island, 316 Eleanor Roosevelt Hall, Kingston, RI 02881, USA.
Email: dhughes@uri.edu

Department ("Monitoring Mechanisms," 2010). The United Nations Office on Drugs and Crime (UNODC; 2010) estimates the market for sexual exploitation in Europe to be €2.5 billion annually.

The EU has acted to combat the problem of trafficking in human beings. In August 2004, the Council Framework Decision 2002/629/JHA (2002) criminalized trafficking in human beings at the EU level. In 2006, the European Commission issued the EU Action Plan (2006), which set combating trafficking in human beings and money laundering as priority areas. In 2011, the European Parliament and the Council issued Directive 2011/36/EU on Preventing and Combating Trafficking in Human Beings and Protecting Its Victims (European Parliament & Council of the European Union, 2011).

In 2012, the European Commission adopted a Communication on the EU Strategy toward the Eradication of Trafficking in Human Beings 2012-2016. One of the strategic actions was to collect and publish data on trafficking disaggregated by age and gender.

The collection of data on the number of victims of trafficking in human beings has been a challenge for organizations, agencies, and governments (Laczko & Gozdziak, 2005). Although no methods of data collection are without flaws, analyses must be made on the best existing data. Researchers and The International Organization for Migration (David, 2007) have made the case that data collection on trafficking in human beings needs to constantly be improved, but data on the number and type of victims are an essential tool for combating trafficking.

In 2013, the European Commission issued the first EU statistical report on the trafficking in human beings in the member states and affiliated countries. The Eurostat report includes data from all 27 of the EU member states and 7 acceding, candidate or associated countries, but the total number and percentages are based on data from the EU member states. The report included data on victims and traffickers. This is the first collection of EU-wide data. Not all countries had complete data in each category for each year, so interpretation has to be done with caution. Still, preliminary findings show that there are differences and trends for types of trafficking, the gender of victims, and source and destination of victims. These new data provide bases for further analysis of trafficking in human beings.

The quantitative report on the trafficking victims in member states of the EU from 2008 to 2010 reported 23,632 identified or presumed victims of human trafficking in the reporting member states (Eurostat, European Commission, 2013). An "identified" victim is defined as "a person who has been formally identified as a victim of trafficking in human beings according to the relevant authority in Member States." A "presumed" victim of trafficking is defined as

a person who has met the criteria of EU regulations and international Conventions but has not been formally identified by the relevant authorities (police) as a trafficking victim or who

has declined to be formally or legally identified as trafficked. (Eurostat, European Commission, 2013, p. 20)

Data on "identified" victims mostly came from the police, whereas data on "presumed" victims mostly came from national rapporteurs, victim assistance services, immigration services, labor inspections, and border guards (Eurostat, European Commission, 2013, p. 23). Over the 3-year period of data collection (2008-2010), the number of victims increased by 18%. The authors of the report urge caution in assuming that the 18% increase in victims over the 3-year period from 2008 to 2010 means an actual increase in the number of victims. They suggest that the increase in numbers may reflect an increase in identifying victims (Eurostat, European Commission, 2013). This cautionary note is reasonable considering that this is the first report done for the EU. Still, the data indicate an increase in trafficking in human beings or new awareness of a serious crime. The goal of this article is to look at the intersection of type of trafficking, gender, and the use of communication technologies. The new EU report provides a basis for this analysis.

Type of Trafficking

Of the total number of victims, 62% were trafficked for sexual exploitation and 25% were trafficked for forced labor. The remaining approximately 14% were trafficked for begging, organ removal, criminal activities, forced marriages, or selling of children (Eurostat, European Commission, 2013). Over the 3-year period, the percentage of victims of trafficking for sexual exploitation increased each year and the percentage of victims of forced labor decreased from 28% in 2008 to 23% in 2009 and 2010 (Eurostat, European Commission, 2013).

Another source of data lends support for the findings in the EU that trafficking for sexual exploitation is the dominant form of trafficking. The UNODC (2012) found that trafficking for sexual exploitation accounted for 58% of trafficking cases globally, but when world regions were compared, Europe had the highest percentage—62%—of trafficking for sexual exploitation, and the lowest—31%—for forced labor. From these two sources of data, the percentage of victims trafficked for sexual exploitation appears to be higher in Europe than in other regions of the world.

Recently, there have been campaigns for increased awareness of forced labor and other types of trafficking, and it is likely that forced labor is still a largely unrecognized crime. Still, these statistics indicate that trafficking for sexual exploitation is the dominant type of trafficking in the EU, and it is likely increasing according to findings of this first collection of data.

Gender and Age of Victims

The EU recognizes that trafficking in human beings is gendered. Directive 2011/36/EU noted that there is "gender-specific" trafficking, and the women and men are often trafficked for different reasons. A goal of the European Commission in 2013 was to

> develop knowledge of the gender dimensions of human trafficking, including the gender specificities of the way men and women are recruited and exported, the gender consequences of the various forms of trafficking and potential differences in the vulnerability of men and women to victimization and its impact on them. (European Commission, 2012)

The EU data found that a majority of victims were women or girls. Of the total number of identified or presumed victims, 80% were female (68% women, 12% girls) and 20% were male (17% men, 3% boys). Children were 15% of the total number of victims of trafficking in the EU (Eurostat, European Commission, 2013, p. 10).

Of the victims of trafficking for sexual exploitation, 96% of them were women and girls in 2010. Of the victims of trafficking for forced labor or services, 77% were male in 2010. The majority of victims of forced begging, organ removal, criminal activity, and selling of children were female, although there was a gradual increase in the number of males over the 3-year period (Eurostat, European Commission, 2013).

The findings from the first EU-wide collection of data on trafficking in human beings reveal that trafficking is a gendered phenomenon. The majority of victims are women and girls (the predominant type of trafficking is for sexual exploitation, and the majority of victims of trafficking for sexual exploitation are women and girls). Also, trafficking for sexual exploitation increased every year over the 3-year time frame (58% in 2008, 60% in 2009, and 66% in 2010). In addition, UNODC found that the European region ranked the highest of all regions in the world for trafficking for sexual exploitation in the world (62%).

Women and girls are also the majority (72%) of victims in the "Other" category, which includes forced begging, criminal activities, removal of organs, forced marriages, and selling of children. Although the majority of victims of forced labor were male, 25% of the victims were female.

These data indicate that trafficking in human beings in the EU is largely comprised of exploitation of women and girls.

Although there may be less awareness and identification of male victims or victims of forced labor and other types of trafficking, this is still an overwhelming finding from the first round of data collection in the EU.

Country of Origin of Victims

The majority of trafficking victims in the EU come from other EU member states. If victims are exploited within their own borders, this is called internal trafficking. As a result of the Schengen Agreement, 22 of the 28 EU member states are open to travel with each other without border controls. Of the total number of victims, the majority, 61%, of victims came from other EU member states. According to Europol's (2013) EU Serious and Organized Crime Threat Assessment, the levels of intra-EU trafficking (internal trafficking) are increasing due to freedom of movement throughout the EU.

There was also a gender difference for victims of internal trafficking. More male victims (74%) were internally trafficked than female victims (66%). This means that female victims are more likely to be recruited outside the EU and transported into the EU (Eurostat, European Commission, 2013).

From 2008 to 2010, the percentage of the identified and presumed victims from outside the EU increased. For male victims, the percentage from outside the EU increased from 12% to 37%. And for the female victims, the percentage from outside the EU increased from 18% to 39%. These are dramatic increases in the number of female victims coming from outside the EU (Eurostat, European Commission, 2013).

One of the weaknesses of the Eurostat report is that some member states were not adequately collecting data on the number of victims by country of recruitment (Eurostat, European Commission, 2013). This indicates that although there were some countries that were strong sending countries, more work needs to be done to ensure that all member states give more precise information on the source of victims so more robust analysis can be done.

In the next section of this article, the role that digital communication technologies play in human trafficking, particularly in the EU, will be examined.

Use of Digital Communication Technologies for Human Trafficking

Since the commercial use of Usenet and the Internet began, they have been used for sexual exploitation (Taylor, Quayle, & Holland, 2001). For example, one newsgroup was called alt.bin.pictures.child.pornography (Lanchet & Hornat, 2008). They have been used to transmit pornography and child sexual abuse material, previously called child pornography (Taylor et al., 2001). These forums were also used to transmit information on venues for commercial sex acts and sex tours (Hughes, 1996, 1999a). With the commercial development of the Internet, it quickly became the site to purchase pornography, exchange or purchase child sexual abuse images (Quayle & Taylor, 2002; Taylor & Quayle, 2003). The Internet has been used to advertise sex tours. In the early to mid-1990s, websites for sex tours could be found that openly advertised the availability of children for sexual exploitation (Hughes, 1999a). And the Internet has been used to advertise numerous locations and services for commercial sex acts (Council of Europe, 2003; Hughes, 2004a).

As a venue for sexual exploitation, sexual predators using the Internet were ahead of laws and law enforcement. Often laws had to be revised to apply to digital material, storage and transmission of illegal images. In addition, law enforcement needed training and tools to investigate crimes facilitated by the Internet or digital technology. With the recognition of the existence of human trafficking, almost every country has passed a law against human trafficking for forced labor and sexual exploitation, and researchers are documenting how digital technologies are used by traffickers.

As social forums on the Internet grew, they became sites for traffickers to contact and recruit victims. Traffickers could place false advertisements on employment sites, offer young women jobs as waitresses or nannies, careers as models or dancers, and present themselves as boyfriends. Later, the traffickers would force victims into prostitution (Sykiotou, 2007). Also women who signed up at marriage agencies could be deceived (Hughes, 2004b).

Although anti–human trafficking laws are more than a decade old, as the EU statistics show, trafficking is a continuing, even growing, crime. The Internet is playing a major role in facilitating these crimes. There are now 2 billion Internet users worldwide (International Telecommunications Union, 2010). According to the World Bank (2012), 75% of the world's population now has a mobile phone, and access is expanding into rural areas. As more communication technologies are available, traffickers will quickly adopt them. This growing access and use of digital communication technologies will increasingly become the way that traffickers contact victims and how victims of sex exploitation are offered to sex buyers.

What we call "online" no longer means using a desktop or laptop computer or Internet cafe. As new communication technologies continue to develop, they include wireless devices, such as smart phones, which are small computers, with access to the Internet through telecommunications companies. These sophisticated, hand-held digital devices have created a "mobile revolution" and increase the capacity of criminals to engage in all aspects of human trafficking.

As the growth and development of digital communi-cation technologies and devices continues, this area will continue to be important for research, investigation, and action for prevention and prosecution of human trafficking.

Cases of Trafficking in Human Beings Involving the Internet in EU Member States

Digital communication technologies are used for trafficking victims for forced labor and sexual exploitation. There are more documented cases of technologies being used for recruiting, controlling, and advertising victims of sexual exploitation than for forced labor. Here are a few typical examples of cases of trafficking in human beings involving digital technologies in EU member states.

- United Kingdom, 2008: Police arrested three men and one woman who trafficked women from Thailand. The perpetrators' network used an escort agency site on the Internet as a front for their activities. Thirty women were recovered and taken to a victim support center. The offenders were charged with controlling prostitution for gain, trafficking, and money laundering ("15 Arrests as Internet Vice Ring Smashed by Police," 2008).
- Czech Republic, 2010: Two perpetrators used the Internet to advertise underage girls for prostitution. They were convicted of human trafficking (UNODC Case Law Database, Czech Republic, CZE0202).
- Germany, 2001: A 16-year-old Polish girl was transported to Germany and used for prostitution in a brothel. One of the perpetrators took photos of her, which were used for an advertisement on the Internet that read "girl for sale" (UNODC Case Law Database, Poland, POL007)
- Romania, 2003-2007: A trafficker contacted high school girls between ages 14 and 17. His apartment was equipped with computers, cameras, and video recorders used for producing pornography. He used the images to coerce the girls into sex acts. He threatened to publicly expose the images on the Internet or to their parents and friends if they did not comply with his orders of exploitation (UNODC Case Law Database, Romania, ROU003).
- Sweden: A man met a woman who was diagnosed with a mental disability on the Internet. He used her vulnerability to exploit her. Another man assisted the first to take explicit photos of the woman. They designed a website for the purpose of making money from the photos (UNODC Case Law Database, Sweden, SWE021).

Although the focus of this article is on relatively new communications technologies, "low" technologies are still used as well, and may be the choice of some criminals because they can avoid detection that newer digital devices allow.

- Czech Republic, 2010: Perpetrators forced women into "window prostitution." The victims were provided with walkie-talkies and had to report to the traffickers about the sex buyers and the earnings. The victims had to hand over most of the money they received to the defendants (UNODC Case Law Database, Czech Republic, CZE022).

A combination of high technology and old means of control is also used to coerce and control victims, such as this case involving voodoo religion.

- Sweden, 2010: Two women from Cameroon coerced women from Rwanda, Nigeria, Uganda, and Cameroon into prostitution. The traffickers used religious rituals and voodoo to control a dozen African women. Medicine Men used rituals, slaughtered animals, and made the women take an oath to reimburse travel costs. They told the women that they would die if the debt was not repaid. Advertisements were placed on the Internet for the women in Sweden. The victims were given mobile phones. They had to respond to calls from sex buyers ("Traffickers Intimidated Victims," 2010; UNODC Case Law Database, Sweden, SWE014).

Use of Digital Technologies for Trafficking for Forced Labor

Traffickers appear to make less use of digital technologies for forced labor compared with sexual exploitation. There are many documented cases of the use of digital technologies for sexual exploitation, but fewer documented cases for cases of forced labor.

Research on the use of digital technology for forced labor in the United States found that traffickers did not rely on technology other than pay-as-you-go cell phones. Victims of forced labor in the United States were recruited by word of mouth from impoverished villages in Latin American countries. Once the victims were trafficked, they had little or no access to technology (Latonero, 2011). An Organization for Security and Co-Operation in Europe (OSCE; 2009) report on human trafficking for labor exploitation in the agricultural sector in European countries reported that people were recruited from newspaper advertisements and by word of mouth. In one large case of forced labor for vegetable picking in Italy, workers were recruited from Poland by newspaper and website advertisements (OSCE, 2009).

These findings are from a limited number of cases and from research done in the United States. It appears that traffickers use job advertisement on the Internet to recruit some workers, but do not use digital technologies to maintain control of victims of work in agriculture, construction, or manufacturing. It is more likely that traffickers contact each other and employers with digital communication technologies. The researchers concluded that how forced labor traffickers operate makes it harder to track them through digital technology (Latonero, 2011).

It is likely that traffickers for forced labor use digital technologies to transfer and launder profits, as they do with trafficking for sexual exploitation, but more research is needed in this area.

In cases of trafficking of domestic workers, victims may be recruited through employment agencies or exploiters make private arrangements with contacts they know in source countries. The situation probably depends widely on the level of available technology in the sending region, particularly outside the EU. It seems likely that some victims of domestic servitude have responded to advertisements online. One of the ways that exploiters of domestic workers maintain control is to prevent the victims from having contact with others and monitoring victims' communications.

Domestic workers are increasingly remotely monitored by their employers. Following several cases in which domestic workers were accused of harming or killing children in the Gulf States, some employers are installing video cameras to monitor workers (Many Installing Cameras at Home to Monitor Maids, 2012). The activities of the domestic worker can be viewed on a smartphone or other remote computer. This surveillance is also a way to monitor and control a worker who is being abused and exploited (Home Surveillance, 2012). The monitoring of domestic workers is being discussed as an issue of privacy and worker rights, but video monitoring of workers is increasingly being recognized as a way that traffickers monitor and control victims (Immigration and Custom Enforcement, 2013).

Use of Digital Technologies for Trafficking for Sexual Exploitation

There are many documented cases of the use of digital communication technologies for trafficking for sexual exploitation. In a recent report, Europol (2011) emphasized the "key role that the Internet was playing in recruiting victims [of sexual exploitation] and advertising their services" (p. 25).

Mobile devices and technologies create a more fluid environment for traffickers, victims, and sex buyers. All of the concerned parties can be in motion, with real-time communication among all of them. Traffickers can engage in real-time communication, such as voice messages, videos, and texting, with the victim. Traffickers can pose as the woman in the advertisement and set up an appointment with a sex buyer. For online "adult entertainment" advertising sites, the ads can be changed and updated throughout the time that sex buyers are most likely to be actively looking for appointments. Sex buyers can search for and make arrangements for sex acts from almost anywhere.

The following activities can be carried out online:

- Recruitment of victims with false employment advertisements
- Contact and groom victims in online forums or dating sites
- Capture images and videos of victims that will be used in advertisements or to threaten the victims with exposure to their families or friends
- Upload text advertisements, images, and videos to brothels, entertainment businesses, or prostitution businesses that operate only from the Internet, such as escort services and online advertisement sites for prostitution
- Arrange meetings between sex buyers and victims

- Communicate with victims to monitor their activities, give them orders, threaten them, and control them
- Make business arrangements with criminal colleagues or legitimate businesses
- Transfer money

Previously, these activities could be done over the Internet with a laptop or a desktop computer, but now, they can be carried out with a mobile, wireless device, enabling criminals to be more mobile.

There are many documented cases of human traffickers using social networking media, such as Facebook, and online advertisement sites, such as Craigslist, and micro-blogging services, such as Twitter (Federal Agents Arrest Twitter Pimp for Sex Trafficking of Child, 2012; Latonero et al., 2012). Many of them are mainstream services commonly used by adults and children. Online, traffickers often commit serious crimes such as human trafficking in public or semi-public spaces.

Organized crime groups involved in human trafficking are flexible and quickly change their tactics following changes in laws and law enforcement investigations (Europol, 2013). New devices, media forums, and mobility increase the flexibility for criminals.

On the "adult entertainment" websites, as they are called, the images of women available for commercial sex acts are displayed. Some of the images are made while the victim is engaged in sex acts. The Dutch Rapporteur on Trafficking in Human Beings has noted that these images "constitute a new dimension to victimhood" ("Monitoring Mechanisms," 2010).

Traffickers for sexual exploitation can be flexible in how they advertise victims to sex buyers. They may post advertisements on mainstream, legitimate, public forums or post to a more marginal site that advertises riskier, rougher sex acts. A website that offers women for prostitution may be run by a criminal gang. A website like this can offer many women for sex acts and also include a comments section, so men can write reviews of the women's performances (Gray, 2010). Digital technology–assisted sexual exploitation is widespread in the EU, particularly in countries where prostitution is legal or tolerated.

There are some positive aspects to the widespread use of digital technology by traffickers. A record is made of all transactions and communications. A smartphone holds a large amount of data that law enforcement can use for investigation and prosecution of traffickers. "The internet has been a good thing for police officers—it has brought all these worms to the surface. We can now identify them and track them down," said Jonathan Rouse, Detective-Inspector with the Queensland Police in Australia in charge of a taskforce on computer-facilitated crimes against children (UNODC, 2013).

Technologies are used for every aspect of sexual exploitation, from recruitment of victims, advertising the victims to sex buyers, coercing them with digital images, to monitoring their behavior. In addition, the financial management of the criminal business is often done online.

The Nexus of Gender, Trafficking for Sexual Exploitation, and Digital Technologies in the EU

According to UNODC (2012), the European region leads the world in trafficking for sexual exploitation. This specific type of crime and human rights violation is the result of factors coming together to enable the criminal victimization of thousands of women and girls.

This article reviewed the first statistical report on EU data on trafficking in human beings by gender and type of exploitation. Then, the article reviewed the use of digital technologies for trafficking for sexual exploitation and forced labor. Three factors emerged—gender, trafficking for sexual exploitation, and digital technologies—to create a nexus of exploitation of women and girls facilitated by digital communication technologies. The nexus of these factors is contributing to gender inequality in the EU, which all EU governmental bodies are committed to eliminating.

Police and governmental bodies in the EU recognize the serious nature of trafficking of human beings and combating it is a priority. In 2013, Europol recommended a high-level response to trafficking in human beings (Europol, 2013).

EU governmental bodies also recognize the seriousness of trafficking in human beings. For over a decade, they have issued directives and recommendations on combating trafficking in human beings. Reports by EU committees, police organizations, and rapporteurs call for making combating trafficking in human beings a top priority, and almost all of them emphasize the importance of understanding and combating the use of new technologies, particularly the Internet. There have been calls for more research on technologies used for trafficking to determine an appropriate response. The Dutch National Rapporteur noted that the "Internet is becoming more anonymous and more accessible at the same time" ("Monitoring Mechanisms," 2010) and called for more research on the use of the Internet for trafficking.

The important role that digital communications is playing in the trafficking of human beings has been recognized as well by the European Commission. In 2014, the European Commission intends to fund programs and support projects that "aim to increase knowledge of recruitment over the Internet and via social networks" (European Commission, 2012).

All aspects of trafficking in human beings need to be researched, but when some factors, particularly gender, sexual exploitation, and digital technologies, converge to create enhanced victimization, special attention is needed to look at the nexus of the problem and not just the separate elements. Trafficking of women and girls for sexual exploitation facilitated by digital technologies has contributed to making

Europe into the world region with the largest amount of trafficking for sexual exploitation.

More research needs to be done on all aspects and categories of trafficking in human beings. To date, most of the research and awareness raising has focused on trafficking for sexual exploitation. More research is needed on forced labor to better understand this type of trafficking in the EU. More research is particularly needed on how digital communication technologies are used to traffic women and girls for sexual exploitation in the EU.

This article looked at the data on gender and type of trafficking for all of the EU. It did not look at inter-EU regions or states or the intersection of gender, type of trafficking, and trafficking in human beings from regions and countries outside the EU. Further research and analysis is needed to look at the dynamics of sending and receiving countries, gender, type of trafficking, and the use of digital communication technologies.

Declaration of Conflicting Interests

The author declared no potential conflicts of interest with respect to the research, authorship, and/or publication of this article.

Funding

The author(s) received no financial support for the research and/or authorship of this article.

References

15 arrests as Internet vice ring smashed by police. (2008, April 21). *Yorkshire Post*. Retrieved from http://www.yorkshirepost.co.uk/news/main-topics/local-stories/15-arrests-as-internet-vice-ring-smashed-by-police-1-2498105

Council Framework Decision 2002/629/JHA. (2004, August).

Council of Europe. (2003, September 16). *Group of specialists on the impact of the use of new information technologies on trafficking in human beings for the purpose of sexual exploitation* (Final Report, EG-S-NT (2002) 9 rev). Strasbourg, France: Author. Retrieved from http://ec.europa.eu/anti-trafficking/download.action?nodePath=/Publications/Group+of+specialists+on+the+impact+of+the+use+of+new+information+technologies.pdf&;fileName=Group+of+specialists+on+the+impact+of+the+use+of+new+information+techn ologies.pdf&fileType=pdf

David, F. (2007). *ASEAN and trafficking in persons: Using data as a tool to combat trafficking in persons*. International Organization for Migration. Retrieved from http://www.human-trafficking.org/uploads/publications/2007/0806_154022_lowres_20asean_20report-complete.pdf

European Commission. (2006, August 7). *Communication from the Commission to the European Parliament, the Council and the European Economic and Social Committee—Developing a comprehensive and coherent EU strategy to measure crime and criminal justice: An EU Action Plan 2006-2010, COM (2006)*. Retrieved from http://europa.eu/legislation_summaries/justice_freedom_security/judicial_cooperation_in_criminal_matters/l33264_en.htm

European Commission. (2012). *The EU Strategy towards the eradication of trafficking in human beings, 2012-2016*. Retrieved from http://ec.europa.eu/anti-trafficking/Publica-tions/Ebook_Strategy

European Parliament & Council of the European Union. (2011, April 5). *Directive 2011/36/EU on preventing and combating trafficking in human beings and protecting its victims, and replacing Council Framework Decision 2002/629/JHA*. Strasbourg, France: Author.

Europol. (2011). *EU organized crime threat assessment*. The Hague, The Netherlands: European Police Office.

Europol. (2013). *EU serious and organized crime threat assessment*. The Hague, The Netherlands: European Police Office.

Eurostat, European Commission. (2013). *Trafficking in human beings*. Retrieved from http://ec.europa.eu/dgs/home-affairs/what-is-new/news/news/2013/docs/20130415_thb_stats_report_en.pdf

Federal agents arrest Twitter pimp for sex trafficking of child. (2012, April 13). *The smoking gun*. Retrieved from http://www.thesmokinggun.com/buster/twitter-pimp-arrested-758490

Gray, R. (2010, February 15). Fears over rise in online brothels. *The Herald*, p. 9.

Home Surveillance. (2012, November 25). *Computing and society*. Retrieved from http://blog.nus.edu.sg/netikit/2012/11/25/home-surveillance/

Hughes, D. M. (1996). Sex tours via the Internet. *Agenda*, *12*(28), 71-76.

Hughes, D. M. (1999a). The Internet and sex industries: Partners in global sexual exploitation. *IEEE Technology and Society Magazine*, *19*, 35-42.

Hughes, D. M. (1999b). *Pimps and predators on the Internet: Globalizing the sexual exploitation of women and children*. Kingston, Rhode Island: Coalition Against Trafficking in Women.

Hughes, D. M. (2004a). Prostitution online. *Journal of Trauma Practice*, *2*, 115-131.

Hughes, D. M. (2004b). The role of "marriage agencies" in the sexual exploitation and trafficking of women from the former Soviet Union. *International Review of Victimology*, *11*, 49-71.

Immigration and Custom Enforcement. (2013, January 16). *Fact sheet: Human trafficking and smuggling*. Retrieved from http://www.ice.gov/news/library/factsheets/human-trafficking.htm

International Telecommunications Union. (2010). *The world in 2010, ICT facts and figures*. Geneva, Switzerland: Author.

Laczko, F., & Gozdziak, E. (2005). *Data and research on human trafficking: A global survey* (Offprint of the Special Issue of the International Migration, Vol. 43, No. 1/2). Geneva, Switzerland: International Organization for Migration.

Lanchet, L., & Hornat, C. A. (2008). *Forensic primer for usenet evidence*. SANS Institute InfoSec Reading Room. Retrieved from http://www.sans.org/reading-room/whitepapers/forensics/forensic-primer-usenet-evidence-32829

Latonero, M. (2011, September). *Human trafficking online: The role of social networking sites and online classifieds*. Center on Communication Leadership and Policy, University of Southern California. Retrieved from https://technologyandtraf-ficking.usc.edu/files/2011/09/HumanTrafficking_FINAL.pdf

Latonero, M., Musto, J., Body, Z., Boyle, E., Bissell, A., Kim, J., & Gibson, K. (2012, November). *The rise of mobile and the diffusion of technology-facilitated trafficking*. USC Annenberg Center on Communication Leadership and Policy, University of Southern California. Retrieved from https://technologyandtrafficking.usc.edu/files/2012/11/HumanTrafficking2012_Nov12.pdf

Many installing cameras at home to monitor maids. (2012, April 3). *The Peninsula*. Retrieved from http://thepeninsulaqatar.com/news/qatar/189438/many-installing-cameras-at-home-to-monitor-maid0073

Monitoring mechanisms in the fight against human trafficking. (2010, October 15). The Hague, Netherlands. Retrieved from http://lastradainternational.org/lsidocs/Final%20Report%20Monitoring%20mechanisms%20in%20the%20fight%20against%20human%20trafficking.pdf

Organization for Security and Co-Operation in Europe. (2009, April 27, 28). *A summary of challenges on addressing human trafficking for labour exploitation in the agricultural sector in the OSCE region* (Background paper for the Alliance Against Trafficking in Persons Conference: Technical Seminar on Trafficking for Labour Exploitation, Focusing on the agricultural sector, Occasional Paper Series No. 3). Vienna, Austria: Author.

Quayle, E., & Taylor, M. (2002). Child pornography and the Internet: Perpetuating a cycle of abuse. *Deviant Behavior, 23,* 331-361.

Sykiotou, A. P. (2007). *Trafficking in human beings: Internet recruitment—Misuse of the Internet for the recruitment of victims of trafficking in human beings* (Council of Europe EG-S-NT: Group of Specialists on the Impact of the Use of New Information Technologies on Trafficking in Human Beings for the Purpose of Sexual Exploitation). Strasbourg, France: Council of Europe.

Taylor, M., & Quayle, E. (2003). *Child pornography: An internet crime*. East Sussex, UK: Psychology Press.

Taylor, M., Quayle, E., & Holland, G. (2001, Summer). Child pornography: The Internet and offending. *Isuma Canadian Journal of Policy Research, 2*(2), 94-100.

Traffickers intimidated victims with voodoo 25. (2010, August). *The Local*. Retrieved from http://www.thelocal.se/20100825/28572

United Nations Office on Drugs and Crime. (2010) *Trafficking in persons to Europe for sexual exploitation*. Retrieved from http://www.unodc.org/documents/publications/TiP_Europe_EN_LORES.pdf

United Nations Office on Drugs and Crime. (2012, December). *Global report on trafficking in persons* (Sales No. E.13.IV.1). Vienna, Austria: United Nations Publications.

United Nations Office on Drugs and Crime. (2013, September 27). *UN crime body to combat online child abuse*. Retrieved from http://www.unodc.org/unodc/en/press/releases/2013/September/un-crime-body-to-combat-online-child-abuse.html

UNODC Case Law Database, Czech Republic, No. CZE020. Retrieved from http://www.unodc.org/cld/case-law-doc/traffickingpersonscrimetype/cze/2011/48_t_32010.html?tmpl=old

UNODC Case Law Database, Czech Republic, No. CZE022. Retrieved from http://www.unodc.org/cld/case-law-doc/traffickingpersonscrimetype/cze/2011/4_t_52010.html?tmpl=old

UNODC Case Law Database, Poland, No. POL007. Retrieved from http://www.unodc.org/cld/case-law-doc/traffickingpersonscrimetype/pol/2007/ii_aka_32505_.html?tmpl=old

UNODC Case Law Database, Romania, No. ROU003. Retrieved from http://www.unodc.org/cld/case-law-doc/traffickingpersonscrimetype/rou/2010/3809892007.html?tmpl=old

UNODC Case Law Database, Sweden, No. SWE014. Retrieved from http://www.unodc.org/cld/case-law-doc/traffickingpersonscrimetype/swe/2011/case_no_b_87-11.html?tmpl=old

UNODC Case Law Database, Sweden, No. SWE021. Retrieved from http://www.unodc.org/cld/case-law-doc/traffickingpersonscrimetype/swe/2008/case_no_b_5886-08.html?tmpl=old

The World Bank. (2012, July 17). *Mobile phone access reaches three quarters of the planet's population*. Retrieved from http://www.worldbank.org/en/news/press-release/2012/07/17/mobile-phone-access-reaches-three-quarters-planets-population

Author Biography

Donna M. Hughes is a professor and holds the Eleanor M. and Oscar M. Carlson Endowed Chair in gender and women's studies at the University of Rhode Island. She researches trafficking of human beings for sexual exploitation. She teaches courses in human trafficking, slavery, and sexual violence and exploitation.

13

Investigating the Factors Affecting Residential Consumer Adoption of Broadband in India

Amir Manzoor[1]

Abstract

This study aims to explore in detail the factors that affect the consumer behavioral intention to adopt broadband Internet in a developing country perspective. Various attitudinal, normative, and control constructs were identified and investigated for their possible influence on broadband Internet adoption. The empirical data for this study were collected using a self-administered questionnaire that included items related to various attitudinal, normative, and control constructs. Descriptive statistics and regression analysis were used to test these constructs for their possible influence on Indian consumers' adoption of broadband Internet. The findings suggest that perceived ease of use (PE), social outcomes (SO), hedonic outcomes (HO), service quality (SQ), facilitating conditions resources (FCR), and self-efficacy (SE) were very significant predictors of Indian consumers' behavioral intention to adopt broadband Internet. This study has multifold significance. The integrated research framework used in this study is an extension of previous well-established research models (such as Model of Adoption of Technology in Households [MATH], Diffusion of Innovation [DOI], and Theory of Planned Behavior [TPB]) and provides an enhanced comprehension of broadband Internet by the Indian household consumers.

Keywords

adoption, broadband, India, survey, consumer behavior, household

Introduction

The Broadband Technology

Papacharissi and Zaks (2006, p-2) have defined broadband access as "all flavors of high-speed digital voice, data and video services, as well as the underlying infrastructure, clients and technologies that enable these services." (p. 2). More specifically, speeds of at least 384 Kbps and packet-switched technology are used at the level of interactivity when adopting broadband access, and these can be allowed for the control and selection of content. Broadband is the new innovation high-speed Internet access technology that provides not only economic and public benefits but also improves people's lives (B. Anderson & Tracey, 2001; Choudrie & Dwivedi, 2006a). Broadband Internet diffusion not only increases national competitiveness (thus increases Foreign Direct Investment [FDI] in the country; Bankole, Osei-Bryson, & Brown, 2013; Manzoor, 2012) but also helps increase international competitiveness by leveraging e-commerce activities (Armenta, Serrano, Cabrera, & Conte, 2012; Ayanso, Cho, & Lertwachara, 2014; Choudrie & Dwivedi, 2004; Dwivedi & Irani, 2009; Johnson, 2010; Wambogo Omole, 2013; Qureshi, 2012; Wei, 2004).

One of the significant changes in Internet technologies has been the replacement of narrowband (dial-up) connection by high-speed broadband connection (Choudrie & Dwivedi, 2004). Broadband is considered a technology that provides high-speed broadcast with greater bandwidth, higher data transmission rates, and always-on access to applications, services, and content (Qureshi, 2010; Sawyer, Allen, & Lee, 2003). The exponential growth of Internet technology has given rise to increased customer expectations (Wei, 2004). Many research studies have been conducted to look into various applications of broadband that have huge potential of increased revenues for communications industries (Crandall, Lehr, & Litan, 2007; Jung, Perez-Mira, & Wiley-Patton, 2009; Katz, Vaterlaus, Zenhäusern, & Suter, 2010; Kolko, 2012; Koutroumpis, 2009). Broadband, therefore, has a key role to provide an ever more-connected world to the consumer communities (Bell, 2008).

Broadband in India

First broadband connection in India was given in 1998. The projected number of broadband subscribers in India by 2007 was 9 million and it was expected that the number will cross

[1]Bahria University, Karachi

Corresponding Author:
Amir Manzoor, Management Sciences Department, Bahria University, 13 National Stadium Road, Karachi, Pakistan 75300.
Email: amirmanzoor@yahoo.com

15 million by 2012. India is considered one of the largest and fastest growing major telecom markets in the world. Broadband Internet in India is growing mainly due to the rapid growth in Indian telecommunication sector. India's growing economy, rapidly expanding middle class, low tariffs, and highly competitive market have further supported the continuous expansion of the broadband sector (ABIresearch, 2012).

Successive Indian governments introduced comprehensive reforms over the last decade that changed the landscape of telecommunication in India. Indian government has taken many initiatives, such as Indian PC program, e-governance, e-learning, and e-healthcare services, to spur the growth of broadband Internet. Broadband network expansion and upgrades are underway in India to match up with comparable Asian broadband markets. Indian government has made three priorities for its plans to increase broadband penetration and usage: lower cost-per-engagemen pricing, affordable broadband software and hardware, and cheap, consistent broadband interconnectivity among villages (Pereira, 2011). The Indian government has continued its policy of an open and competitive broadband market. Some of the key steps in this regard included issuing licenses to new telecom operators, allowing global operators to work with local companies, privatization of the country's long-distance market, bandwidth sharing among Internet service providers (ISPs), allowing the use of Ku-band in both Indian and foreign satellites, and launch of mobile number portability (MNP; the Telecom Regulatory Authority of India [TRAI], 2010). In 2011, the fixed line penetration in India was 3% and around 98% of the Indian population had some form of access to a telephone (BusinessWire.com, 2011). During the last decade, heavy investment was made in telecom infrastructure. The Indian government continued to force completion of restructuring of telecommunications regulatory regime. The TRAI continued its structural reforms such as adoption of unified licensing, increased infrastructure sharing, and revised FDI policy, increasing foreign ownership limit from 49% to 74% (Press Trust of India, 2012). In 2012, Indian government approved a national broadband plan. The objective of the plan was to provide high-speed Internet access and e-government services to more than 160 million people by 2014. Under this plan, an open access countrywide optical-fiber-based broadband network will be established, which will provide connectivity to all areas with population of 500 and above by 2013 (TRAI, 2010).

Broadband Internet and Economic Development

There are many factors that boost economic growth, including product, process, and organizational innovations, and the distribution and generation of information and ideas. Technological changes result from small incremental improvements over a period of time. However, even a small improvement in technology can alter how and where economic activity is managed (Czernich, Falck, Kretschmer, & Woessmann, 2011; Majumdar, Carare, & Chang, 2010; Organisation for Economic Co-operation and Development [OECD], 2008).

OECD (2008) defines broadband as a general-purpose technology that when incorporated with other information communication technologies (ICTs) can change how and where economic activities are managed. This change is affected through many channels. A clear impact occurs from investing in the infrastructure itself and related ICTs. Indirect impacts come from all aspects of economic activity affected by broadband and which drive economic growth (Collins, Day, & Williams, 2007).

Ample academic literature is available that has examined the impact of broadband on firm-level productivity (Holt & Jamison, 2009). A central theme of the academic literature is the ability of broadband to create new business models, processes, and innovation. It helps in increasing the productivity and efficiency of the firm. The study of Allen Consulting Group (2002) reported results from a survey of Australian business on the cost saving obtained from using broadband Internet. The results of the survey showed that around 6.3% businesses experienced cost saving from the utilization of broadband Internet as compared with the 1.5% from the utilization of dial-up Internet. The research declared that for Australian businesses, the average cost saving would result in an overall output gain of approximately 0.32%. These results are consistent with the most recent research by the Australian Industry Group (2008) who claimed that more than 93% of the firms investigated showed that broadband has a positive influence on the efficiency and output.

A panel of 6,060 New Zealand firms was analyzed by Grimes, Ren, and Stevens (2012) in a study to find the effect that various forms of Internet access have on the firm's output. The information collected from the surveys conducted by Statistics New Zealand permitted the authors to control for a range of firm characteristics including those factors that may determine a firm's choice of Internet access. The two estimation approaches used were an instrument variable (IV) estimator and propensity score matching (PSM) estimator. The (IV) estimation results showed higher productivity impacts from the utilization of broadband. The utilization of broadband increased the productivity of firms by 21% to 25%. The PSM results showed that productivity increased by 6.9 to 9.7 due to broadband utilization. These effects were consistent across different types of firms with no significant differences across an urban versus rural split or across high versus low knowledge industries. Although the results firm IV estimation were greater than those gathered from PSM, the authors of the study favored PSM estimation due to their low confidence on IV results. This low confidence resulted from their lack of specific knowledge about the correct functional specification in relation to a firm's labor productivity relative to its sector.

In addition to examination of broadband Internet on firm-level productivity, there is growing academic literature that examines the association between broadband and macroeconomic-level indicators such as economic growth and employment. Czernich et al. (2011) carried out a panel analysis of 25 OECD countries, using an instrumental variable model, to estimate the impact of broadband infrastructure on economic growth. Czernich et al. concluded that the broadband introduction and diffusion had a significant impact in the growth of gross domestic product (GDP) in the OECD countries included in the panel study. After the introduction of broadband, the per capita GDP of a country was on average 2.7% to 3.9% greater than before the introduction of broadband, controlling for country and year fixed effects. It was further concluded that an increase in 10% points in the diffusion of broadband raised annual growth per capita by between 0.9% and 1.5% points annually.

The effect of broadband penetration on the employment growth, rents, wages, business growth, and industry structure at the community, industry, and state level was analyzed in the study by (Lehr, Bauer, Heikkinen, & Clark, 2011)). Lehr et al. concluded that the take-up of broadband promotes the economic activity with major effects in the growth of job and business, especially for larger business and business in IT-intensive sectors. It was further concluded that the take-up of broadband had no substantial effect on the wages but that there was an important link between residential property values and broadband take-up.

Qiang, Rossotto, & Kimura (2009) used the Barro cross-sectional and endogenous growth model by using the data from 120 developing and developed countries to evaluate the effect that broadband has had on the long-term economic growth rates over the years 1980 to 2006. The findings from the empirical analysis recommend a strong and substantial growth dividend from broadband access in developed countries. Keeping other factors constant, in a developed country, a 10% increase in the penetration of broadband would result in 1.21% increase in the growth of economy. It was found that keeping other factors constant, a 10% increase in the diffusion of broadband in developing countries would cause a 1.38% increase in the growth of an economy.

The research by Koutroumpis (2009) used a macroeconomic production function with microeconomic model for broadband investment to estimate how broadband infrastructure investment contributed in economic growth in 22 OECD countries over the period of 5 years (i.e., 2002-2007). Findings showed a strong causal relationship between economic growth and broadband. The results also suggested that there are increasing returns to investment in broadband infrastructure. Countries with broadband penetration rates of more than 30% gained higher returns from broadband investments relative to those countries with lower broadband penetration rates.

The study by Crandall et al. (2007) assessed the impacts of broadband penetration on economic output and employment, in aggregate and by industry sector, for 48 states of the United States over the period 2003-2005. Crandall et al. concluded that non-farm employment in several industries was positively associated with broadband use. Particularly, for every 1% increase in broadband penetration in a state, employment increased by 0.2% to 0.3% each year. The study also concluded that employment in both manufacturing and service industries (especially education, finance, and health care) was positively associated with broadband penetration. Moreover, the state output of goods and services was positively associated with the utilization of broadband.

The Broadband Adoption

There exist both macro- (Choudrie & Lee, 2004; Feijoo, Gomez-Barroso, Ramos, & Rojo-Alonso, 2006; Garfield & Watson, 1997; Giovanis, 2011; Han, 2003; Hargittai, 1999; Nam, Kim, Lee, & Duan, 2009; Ooi, Sim, Yew, & Lin, (2011) ; Sim, Kong, Lee, Tan, & Teo, 2012) and micro-level (Chia, Lee, & Yeo, 1998; Choudrie & Dwivedi, 2006a, 2006b; Dwivedi, Lal, & Williams, 2009; Irani, Dwivedi, & Williams, 2009; OECD, 2001, 2008) studies on broadband Internet adoption and diffusion. Most studies have been conducted in the context of developed countries. There exist few studies in the context of developing countries (Dwivedi, Choudrie, & Brinkman, 2006; Dwivedi, Williams, Lal, & Bhatt, (2007) ; Dwivedi, Khan, & Papazafeiropoulou, 2007; Khoumbati, Dwivedi, Lal, & Chen, 2007; Ooi, Sim, et al., 2011).

There exists room for further research to better understand the current state of broadband Internet deployment, uptake, and diffusion in developing world and a research that may assist in explaining the factors that may hasten the process of broadband adoption. It is believed that broadband Internet can significantly boost Indian economy by enhancing the well-being of its people by assisting in the release of education, health, and telecommunications services at a more affordable rate to the masses (Dwivedi, Khan, et al., 2007). Given such perceived benefits, it is pertinent to gain an understanding of the factors affecting the consumption, acceptance, and usage of broadband among Indian consumers.

Many research studies have been conducted on broadband adoption in the developing countries such as Bangladesh, India (Dwivedi, Williams, Lal, & Bhatt, 2007)), the Kingdom of Saudi Arabia (KSA; Dwivedi & Weerakkody, 2007), and Malaysia (Ooi, Sim, et al., 2011). These studies identified different factors significant in determining the behavioral intentions (BIs) of consumers to adopt broadband. Table 1 lists the factors.

In the era of continued development of ICTs and emerging network economy, developing countries, such as India, are faced with various challenges as well as opportunities (Kapur, 2001). Inadequate broadband access may result in negative impact on Indian productivity and is likely to result in higher operational costs for a number of businesses. This not only affects the performance of existing Indian firms but also hinders the attractiveness of India as a potential investment place. Broadband is immensely important to India

Table 1. Factors Affecting Broadband Adoption in Different Developing Countries.

Country	Significant factors in broadband adoption
Bangladesh	Attitude, primary influence, secondary influence, and facilitating conditions resources (Dwivedi, Khan, & Papazafeiropoulou, 2007)
Kingdom of Saudi Arabia	Usefulness, service quality, age, usage, type of connection, and accommodation (Dwivedi, Williams, Weerakkody, Lal, & Bhatt, 2008)
Pakistan	Primary influence, facilitating conditions resources, cost, and perceived ease of use (Khoumbati, Dwivedi, Lal, & Chen, 2007)
India	Relative advantage, hedonic outcomes, and cost (Dwivedi et al., 2008)
Malaysia	Primary influences (PI), secondary influences, relative advantage (RA), utilitarian outcomes (UO), facilitating condition resources (FCR), and self-efficacy (SE; Ooi, Sim, Yew, & Lin, 2011)

because it can accelerate the economic growth through ICT, which has been identified as a key factor to improve productivity performance. Considering the relatively low levels of broadband adoptions in developing countries, including India, a research in this area may be helpful in understanding and accelerating the process of broadband Internet adoption by consumers in India. It is essential to conduct an empirical study in India, given the slow rate of broadband adoption. This study extends the research conducted by Khoumbati et al. (2007) and Dwivedi et al. (2008) to gain an up-to-date and deeper comprehension of consumers' perception and attitude toward the usefulness of broadband technology.

After witnessing the Korean broadband success and the resultant economic development (Jin, 2005), academicians, government telecom officials, and information technology experts around the world are interested in knowing how the Korean success story can be replicated by gaining an understanding of the factors that influenced such a quick embrace of broadband technology by Korea. This research would be a valuable addition to existing literature as the reasons for adoption of consumers can be understood in a more advanced level from a developing country that is situated in a strategic location neighboring China. The knowledge of such reasons would provide a good foundation for other developing countries that plan to deploy broadband services and to reap associated economic and development benefits.

Research Questions and Theoretical Model

This study addresses two research questions:

Research Question 1: What are the decisive factors in consumers' BI to adopt broadband?
Research Question 2: Which of the factor(s) most significantly influence consumer's BI to adopt broadband?

The factors to be studied in the theoretical model of this research have been derived from theoretical models of Diffusion of Innovation (DOI), Theory of Planned Behavior (TPB), and Model of Adoption of Technology in Households (MATH). Other factors might have an effect on the adoption

of broadband and these factors have been taken into consideration to predict the intention to adopt broadband Internet. There are five sections of this study. The first section provides an overview of the research undertaken. The next section provides a discussion of the theoretical background and research model. This is followed by the discussion of methodology. After that, a discussion and detailed interpretation of findings will be presented. The last section provides a conclusion with implications, limitations, and suggestions for future research.

Theoretical Background and the Research Model

Theoretical Basis

This study utilizes some important theoretical models, such as TPB, DOI, and MATH, to examine broadband diffusion. According to TPB, the IT usage among individuals is driven by their BIs. These BIs comprised of Attitudes (personal belief about expected outcomes of the usage of IT), Subjective Norm (belief about personal behavior regarding IT usage with respect to other people expectations), and Perceived Behavioral Control (personal belief about degree of usage capability of IT). Many researches have used and adopted TPB to study the adoption and implementation of information technology (Aboelmaged, 2010; Ajzen, 1985, 1988, 1991; Ajzen & Madden, 1986; C. L. Anderson & Agarwal, 2010; Hernández, Jiménez, & Martín, 2010, 2011; M.-C. Lee & Tsai, 2010; Ortiz de Guinea & Markus, 2009; San Martín & Herrero, 2012; Schifter & Ajzen, 1985; Venkatesh, Thong, & Xu, 2012; Yousafzai, Foxall, & Pallister, 2010). Many studies have extended TPB to fit varying needs of researches such as the Decomposed Theory of Planned Behavior (Decomposed TPB; Cheon, Lee, Crooks, & Song, 2012; Choudrie & Dwivedi, 2006b; Hartshorne & Ajjan, 2009; Lin, 2010; Pham, Pham, & Nguyen, 2011; Püschel, Mazzon, & Hernandez, 2010; Ramayah, Rouibah, Gopi, & Rangel, 2009; Venkatesh & Brown, 2001; Yaghoubi, 2010). The MATH framework is based on three main elements, namely, attitudinal belief (consisting of utilitarian outcomes [UO], hedonic outcomes [HO], and social outcomes [SO]), normative belief (consisting of primary and secondary influences), and control belief (consisting of perceived knowledge [K], perceive ease of use [PE], and perceived cost). This framework has also been used in research considering it to be more appropriate and useful to study broadband adoption (Venkatesh & Brown, 2001). The Rogers's DOI has been used to study the pattern of diffusion and adoption of new technology to predict the success of new technology invention. This model focuses on both the usage and subsequent usage aspect (Rogers, 1995). One element of DOI, relative advantage (RA), has been found to be easily applied and integrated with other constructs used to examine diffusion of broadband (Choudrie & Dwivedi, 2004). Against this backdrop, this research uses a modified framework that extends existing models to investigate the factors that affect the BI of consumers to adopt broadband Internet. The definitions of each variable are shown in Table 2.

It was found that constructs such as RA, UO, HO, primary influence, secondary influence, self-efficacy (SE), facilitating conditions, overall service quality (SQ), and service value and satisfaction significantly influenced BI to adopt broadband (Choudrie & Dwivedi, 2005; Dwivedi, 2005; Dwivedi et al., 2006; Dwivedi, Khan, & Papazafeiropoulou, 2007); Rogers, 1995). However, other constructs are yet to be applied to examine broadband adoption. For this study, we have divided the facilitating conditions variable into two variables, namely, facilitating conditions resources (FCR) and facilitating conditions technology (FCT). The purpose is to look separately at the impact of general facilitating conditions and facilitating conditions related to technology. This study postulates that BI to adopt broadband is influenced by several independent variables, which can be categorized into three broad groups (Choudrie & Dwivedi, 2004, 2005; Dwivedi et al., 2006; Dwivedi, Khan, et al., 2007; Dwivedi, Khoumbati, Williams, & Lal, 2007). These are as follows:

1. **Attitudinal factors**, which describe the individual's perception toward broadband technologies (RA, UO, HO, SO, SQ);
2. **Normative factors**, which describe the social influences that may affect the intention to adopt broadband (referents influence [RI], social influence [SI]); and
3. **Control factors**, which control or influence the ability to initiate and maintain a broadband subscription (K, skills [SK], FCT, FCR, PE, SE).

RA and UO have been found to be significant factors in determining broadband adoption (Dwivedi & Irani, 2009; Dwivedi, Khan, & Papazafeiropoulou, 2007; Venkatesh & Brown, 2001). Figure 1 shows the conceptual framework used by this research. This framework is adapted and modified from Dwivedi (2005) and Dwivedi, Choudrie, & Brinkman (2006). This framework takes broadband adoption

as an independent variable that is affected by a number of independent variables that can be categorized as attitudinal, normative, and control factors. Because very few studies have examined the broadband adoption in developing countries (especially Southeast Asian countries), it was decided that all the possible and appropriate constructs (within the Indian context) from previous studies, as described in Table 2, would be included in this study.

Hypotheses

Based on the literature review and the developed conceptual model, following hypotheses were developed for this study.

> **Hypothesis 1.** Attitudinal, normative, and control factors significantly influence consumers' BI to adopt broadband (Choudrie & Dwivedi, 2005; Dwivedi, 2005; Dwivedi, Khan, et al., 2007).
>
> **Hypothesis 2.** The proposed conceptual model of broadband adoption provides an appropriate level of explanation of variance in the consumer's BI to adopt broadband (Dwivedi, Khan, et al., 2007; Dwivedi & Weerakkody, 2007).

Research Method

Sampling and Data Collection

For the purpose of exploratory research, such as the current study, the survey is considered a suitable instrument for primary data collection (Choudrie & Dwivedi, 2005). A self-administered questionnaire was developed and used. Literature review provided the foundation for development of questionnaire. First, a draft questionnaire was prepared. The final questionnaire consisted of 21 questions. The Likert-type scale questions were adapted from Dwivedi (2005) and Dwivedi et al. (2006) and the demographic categories were adapted from Choudrie and Dwivedi (2006a). Each question was a statement followed by a 5-point Likert-type scale ranging from "strongly disagree" through "neither agree nor disagree" to "strongly agree." Questionnaires were distributed both in hard copy format and via e-mails.

Profile of Respondents

Table 3 provides the breakdown of various residential consumers in our sample.

The Sample Plan

Due to the uncertainty regarding the identity of consumers currently using the broadband facility and the nomadic nature of access, the snowballing sampling technique (Dwivedi, Khan, et al., 2007; Dwivedi, Khoumbati, et al., 2007) was used. The researcher sought help from his friends

Table 2. Constructs Related to Broadband Internet Adoption.

Constructs	Definition	Source
Behavioral intention	The consumer's intention to adopt and use broadband Internet	Brown and Venkatesh (2005); Venkatesh and Brown (2001)
Relative advantage	Degree to which the consumer perceives broadband Internet beneficial	Rogers (1995)
Utilitarian outcomes	Degree to which broadband Internet is beneficial in enhancing the effectiveness of typical daily activities of the consumer	Brown and Venkatesh (2005); Venkatesh and Brown (2001)
Hedonic outcomes	The degree of pleasure that the consumer derives from the use of broadband Internet	Brown and Venkatesh (2005); Venkatesh and Brown (2001)
Social outcomes	The enhancement of social status that the consumer derives from the use of broadband Internet	Venkatesh and Brown (2001)
Service quality	The perceived quality of service of Internet service provider	Brady and Cronin, (2001); Cronin, Brady, and Hult (2000); DeLone and McLean (2003); Parasuraman, Berry, and Zeithaml (1991); Sweeney, Soutar, and Johnson (1999); Szymanski and Henard (2001); Teas and Agarwal (2000)
Satisfaction	The satisfaction a consumer derives from the service of the Internet service provider	Brady and Cronin (2001); Brady et al. (2001); Cronin et al. (2000); Sweeney et al. (1999); Szymanski and Henard (2001); Teas and Agarwal (2000)
Service value	The consumer perceived value of service of Internet service provider	Bolton and Drew (1991); Brady and Cronin (2001); Brady et al. (2001); Cronin et al. (2000); Dodds, Monroe, and Grewal (1991); Sweeney et al. (1999); Szymanski and Henard (2001); Teas and Agarwal (2000); Zeithaml (1988)
Primary influences	Influences from the consumer's family and friends to use or not to use broadband Internet	Brown and Venkatesh (2005); Venkatesh and Brown (2001)
Workplace referents' influences	The degree of peer influence on the consumer's use of broadband Internet	Brown and Venkatesh (2005)
Secondary influences	The degree of influence from the secondary sources of information (such as newspapers, advertisement) on consumer's adoption of broadband Internet	Brown and Venkatesh (2005); Rogers (1995); Venkatesh and Brown (2001)
Perceived knowledge	The consumer's perceived knowledge about the broadband Internet	Rogers (1995); Venkatesh and Brown (2001)
Self-efficacy	The consumer's skill to use broadband Internet without assistance	Dwivedi (2005)
Perceived ease of use	The perceived degree of ease in using the computer	Venkatesh and Brown (2001)
Facilitating conditions	How resourceful a consumer feels when he or she subscribes to broadband Internet	Venkatesh and Brown (2001)
Cost	The cost of current broadband subscription	Venkatesh and Brown (2001)
Declining cost	Extent to which the falling cost of broadband Internet access inhibits its adoption	Venkatesh and Brown (2001)

in India. Initial respondents from academia, the private sector, government, students, and the public within Mumbai city were first identified. They in turn referenced their friends and colleagues from other cities of India who utilized broadband. This progressively increased the sample size (Selamat et al., 2008; Ooi, Sim, Yew, & Lin, 2011)). This strategy led to the questionnaire being administered to 600 broadband users during the period of June to December 2013. Of the 600 questionnaires administered, 203 respondents returned completed and usable questionnaires. Thus, a response rate of

33% was achieved. This response rate is comparable with response rates in recent studies on broadband adoption (Dwivedi, Khan, et al., 2007; Dwivedi, Khoumbati, et al., 2007; Mugeni, Wanyembi, & Wafula, 2012; Ooi, Lin, et al., 2011).

Data Analysis

In the initial phase of data analysis, responses were checked and assigned a unique ID number (Fowler, 2002; Holmström,

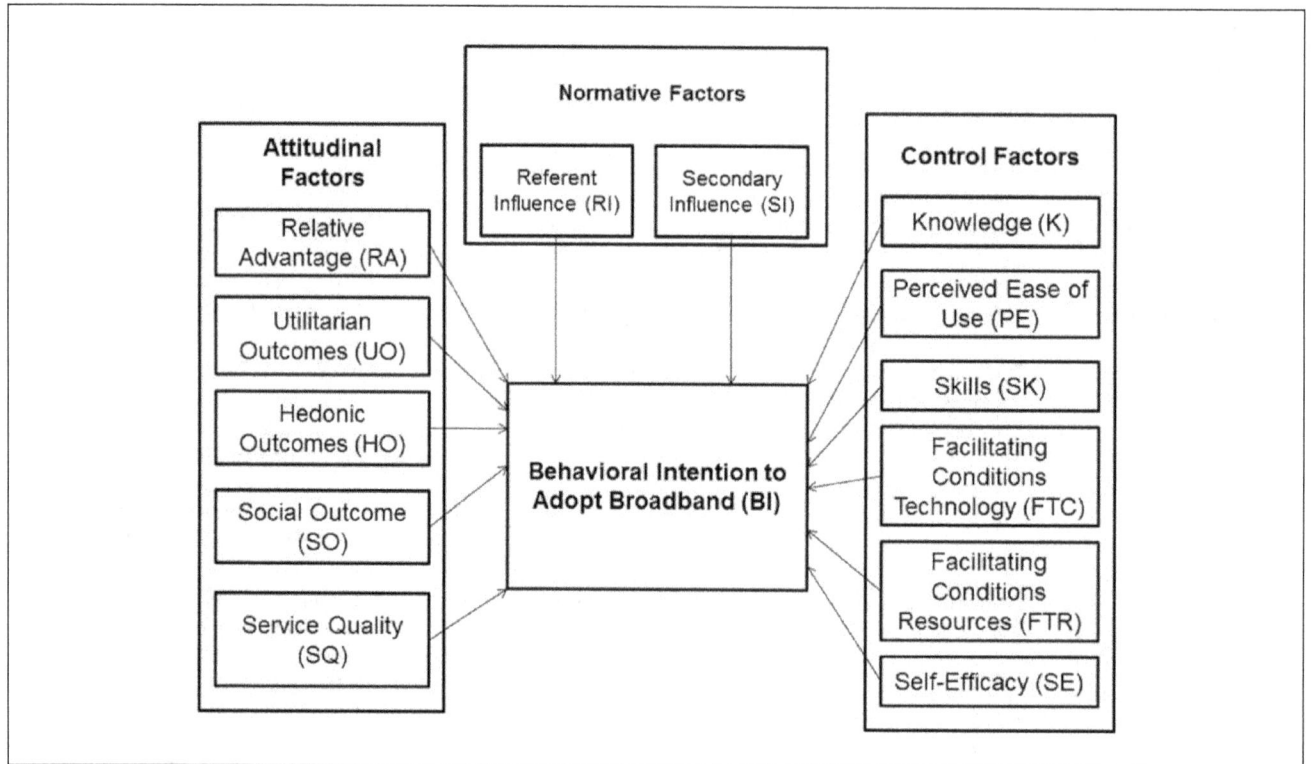

Figure 1. Research model.
Note. Adapted from Dwivedi (2005) and (Dwivedi, Williams, Lal, & Bhatt, 2007).

Table 3. Frequency Distributions of Respondents' Demographics.

	n (%)
Gender	
Male	390 (65)
Female	210 (35)
Age (years)	
Below 20	25 (4.16)
21-30	160 (26.66)
31-40	210 (35)
41-50	150 (25)
51 and older	55 (9.16)
Income (INR)	
<10,000	40 (6.66)
10,000-19,999	15 (2.5)
20,000-29,999	95 (15.83)
30,000-39,999	178 (29.66)
40,000-49,999	90 (15)
50,000-59,999	100 (16.66)
60,000-69,999	52 (8.66)
>70,000	30 (5)
Education level	
High school	178 (29.66)
Technical college	155 (25.83)
4-year degree	200 (33.33)
Graduated degree	67 (11.16)
Total	600

Ketokivi, & Hameri, 2009). SPSS software (Version 21) was used to calculate descriptive statistics. The same software was used for reliability testing and regression analysis of research data collected through questionnaires (Hinton, Brownlow, McMurray, & Cozens, 2004; Straub, Boudreau, & Gefen, 2004).

Reliability Test

Reliability of constructs was estimated using Cronbach's coefficient (α; Table 4).

Higher values of Cronbach's alpha represent the higher internal consistency of the construct. Values of alpha greater than .9 represent very high consistency whereas values between .5 and .7 represent moderate consistency (Hinton et al., 2004). In this study, Cronbach's alpha varied between .879 for SO and .911 for BI. The values obtained suggest that all constructs were internally consistent.

Testing of Multivariate Assumptions

Before data analysis, we checked the multivariate assumptions (such as multicollinearity; Fotopoulos & Psomas, 2009; Hair, Tatham, Anderson, & Black, 2005; V.-H. Lee, Ooi, Tan, & Chong, 2010). Table 5 shows the variance inflation factor (VIF) values of various independent variables. The values ranged from 1.120 to 1.914 and all values were less than 10

Table 4. Reliability Values (N = 203).

Construct	Number of items	Cronbach's α	Type
BI: Behavioral intention	3	.911	Excellent reliability
RA: Relative advantage	7	.884	High reliability
UO: Utilitarian outcomes	11	.892	High reliability
HO: Hedonic outcome	4	.897	High reliability
SQ: Service quality	4	.898	High reliability
RI: Referents influence	8	.886	High reliability
SI: Social influence	2	.898	High reliability
K: Knowledge	3	.883	High reliability
SK: Skills	6	.884	High reliability
FCT: Facilitating conditions technology	6	.883	High reliability
FCR: Facilitating conditions resources	7	.893	High reliability
SO: Social outcome	4	.879	High reliability
PE: Perceived ease of use	2	.887	High reliability
SE: Self-efficacy	3	.889	High reliability

Table 5. Multicollinearity Statistics.

Model		Collinearity statistics	
		Tolerance	VIF
I	(Constant)		
	BI	.701	1.428
	UO	.815	1.226
	SO	.893	1.120
	RA	.621	1.610
	HO	.504	1.985
	SQ	.816	1.226
	SK	.574	1.742
	FCT	.634	1.578
	PE	.598	1.673
	K	.585	1.708
	SE	.707	1.414
	FCR	.522	1.914
	SI	.623	1.604
	RI	.692	1.446

Note. VIF = variance inflation factor; BI = behavioral intention to adopt broadband; UO = utilitarian outcomes; SO = social outcomes; RA = relative advantage; HO = hedonic outcomes; SQ = service quality; SK = skills; FCT = facilitating conditions technology; PE = perceived ease of use; K = knowledge; SE = self-efficacy; FCR = facilitating conditions resources; SI = social influence; RI = referents influence. Dependent variable: SNO = Serial Number.

Table 6. Descriptive Statistics for Behavioral Intention (N = 203).

Factors	Detailed factors	M	SD
BI (behavioral intention)	Scale-BI	4.96	1.44
	BI1	5.54	1.26
	BI2	4.20	1.69
	BI3	5.16	1.36

Descriptive Statistics

In this study, we used 14 constructs to measure the factors affecting broadband adoption in India. In this section, the means and standard deviations of the dependent variable BI and the items related to the 14 constructs are described.

Descriptive statistics for BI. Three questions were used to measure the consumer's BI to subscribe to broadband (Azab, 2009). Table 6 shows the mean and standard deviations of aggregate measures for the three constructs used to measure BI. Respondents showed strong agreement for all the items of the BI with average score of aggregate measure (M = 4.96, SD = 1.44). Item BI1 scored the maximum (M = 5.54, SD = 1.26) and Item BI2 scored the minimum (M = 4.20, SD = 1.69).

Descriptive statistics for attitudinal factors. Table 7 shows the mean and standard deviations of aggregate measures for the five constructs used to measure attitudinal factors. Respondents showed strong agreement for all the constructs as evident from their mean and standard deviation values. RA showed the strongest agreement (M = 5.78, SD = 1.24) whereas SQ showed the least agreement (M = 5.29, SD = 1.47).

with tolerance value greater than 0.1. Thus, multicollinearity of data was not shown (Hair et al., 2005). There were no independent variables that had condition indexes above 30 coupled with 2 variance proportions greater than .50 (Tabachnick & Fidell, 2007). Hence, based on these basic assumptions of the multivariate model, it implies that there were no statistically significant violations (V.-H. Lee et al., 2010; Sit, Ooi, Lin, & Chong, 2009).

Table 7. Descriptive Statistics for Attitudinal Constructs (N = 203).

Factors	Detailed factors	M	SD	Rank
UO (utilitarian outcomes)	**Scale-UO**	**5.64**	**1.24**	**2**
	UO1	5.70	2.60	
	UO2	5.79	1.16	
	UO3	5.75	1.30	
	UO4	6.00	1.07	
	UO5	5.23	1.51	
	UO6	5.56	0.99	
	UO7	5.53	0.92	
	UO8	5.31	1.05	
	UO9	5.34	1.06	
	U10	5.80	0.77	
	UO11	6.01	1.18	
RA (relative advantage)	**Scale-RA**	**5.78**	**1.24**	**1**
	RA1	5.88	0.98	
	RA2	6.09	1.12	
	RA3	5.73	1.29	
	RA4	5.55	1.19	
	RA5	6.02	1.17	
	RA6	5.54	1.40	
	RA7	5.57	1.56	
SO (social outcomes)	**Scale-SO**	**5.56**	**1.36**	**3**
	SO1	5.70	1.31	
	SO2	6.11	1.17	
	SO3	5.79	1.14	
	SO4	5.89	1.18	
HO (hedonic outcomes)	**Scale-HO**	**5.42**	**1.47**	**4**
	HO1	5.45	1.53	
	HO2	5.79	1.01	
	HO3	5.29	1.36	
	HO4	4.64	1.96	
SQ (service quality)	**Scale-SQ**	**5.29**	**1.47**	**5**
	SQ1	5.82	1.33	
	SQ2	5.57	1.48	
	SQ3	5.40	1.70	
	SQ4	5.52	1.40	

Table 8. Descriptive Statistics for Control Constructs (N = 203).

Factors	Detailed factors	M	SD	Rank
SK (skills)	**Scale-SK**	**6.09**	**1.09**	**2**
	SK1	5.89	1.24	
	SK2	6.14	1.34	
	SK3	6.31	0.93	
	SK4	6.17	1.02	
	SK5	6.07	1.06	
	SK6	5.95	0.98	
K (knowledge)	**Scale-K**	**5.78**	**1.22**	**5**
	K1	5.71	1.21	
	K2	5.85	1.24	
	K3	5.97	1.16	
FCT (facilitating conditions technology)	**Scale-FCT**	**6.22**	**1.05**	**1**
	FCT1	6.29	1.06	
	FCT2	6.16	1.05	
	FCT3	6.19	1.07	
	FCT4	6.00	0.92	
	FCT5	5.68	1.19	
	FCT6	5.56	0.97	
FCR (facilitating conditions resources)	**Scale-FCR**	**5.37**	**1.61**	**6**
	FCR1	5.79	1.31	
	FCR2	4.95	1.90	
	FCR3	5.58	1.15	
	FCR4	5.96	1.03	
	FCR5	5.73	1.42	
	FCR6	5.22	1.90	
	FCR7	5.48	1.20	
PE (perceived ease of use)	**Scale-PE**	**6.02**	**1.02**	**3**
	PE1	5.84	1.11	
	PE2	6.20	0.92	
SE (self-efficacy)	**Scale-SE**	**5.82**	**1.14**	**4**
	SE1	5.54	1.22	
	SE2	6.10	1.05	
	SE3	5.99	1.24	

Descriptive statistics for control factors. Table 8 shows the mean and standard deviations of aggregate measures for the eight constructs used to measure control factors. Respondents showed strong agreement for all the constructs as evident from their mean and standard deviation values. FCT showed the strongest agreement (M = 6.22, SD = 1.05) whereas FCR showed the least agreement (M = 5.37, SD = 1.61).

Descriptive statistics for normative factors. Table 9 shows the mean and standard deviations of aggregate measures for the two constructs used to measure normative factors. Respondents showed strong agreement for all the constructs as evident from their mean and standard deviation values. RI showed the strongest agreement (M = 5.17, SD = 1.39) whereas SI showed the least agreement (M = 5.03, SD = 1.1.68).

Regression Analysis: Influence of Independent Variables on Broadband Intention

The study used ordinary least square regression analysis. BI was taken as dependent variable and 14 constructs as predictor variables. The adjusted R^2 was .555 (F = 20.387, p < .001; see Tables 10 and 11).

The conceptual model of broadband Internet adoption presented in this study performed well as compared with previous studies on broadband Internet adoption. The adjusted R^2 for previous behavioral models was .32 (Davis, Bagozzi, & Warshaw, 1989)), .51 (Davis, 1989), and .435 (Dwivedi, 2005). The adjusted R^2 for the model presented in this study was found to be .555. It shows that the model presented in

Table 9. Descriptive Statistics for Normative Construct (N = 203).

Factors	Detailed factors	M	SD	Rank
RI (referents influence)	**Scale-RI**	**5.17**	**1.39**	**1**
	RI1	5.78	1.20	
	RI2	5.26	1.23	
	RI3	5.44	1.38	
	RI4	5.09	1.89	
	RI5	5.96	1.19	
	RI6	3.35	1.31	
	RI7	5.26	1.47	
	RI8	5.24	1.47	
SI (social influence)	**Scale-SI**	**5.03**	**1.68**	**2**
	SI1	5.29	1.65	
	SI2	4.76	1.71	

Table 10. Model Summary.

Model	R	R^2	Adjusted R^2	Standard error of the estimate
1	.764[a]	.584	.555	0.640036226698565

Note. Dependent variable: BI. UO = utilitarian outcomes; SO = social outcomes; RA = relative advantage; HO = hedonic outcomes; SQ = service quality; SK = skills; FCT = facilitating conditions technology; PE = perceived ease of use; K = knowledge; SE = self-efficacy; FCR = facilitating conditions resources; SI = social influence; RI = referents influence; BI = behavioral intention to adopt broadband.
[a]Predictors: (Constant), SE, FCR, HO, UO, SQ, PE, SI, RI, RA, FCT, K, SO, SK.

Table 11. ANOVA.

Model		Sum of squares	df	M^2	F	Significance
1	Regression	108.568	13	8.351	20.387	.000[a]
	Residual	77.423	189	0.410		
	Total	185.991	202			

Note. Dependent variable: BI. UO = utilitarian outcomes; SO = social outcomes; RA = relative advantage; HO = hedonic outcomes; SQ = service quality; SK = skills; FCT = facilitating conditions technology; PE = perceived ease of use; K = knowledge; SE = self-efficacy; FCR = facilitating conditions resources; SI = social influence; RI = referents influence; BI = behavioral intention to adopt broadband.
[a]Predictors: (Constant), SE, FCR, HO, UO, SQ, PE, SI, RI, RA, FCT, K, SO, SK.

this study was able to explain an appropriate and comparable level of variance in BI. That means that predictor variables considered in the model of this study are significant in understanding the consumers' BI to adopt broadband.

In this validated model, six predictor variables were found significant for explaining the variation in BI (Table 12). These predictor variables include HO (β = .249, p < .001), SQ (β = −.506, p < 0.001), FCR (β = .333, p < .001), SO (β = −.570, p < .001), PE (β = .922, p < .001), and SE (β = .433,

p < .001). The regression equation for the validated model was found to be as follows:

$$BI = 3.11 + 0.249HO - 0.506SQ + 0.333FCR - 0.570SO + 0.922PE + 0.433SE + 0.552$$

The value of β suggests that the PE had the largest impact in the explanation of variations of consumer BI to adopt broadband, followed by SO, SQ, SE, FCR, and HO. RA, The variables UO, RI, SI, K, SK, and FCT did not significantly affect the BI.

Discussion

The aim of this study is to examine the significant factors that influence the consumer attitude toward broadband adoption and usage in India. India is progressing fast in IT infrastructure development and broadband adoption rate. In comparison with a study on Bangladesh (Dwivedi, Khan, & Papazafeiropoulou, 2007), a previous study on India (Dwivedi et al., 2008), the KSA (Dwivedi & Weerakkody, 2007), Pakistan (Khoumbati et al., 2007), and Malaysia (Ooi, Sim, et al., 2011). Our results show that four constructs—PE FCR, SE, FCT, and HO—have similar results with the previous studies. Meanwhile, the findings on SO, SQ, RI, SI, RA, K, and FCR were found to be contradicted as compared with the same studies. Following is the explanation of the comparison of the differences between our findings and previous studies in other developing countries.

The RI and SI were found to have no significant effect on consumers' intention in adopting broadband. This implies that the influences perceived from social groups (such as family and peers) and secondary sources (news, newspaper, or magazines) have no significant effect on consumers' intention in adopting broadband. This is a significant contrast compared with previous studies. The broadband usage in India is increasing although not at the pace comparable with its neighbor China. With this, the influence from reference groups should be a trusted information source for consumers deciding to adopt and use broadband. However, the results indicate a deficit of consumer trust on information provided by the reference groups. Therefore, the strategy of emphasizing on primary influence as a driver to increase broadband adoption in early stages may require rethinking.

For FCR, our results showed significant effect toward intention to adopt broadband. This may imply that the strategy to increase broadband adoption rate by providing facilitating conditions and supporting facilities (Rogers, 1995) seems to have significant impact. This is in agreement with previous studies of Pakistan (Khoumbati et al., 2007) and Bangladesh (Dwivedi, Khan, et al., 2007). The results suggest that facilitating factors (such as technical and support facilities) are important to facilitate and boost up the broadband uptake in nations. This finding has significant implication for ISPs in India investing significantly on enhancing their technical and support facilities.

Table 12. Regression Analysis: Coefficients (Dependent Variable: Behavioral Intention).

Model		Unstandardized coefficients		Standardized coefficients		
		B	SE	β	t	Significance
I	(Constant)	3.110	0.552		5.629	.000
	RA	1.136	0.120	.922	9.426	.000
	UO	−0.027	0.098	−.024	−0.271	.786
	HO	0.198	0.054	.249	3.688	.000
	SQ	−0.397	0.060	−.506	−6.599	.000
	RI	0.021	0.113	.021	0.184	.854
	SI	−0.019	0.052	−.032	−0.369	.712
	K	0.303	0.102	.343	2.973	.003
	SK	−0.449	0.173	−.402	−2.588	.010
	FCT	0.055	0.126	.045	0.438	.662
	FCR	0.427	0.098	.333	4.378	.000
	SO	−0.520	0.119	−.570	−4.379	.000
	PE	−0.727	0.107	−.664	−6.792	.000
	SE	0.394	0.098	.433	4.023	.000

Note. UO = utilitarian outcomes; SO = social outcomes; RA = relative advantage; HO = hedonic outcomes; SQ = service quality; SK = skills; FCT = facilitating conditions technology; PE = perceived ease of use; K = knowledge; SE = self-efficacy; FCR = facilitating conditions resources; SI = social influence; RI = referents influence.

For FCT, our results showed no significant effect toward intention to adopt broadband. Considering the various factors used to analyze facilitating conditions of technology (availability of electricity, reliability of broadband connection, availability of diverse access technologies, and choice of different service providers), we can explain this finding by looking at the current highly competitive landscape of broadband services in India. With the emergence of many service providers providing various broadband access technologies and low switching costs, consumers have liberty to switch service providers quickly.

From our findings, HO was found to be least significant in influencing broadband adoption, while UO was found to be not significant. It appears that consumers' adoption of broadband is not affected by either hedonic reasons or utilitarian reasons. This finding has significant implication for service providers who invest millions of dollars in broadband services to satisfy the consumers' emotional needs and pleasurable experiences. This finding is in contrast to experiences of developed nations, such as South Koreas and the United Kingdom, that successfully promoted and increased broadband adoption in the country by utilizing online games, real-time sport events, and on-demand video or music to attract consumers to subscribe to broadband (H. Lee, O'Keefe, & Yun, 2003). The authorities or industry players in developing nations may need to revisit this strategy of providing entertainment-based products in their particular national context to find out what application or content consumers prefer to subscribe to broadband.

From our results, SO exhibits significant relationships on intention to use broadband services. No previous study

included this construct in their research model. The finding suggests consumers believe that use of broadband services enhance their social status. Combining this finding with the finding on RI and SI, we could explain by saying that consumers probably do not trust the information provided by reference groups but believe that use of broadband Internet is an aid to enhance their social status. This suggests that broadband Internet has a strong social connection with respect to consumer preferences. This has significant implication for broadband service providers. Service providers may need to strengthen the brand image of their broadband services.

The non-significant relationship between RA and consumers' intention to adopt broadband is a significant shift because previous literature has consistently showed strong positive influence of RA on broadband adoption (Dwivedi, Khan, & Papazafeiropoulou, 2007; Holak & Lehmann, 1990; Tornatzky & Klein, 1982). Broadband service providers in developing countries have been focusing on providing their customers various advantages as compared with traditional dial-up connection (such as faster downloads, instant communication, always-on connectivity). The finding of this research suggests that perception of broadband is becoming less important as an aid to speed up the overall broadband adoption process. Service providers should not put too much emphasis on emphasizing broadband benefits to increase broadband adoption by consumers.

A significant positive relationship of SE with BIs was expected as mentioned by previous literature (Dwivedi et al., 2006; LaRose, Gregg, Strover, Straubhaar, & Carpenter, 2007). With higher level of confidence in technology, consumers are more likely to adopt broadband. The industry stakeholders may have to adopt a segmental approach to identify and provide relevant education and training to various segments of population that do not have access to opportunities to learn computers and Internet.

SK was found to have no significant relationship with BIs. It appears that most consumers already possess the requisite knowledge and skills to adopt broadband services. Currently, the government of India is following a push strategy by providing more facilitating infrastructure, such as skill-oriented courses, to motivate consumers learn and use new technologies such as broadband Internet. The research findings suggest that governments and industry players may need to revisit this strategy to make it meaningful for consumers and to increase broadband adoption.

From our results, knowledge (K) did not exhibit significant relationships on intention to use broadband services. It is in contrast with previous studies (Choudrie & Lee, 2004; Dwivedi, Khoumbati, Williams, & Lal, 2007)) that showed a significant relation of knowledge (K) with BI. PE exhibited significant relationship on intention to use broadband services. It was expected as suggested by previous literature (Dwivedi, Khan, & Papazafeiropoulou, 2007).

For knowledge (K), the consumers appear familiar in using computer and Internet technologies. The government of India has taken many initiatives to boost computer learning. This finding suggests that the computer learning initiatives need to be further emphasized by educational institutions (such as schools and universities) by offering different varieties of computer learning programs that best suit consumer needs. For PE, it could be explained that Indian consumers may lack the knowledge and skill to use Internet technologies. If we combine it with the finding on knowledge (K), a further explanation could be that even if the consumers feel that they have the requisite knowledge to use Internet technologies, they tend to have less desire in adopting new technologies and this may prove to be a barrier in using the particular services. Interestingly, this finding is in contrast with B. Anderson's (2008) study of European broadband consumers. B. Anderson found that those users who had the knowledge and experience of using broadband were most benefited with broadband Internet, whereas those who lack the skills, knowledge, and perhaps self-confidence were left further behind. Therefore, successful broadband adoption requires further motivation of consumers about broadband usage and benefits to develop their positive attitudes toward broadband adoption and use.

From our results, SQ exhibits significant relationships on intention to use broadband services. This is in line with the previous study of the KSA (Dwivedi & Weerakkody, 2007). One reason could be that in the KSA, only one service provider provides broadband services and consumers have no liberty of switching in case of dissatisfaction with the service. In India, many broadband service providers are offering different packages to different consumer segments. The switching costs are low and consumers switch quickly if they are not satisfied with a service provider. In the past 5 years, many service providers, especially in the wireless Internet services, have emerged and created challenges to improve competitiveness among service providers by providing better quality of broadband services. To develop knowledge-based societies in developing countries, it is imperative to improve quality of broadband services.

Conclusion and Implications

Practical Implications

This study contributes in several ways to the research literature. At a methodological level, this study proposes and tests a new model for broadband adoption in a developing country perspective. Furthermore, it contributes to the ongoing search for policies that are more effective and strategies to increase broadband penetration by providing a more detailed examination of factors that possibly influence broadband adoption by household consumers. In line with other recent studies, we find that broadband penetration is positively influenced by broadband SQ, knowledge (K), and RA. In addition, this study identified a new construct SO, which is positively related with broadband adoption by household consumers.

The findings of this study generate a number of issues that may be helpful for various stakeholders in Indian broadband services, namely, service providers, policy makers, regulators, academia, general public, and so on, for informed decision making and developing a strategy for accelerated broadband diffusion in India. This informed decision making is important due to the established positive impact of increased broadband deployment on economic development of a country (Bauer, Madden, & Morey, 2014; Ng, Lye, & Lim, 2013).

The gradual adoption of broadband by the consumers in developing countries is a concern for both the industry and government due to the strong positive impact of broadband deployment on national economic development. Therefore, this research should offer a substantial contribution to all interested stakeholders including the ISPs and government. Practitioners could use the findings of this study when revising and restructuring their marketing strategy. Academicians could use the findings of this research to enhance their knowledge of broadband adoption in a developing but fast emerging economy. Such knowledge can provide a basis for further research in broadband adoption in other emerging economies.

This study provides useful guidelines for both ISPs and policy makers to understand factors that can influence consumers' intention to adopt broadband technology. Service providers and policy makers can use this study to get useful guidelines and improve their strategies of increasing broadband Internet adoption rate. This research identifies many factors that stakeholders need to focus on to attract more broadband consumers.

The findings of this research suggest that RI and SI have no significant relationship with BI. RI has more significance than SI. Taking this into account, service providers need to rethink their customer communication strategy to make it more effective by creating a more effective impression toward consumers. This is essential so that customers are attracted to spread positive word of mouth in their social circles and attract more consumers toward broadband.

The significant influence of self-efficacy on the broadband adoption among Indian consumers suggests that service providers should invest in developing responsive and available technical support (such as in the form of user manuals, live streaming on the company website) to attract consumers who are likely to learn and are confident in exploring the broadband services.

With less significance attached to the RA, service providers need to think about some innovative RAs that broadband can provide, for example, developing content that is specifically available with broadband services. The more innovative benefits customers receive, the more likely they will subscribe or use the broadband service.

The significance of SO suggests that service providers need to think about ways to make their service a tool of social status enhancement, for example, by branding their service. It is similar to what happened with the cellular phone in developing countries. Initially, it was a communication device. Later on with proliferation of cellular phone devices, the prices fell and everyone had access to a cellular phone. The consumers started to make cellular phone a social status symbol, for example, by buying expensive branded cellular phones.

The least significance attached to HO has a significant learning for service providers. They need to look into HO from the consumers' perspective and provide those applications to the consumers that could drive their adoption of broadband. Enhancement and upgradation of broadband service may improve consumer experience with broadband services. Finally, the high significance attached to the FCR suggests the need for service providers to invest more on providing customer-centric services.

Theoretical Contributions

This research, with its advanced and progressive standpoint, differs from previous research on broadband adoption. This research is similar to the work done by Venkatesh and Brown (2001). However, it differs in that it found subjective norm a non-significant determinant of broadband adoption. Tan and Teo (2000) and Anckar (2003) found attitudinal and control factors essential for technology adoption. Both studies found that attitudinal and control factors were highly significant predictors of technology adoption or non-adoption. On similar lines, this research found that subjective norm was a non-significant predictor of broadband adoption by Indian households. Anchor also suggested the differences between

the critical barriers for e-commerce adoption and adoption of computers in U.S. households (Venkatesh & Brown, 2001).

By examining the relationships between BI to adopt broadband and the 14 constructs, this research provides an extended model based on three previous models. The model used in this research provides a greater understanding of consumers' adoption of broadband in a developing country context. India is a fast growing Southeast Asian nation where consumers still value their traditions despite becoming urban and technology savvy.

Limitation and Future Studies

Every research has its limitations and this research is no exception. First, this study used a larger snowball sample size but it does not ensure the homogeneity of target respondents. Therefore, the results obtained may not be generalized for the population of India as a whole. Future research on broadband adoption may emphasize considering a cross-country or cross-cultural survey. Second, the study was conducted in four major cities of India. The findings may not be applicable across geographies and cultures. However, considering the embryonic stage of development of broadband in India, the selected cities still represent a good sample to investigate broadband adoption in India. Third, the study did not supplement the survey-based approach with interviews to have longitudinal data. Future research can use a longitudinal data collection mechanism to generate better statistical results. Fourth, this study did not take into account the correlation of factors and effect on others. Future research may consider including some moderating constructs to examine the inter-relationships among various factors affecting BI.

This research identified five factors—PE, SO, SQ, FCR, and SE—that influence consumers' adoption of broadband in India. Therefore, the first research question has been verified. From the analysis done, it has been proven that PE has the most significant impact on BI to adopt broadband. Therefore, the second research question has been verified.

In summary, this research provides stakeholders of the broadband industry some very useful guidelines that would be helpful when industries plan to expand and provide high-quality beneficial broadband services to the consumers.

Declaration of Conflicting Interests

The author(s) declared no potential conflicts of interest with respect to the research, authorship, and/or publication of this article.

Funding

The author(s) received no financial support for the research and/or authorship of this article.

References

ABIresearch. (2012, February 21). *Indian broadband subscriber base to top 15 million in 2012*. Retrieved from https://www.

abiresearch.com/press/indian-broadband-subscriber-base-to-top-15-million

Aboelmaged, M. G. (2010). Predicting e-procurement adoption in a developing country: An empirical integration of technology acceptance model and theory of planned behaviour. *Industrial Management & Data Systems*, *110*, 392-414.

Ajzen, I. (1985). From Intentions to Actions: A Theory of Planned Behavior. In P. D. J. Kuhl & D. J. Beckmann (Eds.), *Action Control* (pp. 11–39). Springer Berlin Heidelberg. Retrieved from http://link.springer.com/chapter/10.1007/978-3-642-69746-3_2

Ajzen, I. (1988). *Attitudes, personality, and behavior*. Chicago, IL: Dorsey Press.

Ajzen, I. (1991). The theory of planned behavior. *Organizational Behavior and Human Decision Processes*, *50*, 179-211.

Ajzen, I., & Madden, T. J. (1986). Prediction of goal-directed behavior: Attitudes, intentions, and perceived behavioral control. *Journal of Experimental Social Psychology*, *22*, 453-474.

Allen Consulting Group. (2002, November 5). Built for business II: Beyond basic connectivity. *Sydney Morning Herald*. Retrieved from http://www.smh.com.au/articles/2002/11/02/1036027090-623.html

Anckar, B. (2003). Drivers and inhibitors to ecommerce adoption: Exploring the rationality of consumer behavior in the electronic marketplace. ECIS 2003 Proceedings, 24.

Anderson, B. (2008). The social impact of broadband household internet access. *Information, Community & Society*, *11*, 5-24.

Anderson, B., & Tracey, K. (2001). Digital living the impact (or otherwise) of the internet on everyday life. *American Behavioral Scientist*, *45*, 456-475.

Anderson, C. L., & Agarwal, R. (2010). Practicing safe computing: A multimedia empirical examination of home computer user security behavioral intentions. *MIS Quarterly*, *34*, 613-643.

Armenta, A., Serrano, A., Cabrera, M., & Conte, R. (2012). The new digital divide: The confluence of broadband penetration, sustainable development, technology adoption, and community participation. *Information Technology for Development*, *18*, 345-353.

Australian Industry Group. (2008). *National CEO Survey: High speed to broadband: Measuring industry demand for a world class service*. Retrieved from http://www.aigroup.com.au/portal/binary/com.epicentric.contentmanagement.servlet.ContentDeliveryServlet/LIVE_CONTENT/Publications/Reports/2008/7122_CEO_Broadband_web.pdf

Ayanso, A., Cho, D. I., & Lertwachara, K. (2014). Information and communications technology development and the digital divide: A global and regional assessment. *Information Technology for Development*, *20*, 60-77.

Azab, N. A. (2009, December). *Assessing electronic government readiness of public organisations: Effect of internal factors (case of Egypt)* (Doctoral dissertation). Middlesex University, London, England. Retrieved from http://eprints.mdx.ac.uk/6510/

Bankole, F. O., Osei-Bryson, K.-M., & Brown, I. (2013). The Impact of Information and Communications Technology Infrastructure and Complementary Factors on Intra-African Trade. *Information Technology for Development*, *0*(0), 1–17. doi:10.1080/02681102.2013.832128 .

Bauer, J. M., Madden, G., & Morey, A. (2014). Effects of economic conditions and policy interventions on OECD broadband adoption. *Applied Economics*, *46*, 1361-1372.

Bell, R. (2008). Five ways for communities to shape their broadband futures. *Economic Development Journal*, *7*(3).

Bolton, R. N., & Drew, J. H. (1991). A multistage model of customers' assessments of service quality and value. *Journal of Consumer Research*, *17*, 375-384.

Brady, M. K., & Cronin, J. J., Jr. (2001). Some new thoughts on conceptualizing perceived service quality: A hierarchical approach. *Journal of Marketing*, *65*, 34-49.

Brown, S. A., & Venkatesh, V. (2005). Model of adoption of technology in households: A baseline model test and extension incorporating household life cycle. *MIS Quarterly*, *29*, 399-426.

BusinessWire.com. (2011, July 15). *India telecoms, mobile, broadband and forecasts report: India continues to be one of the fastest and largest growing major telecom markets in the world*. Retrieved from http://www.businesswire.com/news/home/20110715005047/en/Research-Markets-2011-India-Telecoms-Mobile-Broadband#.VC7ZsRZCd4k

Cheon, J., Lee, S., Crooks, S. M., & Song, J. (2012). An investigation of mobile learning readiness in higher education based on the theory of planned behavior. *Computers & Education*, *59*, 1054-1064.

Chia, L. T., Lee, B. S., & Yeo, C. K. (1998). Information technology and the Internet: The Singapore experience. *Information Technology for Development*, *8*, 101-120.

Choudrie, J., & Dwivedi, Y. K. (2004). Towards a conceptual model of broadband diffusion. *Journal of Computing and Information Technology-CIT*, *12*, 323-338.

Choudrie, J., & Dwivedi, Y. (2005). Investigating Broadband Diffusion in the Household: Towards Content Validity and Pre-Test of the Survey Instrument. ECIS 2005 Proceedings. Retrieved from http://aisel.aisnet.org/ecis2005/38.

Choudrie, J., & Dwivedi, Y. K. (2006a). Examining the socio-economic determinants of broadband adopters and non-adopters in the United Kingdom. In *Proceedings of the 39th Annual Hawaii International Conference on System Sciences, 2006. HICSS'06.* (Vol. 4, pp. 85a). IEEE.

Choudrie, J., & Dwivedi, Y. K. (2006b). Investigating factors influencing adoption of broadband in the household. *Journal of Computer Information Systems*, *46*, 25-34.

Choudrie, J., & Lee, H. (2004). Broadband development in South Korea: Institutional and cultural factors. *European Journal of Information Systems*, *13*, 103-114.

Collins, P., Day, D., & Williams, C. (2007). *The economic effects of broadband: An Australian perspective*. Retrieved from http://www.oecd.org/dataoecd/29/9/38698062.pdf

Crandall, R. W., Lehr, W., & Litan, R. E. (2007). *The effects of broadband deployment on output and employment: A cross-sectional analysis of US data*. Washington, DC: Brookings Institution Press.

Cronin, J. J., Jr., Brady, M. K., & Hult, G. T. M. (2000). Assessing the effects of quality, value, and customer satisfaction on consumer behavioral intentions in service environments. *Journal of Retailing*, *76*, 193-218.

Czernich, N., Falck, O., Kretschmer, T., & Woessmann, L. (2011). Broadband infrastructure and economic growth. *The Economic Journal*, *121*, 505-532.

Davis, F. D. (1989). Perceived usefulness, perceived ease of use, and user acceptance of information technology. *MIS Quarterly*, *13*, 319-340.

Davis, F. D., Bagozzi, R. P., & Warshaw, P. R. (1989). User acceptance of computer technology: A comparison of two theoretical models. *Management Science, 35*, 982-1003.

DeLone, W. H., & McLean, E. R. (2003). The DeLone and McLean model of information systems success: A ten-year update. *Journal of Management Information Systems, 19*, 9-30.

Dodds, W. B., Monroe, K. B., & Grewal, D. (1991). Effects of price, brand, and store information on buyers' product evaluations. *Journal of Marketing Research, 28*, 307-319.

Dwivedi, Y., & Irani, Z. (2009). Understanding the adopters and non-adopters of broadband. *Communications of the ACM, 52*, 122-125.

Dwivedi, Y. K. (2005). *Investigating consumer adoption, usage and impact of broadband: UK households* (Doctoral dissertation). Brunel University, School of Information Systems, Computing and Mathematics. Retrieved from http://bura.brunel.ac.uk/handle/2438/5335

Dwivedi, Y. K., Choudrie, J., & Brinkman, W.-P. (2006). Development of a survey instrument to examine consumer adoption of broadband. *Industrial Management & Data Systems, 106*, 700-718.

Dwivedi, Y. K., Khan, N., & Papazafeiropoulou, A. (2007). Consumer adoption and usage of broadband in Bangladesh. *Electronic Government: An International Journal, 4*, 299-313.

Dwivedi, Y. K., Khoumbati, K., Williams, M. D., & Lal, B. (2007). Factors affecting consumers' behavioural intention to adopt broadband in Pakistan. *Transforming Government: People, Process and Policy, 1*, 285-297.

Dwivedi, Y. K., Lal, B., & Williams, M. D. (2009). Managing consumer adoption of broadband: Examining drivers and barriers. *Industrial Management & Data Systems, 109*, 357-369.

Dwivedi, Y. K., & Weerakkody, V. (2007). Examining the factors affecting the adoption of broadband in the Kingdom of Saudi Arabia. *Electronic Government: An International Journal, 4*, 43-58.

Feijoo, C., Gomez-Barroso, J. L., Ramos, S., & Rojo-Alonso, D. (2006). Active policy measures against the digital divide based on mobile/wireless connectivity development: The Latvian experience. *International Journal of Mobile Communications, 4*, 727-742.

Fotopoulos, C. B., & Psomas, E. L. (2009). The impact of "soft" and "hard" TQM elements on quality management results. *International Journal of Quality & Reliability Management, 26*, 150-163.

Fowler, J. F. J. (2002). *Survey research methods* (4th ed.). Thousand Oaks, CA: SAGE.

Garfield, M. J., & Watson, R. T. (1997). Differences in national information infrastructures: The reflection of national cultures. *The Journal of Strategic Information Systems, 6*, 313-331.

Giovanis, A. N. (2011). Factors affecting Greek internet users' intentions to adopt online shopping: The perspective of an extended technology acceptance model. *International Journal of Technology Marketing, 6*, 290-304.

Grimes, A., Ren, C., & Stevens, P. (2012). The need for speed: Impacts of internet connectivity on firm productivity. *Journal of Productivity Analysis, 37*, 187-201.

Hair, J. F., Tatham, R. L., Anderson, R. E., & Black, W. (2005). *Multivariate data analysis* (Vol. 6). Upper Saddle River, NJ: Pearson Prentice Hall.

Han, G. (2003). Broadband adoption in the United States and Korea: Business driven rational model versus culture sensitive policy model. *Trends in Communication, 11*, 3-25.

Hargittai, E. (1999). Weaving the Western Web: Explaining differences in Internet connectivity among OECD countries. *Telecommunications Policy, 23*, 701-718.

Hartshorne, R., & Ajjan, H. (2009). Examining student decisions to adopt Web 2.0 technologies: Theory and empirical tests. *Journal of Computing in Higher Education, 21*, 183-198.

Hernández, B., Jiménez, J., & Martín, M. J. (2010). Customer behavior in electronic commerce: The moderating effect of e-purchasing experience. *Journal of Business Research, 63*, 964-971.

Hernández, B., Jiménez, J., & Martín, M. J. (2011). Age, gender, and income: Do they really moderate online shopping behaviour? *Online Information Review, 35*, 113-133.

Hinton, P. R., Brownlow, C., McMurray, I., & Cozens, B. (2004). *SPSS explained* (1st ed.). New York, NY: Routledge.

Holak, S. L., & Lehmann, D. R. (1990). Purchase intentions and the dimensions of innovation: An exploratory model. *Journal of Product Innovation Management, 7*, 59-73.

Holmström, J., Ketokivi, M., & Hameri, A.-P. (2009). Bridging practice and theory: A design science approach. *Decision Sciences, 40*, 65-87. doi:10.1111/j.1540-5915.2008.00221.x

Holt, L., & Jamison, M. (2009). Broadband and contributions to economic growth: Lessons from the US experience. *Telecommunications Policy, 33*, 575-581.

Irani, Z., Dwivedi, Y. K., & Williams, M. D. (2009). Understanding consumer adoption of broadband: An extension of the technology acceptance model. *Journal of the Operational Research Society, 60*, 1322-1334.

Jin, D. Y. (2005). Socioeconomic implications of broadband services: Information economy in Korea. *Information, Community & Society, 8*, 503-523.

Johnson, M. (2010). Barriers to innovation adoption: A study of e-markets. *Industrial Management & Data Systems, 110*, 157-174.

Jung, Y., Perez-Mira, B., & Wiley-Patton, S. (2009). Consumer adoption of mobile TV: Examining psychological flow and media content. *Computers in Human Behavior, 25*, 123-129.

Kapur, S. (2001). Developing countries in the network economy: A blueprint for success. *Network, 19*, 20.

Katz, R. L., Vaterlaus, S., Zenhäusern, P., & Suter, S. (2010). The impact of broadband on jobs and the German economy. *Intereconomics, 45*, 26-34.

Khoumbati, K., Dwivedi, Y. K., Lal, B., & Chen, H. (2007). Broadband adoption in Pakistan. *Electronic Government: An International Journal, 4*, 451-465.

Kolko, J. (2012). Broadband and local growth. *Journal of Urban Economics, 71*, 100-113.

Koutroumpis, P. (2009). The economic impact of broadband on growth: A simultaneous approach. *Telecommunications Policy, 33*, 471-485.

LaRose, R., Gregg, J. L., Strover, S., Straubhaar, J., & Carpenter, S. (2007). Closing the rural broadband gap: Promoting adoption of the Internet in rural America. *Telecommunications Policy, 31*, 359-373.

Lee, H., O'Keefe, R. M., & Yun, K. (2003). The growth of broadband and electronic commerce in South Korea: Contributing factors. *The Information Society, 19*, 81-93.

Lee, M.-C., & Tsai, T.-R. (2010). What drives people to continue to play online games? An extension of technology model and theory of planned behavior. *International Journal of Human-Computer Interaction, 26*, 601-620.

Lee, V.-H., Ooi, K.-B., Tan, B.-I., & Chong, A. Y.-L. (2010). A structural analysis of the relationship between TQM practices and product innovation. *Asian Journal of Technology Innovation, 18*, 73-96.

Lehr, W., Bauer, S., Heikkinen, M., & Clark, D. (2011). Assessing Broadband Reliability: Measurement and Policy Challenges. In Research Conference on Communications, Information and Internet Policy, Arlington, VA.

Lin, H.-F. (2010). Applicability of the extended theory of planned behavior in predicting job seeker intentions to use job-search websites. *International Journal of Selection and Assessment, 18*, 64-74.

Majumdar, S. K., Carare, O., & Chang, H. (2010). Broadband adoption and firm productivity: Evaluating the benefits of general purpose technology. *Industrial and Corporate Change, 19*, 641-674.

Manzoor, A. (2012). Broadband Internet development and economic growth: A comparative study of two Asian countries. *IOSR Journal of Business and Management, 1*, 1-14.

Mugeni, G. B., Wanyembi, G. W., & Wafula, J. M. (2012). Evaluating factors affecting broadband readiness in Kenya: A pilot study. *International Journal of Information and Communication Technology Research, 2*, 491-498.

Nam, C., Kim, S., Lee, H., & Duan, B. (2009). Examining the influencing factors and the most efficient point of broadband adoption in China. *Journal of Research and Practice in Information Technology, 41*, 25-38.

Ng, T. H., Lye, C. T., & Lim, Y. S. (2013). Broadband penetration and economic growth in ASEAN countries: A generalized method of moments approach. *Applied Economics Letters, 20*, 857-862.

Ooi, K.-B., Sim, J.-J., Yew, K.-T., & Lin, B. (2011). Exploring factors influencing consumers' behavioral intention to adopt broadband in Malaysia. *Computers in Human Behavior, 27*, 1168-1178.

Organisation for Economic Co-operation and Development. (2001). *The development of broadband access in the OECD Countries.* Paris, France: Author.

Organisation for Economic Co-operation and Development. (2008). *Broadband growth and policies in OECD countries.* Paris, France: Author.

Ortiz de Guinea, A., & Markus, M. L. (2009). Why break the habit of a lifetime? Rethinking the roles of intention, habit, and emotion in continuing information technology use. *Management Information Systems Quarterly, 33*, 433-444.

Papacharissi, Z., & Zaks, A. (2006). Is broadband the future? An analysis of broadband technology potential and diffusion. *Telecommunications Policy, 30*, 64-75.

Parasuraman, A., Berry, L. L., & Zeithaml, V. A. (1991). Perceived service quality as a customer-based performance measure: An empirical examination of organizational barriers using an extended service quality model. *Human Resource Management, 30*, 335-364.

Pereira, B. (2011, December 14). Government of India shares plans to increase broadband penetration. *InformationWeek.* Retrieved from http://www.informationweek.in/information-week/news-analy-sis/176062/government-india-shares-plans-increase-broadband-penetration

Pham, L., Pham, L. N., & Nguyen, D. T. (2011). Determinants of e-commerce adoption in Vietnamese small and medium sized enterprises. *International Journal of Entrepreneurship, 15*, 45-72.

Press Trust of India. (2012, May 16). *The Times of India.* Retrieved from http://timesofindia.indiatimes.com/tech/tech-news/Trais-proposal-can-lead-to-hike-in-mobile-tariff-Uninor/article-show/13163405.cms

Püschel, J., Mazzon, J. A., & Hernandez, J. M. C. (2010). Mobile banking: Proposition of an integrated adoption intention framework. *International Journal of Bank Marketing, 28*, 389-409.

Qiang, C. Z.-W., Rossotto, C. M., & Kimura, K. (2009). *Economic impacts of broadband in Information and communications for development 2009: Extending reach and increasing impact* (pp. 35-50). Washington, DC: The International Bank for Reconstruction and Development/The World Bank.

Qureshi, S. (2010). Driving development through innovations in information technology and its applications. *Information Technology for Development, 16*, 93-95.

Qureshi, S. (2012). As the global digital divide narrows, who is being left behind? *Information Technology for Development, 18*, 277-280.

Ramayah, T., Rouibah, K., Gopi, M., & Rangel, G. J. (2009). A decomposed theory of reasoned action to explain intention to use Internet stock trading among Malaysian investors. *Computers in Human Behavior, 25*, 1222-1230.

Rogers, E. M. (1995). *Diffusion of innovations* (4th ed.). New York, NY: Free Press.

San Martín, H., & Herrero, Á. (2012). Influence of the user's psychological factors on the online purchase intention in rural tourism: Integrating innovativeness to the UTAUT framework. *Tourism Management, 33*, 341-350.

Sawyer, S., Allen, J. P., & Lee, H. (2003). Broadband and mobile opportunities: A socio-technical perspective. *Journal of Information Technology, 18*, 121-136.

Schifter, D. E., & Ajzen, I. (1985). Intention, perceived control, and weight loss: An application of the theory of planned behavior. *Journal of Personality and Social Psychology, 49*, 843-851.

Selamat, M. H., Dwivedi, Y. K., Wahab, A., Syahir, M., Samsudin, M., Amir, M., . . .Lal, B. (2008). Factors affecting Malaysian accountants' broadband adoption and use behavior.

Sim, J.-J., Kong, F.-M., Lee, V.-H., Tan, G. W.-H., & Teo, A.-C. (2012). Determining factors affecting broadband services adoption: An empirical analysis of Malaysian consumers. *International Journal of Services, Economics and Management, 4*, 236-251.

Sit, W.-Y., Ooi, K.-B., Lin, B., & Chong, A. Y.-L. (2009). TQM and customer satisfaction in Malaysia's service sector. *Industrial Management & Data Systems, 109*, 957-975.

Straub, D., Boudreau, M.-C., & Gefen, D. (2004). Validation guidelines for IS positivist research. *The Communications of the Association for Information Systems, 13*, 63.

Sweeney, J. C., Soutar, G. N., & Johnson, L. W. (1999). The role of perceived risk in the quality–value relationship: A study in a retail environment. *Journal of Retailing, 75*, 77-105.

Szymanski, D. M., & Henard, D. H. (2001). Customer satisfaction: A meta-analysis of the empirical evidence. *Journal of the Academy of Marketing Science, 29*, 16-35.

Tabachnick, B. G., & Fidell, L. S. (2007). *Using multivariate statistics* (5th ed.). Boston, MA: Pearson.

Tan, M., & Teo, T. S. (2000). Factors influencing the adoption of Internet banking. *Journal of the AIS, 1*(1es), 5.

Teas, R. K., & Agarwal, S. (2000). The effects of extrinsic product cues on consumers' perceptions of quality, sacrifice, and value. *Journal of the Academy of Marketing Science, 28*, 278-290.

Telecom Regulatory Authority of India. (2010). *Recommendation on national broadband plan*. Retrieved from http://www.trai.gov.in/Content/TelDis/52_1_1.aspx

Tornatzky, L. G., & Klein, K. J. (1982). Innovation characteristics and innovation adoption-implementation: A meta-analysis of findings. *IEEE Transactions on Engineering Management, 29*, 28-45.

van Welsum, D. (2008). *Broadband and the economy* (Ministerial background report, DSTI/ICCP/IE (2007) 3/FINAL). Report prepared for the OECD Ministerial Meeting on the future of the Internet Economy, Seoul, Korea, June 17-18, 2008.

Venkatesh, V., & Brown, S. A. (2001). A longitudinal investigation of personal computers in homes: Adoption determinants and emerging challenges. *MIS Quarterly, 25*, 71-102.

Venkatesh, V., Thong, J. Y., & Xu, X. (2012). Consumer acceptance and use of information technology: Extending the unified theory of acceptance and use of technology. *MIS Quarterly, 36*, 157-178.

Wambogo Omole, D. (2013). Harnessing information and communication technologies (ICTs) to address urban poverty: Emerging open policy lessons for the open knowledge economy. *Information Technology for Development, 19*, 86-96.

Wei, J. (2004). Worldwide internet usages and online multi-linguistic population comparison study. *Information Systems Education Journal, 2*, 1-16.

Yaghoubi, N.-M. (2010). Factors affecting the adoption of online banking: An integration of technology acceptance model and theory of planned behavior. *International Journal of Business and Management, 5*, P159.

Yousafzai, S. Y., Foxall, G. R., & Pallister, J. G. (2010). Explaining internet banking behavior: Theory of reasoned action, theory of planned behavior, or technology acceptance model? *Journal of Applied Social Psychology, 40*, 1172-1202.

Zeithaml, V. A. (1988). Consumer perceptions of price, quality, and value: A means-end model and synthesis of evidence. *The Journal of Marketing, 52*, 2-22.

Author Biography

Amir Manzoor holds a bachelor's degree in engineering from NED University, Karachi, an MBA from Lahore University of Management Sciences (LUMS), and an MBA from Bangor University, United Kingdom. He has many years of diverse professional and teaching experience working at many renowned national and internal organizations and higher education institutions. His research interests include electronic commerce and technology applications in business.

14

On a Rhetorical Technique in Leopold's *The Land Ethic*: "that Imperial First Word"

Henry St. Maurice[1]

Abstract

Analysis of definite articles in Leopold's essay *The Land Ethic* found evidence of deliberate usage of a rhetorical device for emphasis. The device, a type of definite article, is commonly used and usually abused. Leopold's uses evidently affected his essay's contribution to environmental communications.

Keywords

Aldo Leopold, land ethic, rhetoric

Introduction

For centuries, the value of rhetoric has been vociferously debated (Vickers, 1989). In *Gorgias*, Plato (ca. 380 B.C.E./1871) showed Socrates brilliantly using rhetoric to debate Sophistic rhetoricians (Wardy, 1996). In his definition, Socrates shows his distrust of their persuasive discourse as follows:

> . . . rhetoric is of two sorts: one, which is mere flattery and disgraceful declamation; the other, which is noble and aims at the training and improvement of the souls of the citizens, and strives to say what is best, whether welcome or unwelcome, to the audience . . .

Then he asks, "but have you ever known such a rhetoric . . . ?" (Plato, ca. 380 B.C.E./1871). This article is a response to that question. It is a discussion of an exemplary rhetoric that both ennobles and flatters audiences in the discourse of environmental studies.

Examples of both kinds of rhetoric abound in works of environmental sciences, literature, and criticism. One prominent example is the emergence of the term *sustainability* (Anholt, 2003; Cronon, 2013) as a replacement for terms evoking imminent crises, such as *Earth in the Balance* (Gore, 1992). An Ngram (Figure 1) for *sustainability* shows a steep climb since the 1970s, perhaps due in part to its Socratic "noble aim," but also possibly its power to flatter audiences that policies, practices, or products are sustainable (e.g., BP, 2013). Another example is growing rebranding of *global warming* as *climate change* (Figure 2) following audience responses to weather patterns (Conway, 2008). In these and other examples, persuasive discourse leads to changes in scholarly research, public policy, and cultural practices.

To ask a rhetorical question, how does persuasive discourse work? In seemingly small ways, ideas form, enter discourse, and reproduce in a vehicle that Dawkins (1989) defined as a *meme*, "a unit of cultural transmission" (p. 192). Long before, Hugo (1877) described the effects of persuasive discourse as, "One resists the invasion of armies; one does not resist the invasion of ideas" (Chapter X). In such cases, ideas attached to discourse can conquer and occupy the world.

This essay examines one specific rhetorical device that helped to spread an environmental idea throughout the world. Aldo Leopold's (1949/2013) essay *The Land Ethic* (*TLE*) is regarded as a seminal document in histories of environmental movements (Meine, 1988, 2004). The Ngram (Figure 3) for the phrase, *the land ethic*, shows a steep upward curve starting in the 1970s, long after its initial publication, as this idea spread along with the expansion of environmental studies (Worster, 1994). Previous analyses of Leopold's rhetoric (e.g., Willard, 2007) have looked at Leopold's writings as through a telescope, showing the depth of his invention and disposition. This analysis looks as through a microscope at one technique of his elocution, evidently aimed at both flattering and ennobling his audiences.

"That Imperial First Word"

TLE is about 6,650 words long, including its title and section headings. This essay is focused on Leopold's usage of

[1]University of Wisconsin–Stevens Point, USA

Corresponding Author:
Henry St. Maurice, School of Education, University of Wisconsin–Stevens Point, 2100 Main Street, Stevens Point, WI 54481, USA.
Email: hstmauri@uwsp.edu

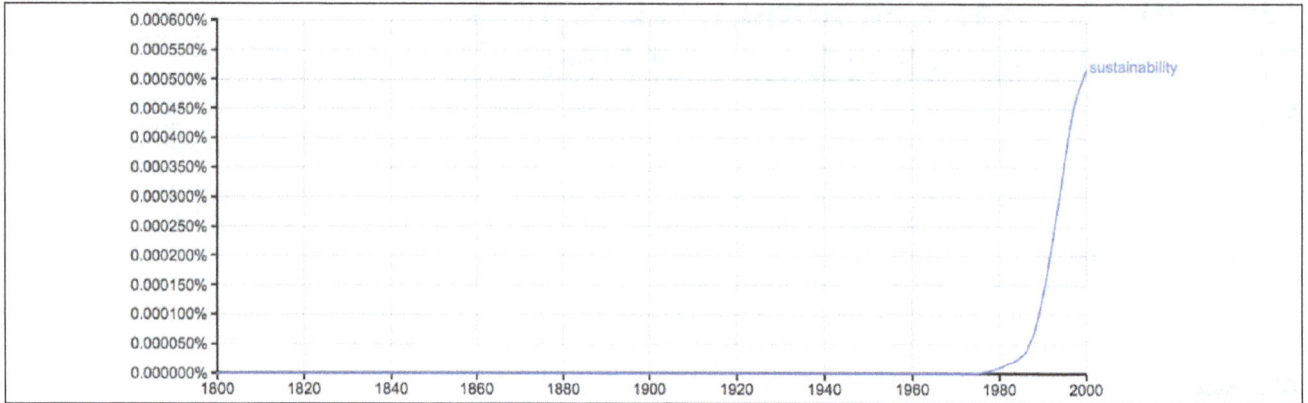

Figure 1. Sustainability Ngram.

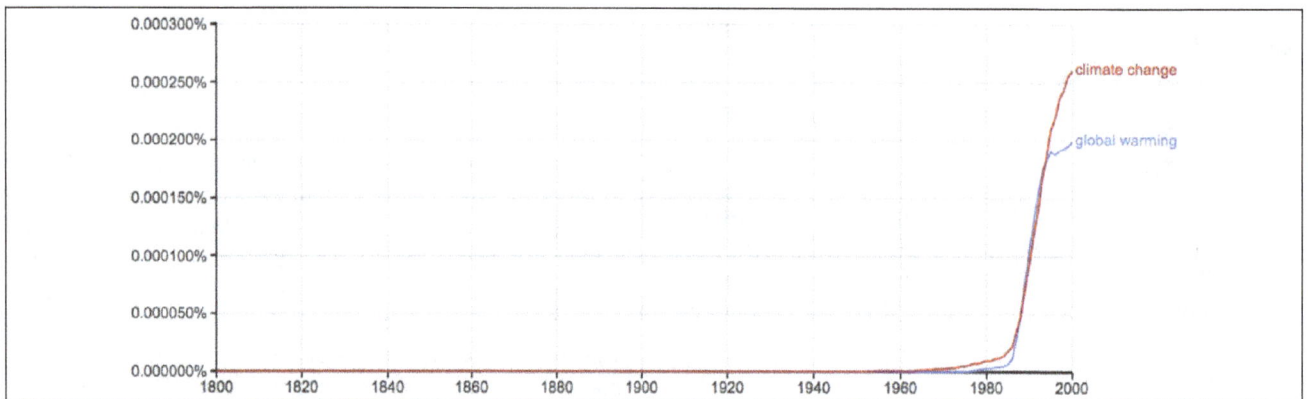

Figure 2. Global warming versus climate change Ngram.

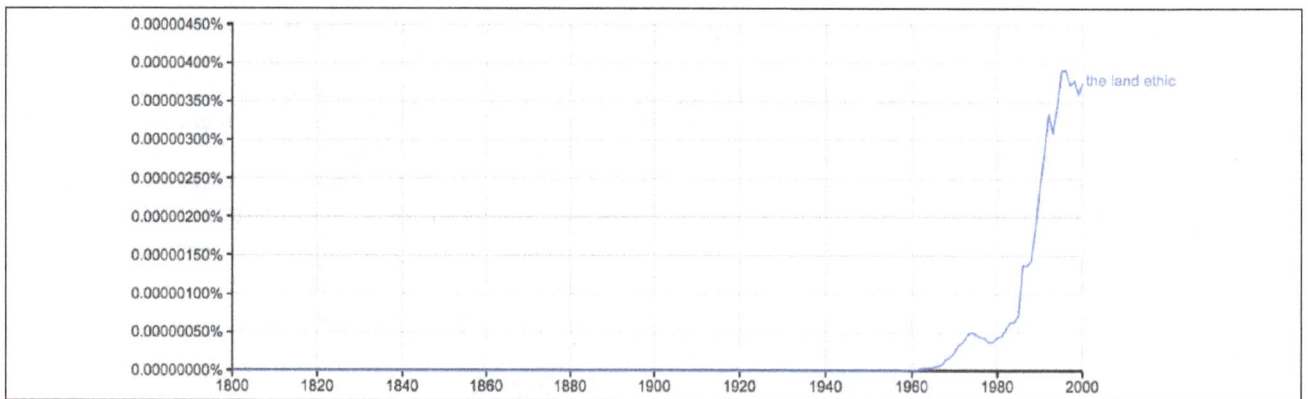

Figure 3. The land ethic Ngram.

one of those words, *the*, a definite article that he uses 477 times in *TLE*. The Oxford English Dictionary (OED) lists 23 various definitions of *the*, one of which is, " . . . Referring to a term used generically or universally" (Item II, Example 19). This study concentrates on Leopold's usage of that specific meaning, making a simple

distinction among three types of usage, as per *The Chicago Manual of Style* (16th ed.):

A definite article points to a definite object that (1) is so well understood that it does not need description (e.g., "the package is here" is a shortened form of "the package that you expected is

Table 1. Examples of Usage.

	Type
The Ethical Sequence: This extension of ethics, so far studied only by philosophers, is actually a process in ecological evolution. Its sequence may be described in ecological as well as in philosophic terms. An ethic, ecologically, is a limitation on freedom of action in **the** struggle for existence. An ethic, philosophically, is a differentiation of social from anti-social conduct. These are two definitions of one thing. **The** thing has its origin in **the** tendency of interdependent individuals or groups to evolve modes of co-operation . . .	2 3 1 3

here"); (2) is a thing that is about to be described (e.g., "the sights of Chicago"); or (3) is important (e.g., "the grand prize").

Examples of these three usages are shown in Table 1, a tabulation of the fourth paragraph of *TLE*, in a section headed "The Ethical Sequence."

This essay will concentrate on usages categorized as Type 3 excluding some singular nouns (i.e., Plato's *the soul*). Most Type 3 usages have grammatical and semantic alternatives. For instance, Leopold's usage of *the tendency* could have been *tendencies*, which would be more grammatically consistent with two plural nouns that it precedes, *individuals or groups*.

Among alternatives in English (Master, 1997), Leopold chose a Type 3 definite article in titling his essay *TLE* as would have been distinct from *a land ethic* or *land ethics*. The former usage is an example of an indefinite article, and the latter use is an example of a null article. A fourth type of article is partitive, as in *some land ethics*.

The incidence of Leopold's usages of Type 3 definite articles in *TLE* is shown in Tables 2 and 3. Among 32 usages counted as Type 3, Leopold uses the phrase *the land ethic* only twice not counting the title; those usages are described below. He uses 75% of these Type 3 articles in three of the eight sections, as will also be discussed below after some general points about their rhetorical functions.

For what rhetorical purposes does Leopold use these words? It seems fair to infer that he is using definite articles to claim universality, a well-known rhetorical technique. Such claims are occasionally overstatements, known to rhetoricians as *hyperbole*, an example of which Mark Twain (1907) colorfully described in his article on The First Church of Christ, Scientist:

"THE" I uncover to that imperial word. . . . It lifts the Mother-Church away up in the sky, and fellowships it with the rare and select and exclusive little company of the THE's of deathless glory—persons and things whereof history and the ages could furnish only single examples, not two: the Saviour, the Virgin, the Milky Way, the Bible, the Earth, the Equator, the Devil, the Missing Link—and now The First Church, Scientist . . .

Twain found bombastic flattery in a usage that literally uplifts an institution's name. In addition to such sublime connotations, definite articles can be used to connote solidity, as Leopold does by invoking pyramids to describe invisible ecological relations. This technique is called *reification* as defined by Whitehead (1925/1997):

. . . a process of constructive abstraction [by which] we can arrive at abstractions [through] simply located bits of material . . .

In short, using Type 3 definite articles can lend both uplift and solidity to abstractions. This usage, however, entails risks along with benefits of accessibility. Whitehead calls one great risk "the fallacy of misplaced concreteness," making false oversimplifications of complex abstractions.

A lexical authority in Leopold's time, Fowler (1926/1983) inveighed against misuses of this type: " . . . inserting *the* where it is indefensible, in the false belief that it is impressive or literary . . . " (p. 642). For example, *the land* is often invoked as a reified abstraction without reference to specific places (Williams, 1985). Despite risks of illogical reification, Type 3 definite articles derive rhetorical force by combining specificity from Types 1 and 2 in a tactic that Burke (1966) called a " . . . terministic screen . . . stretched to cover not just its own special field but a more comprehensive area" (p. 52).

In *TLE*, Leopold uses Type 3 definite articles for specific purposes. First of all, he modulates them in alternating sections that have more frequent usage than sections in which he uses few or even none (Figure 4).

Three sections of *TLE* were modified from articles published in scholarly journals and had been edited for usage according to various styles. Leopold revised them for this essay (Meine, 1988). In the resulting sections that Meine calls "the scientific backbone" of TLE occur the most frequent usages of reified abstractions such as "the land" or "the individual."

Leopold made special use of Type 3 definite articles in two other sections. In the section headed "Land Health and the A–B Cleavage," Leopold identified two types of conservationists: One he called Group A, who use land for "commodity production" contrasted with Group B conservationists who regard land "as a biota, and its function as something broader." In the next section headed "The Outlook," Leopold elevated Group B's ethics to higher status with a canny usage of articles in his definition: "A thing is right when it tends to preserve the integrity, stability, and beauty of *the* biotic community. It is wrong when it tends otherwise" (at Word 6301, emphasis added). He could have used " . . . *a* biotic community" without diminishing the force of that statement, but he used a Type 3 article to infer a universal idea that all living things are joined.

In that concluding section of Leopold's last published work, in what Meine (2004) called " . . . arguably the most important sentence that Aldo Leopold ever wrote" (p. 210), Leopold said, "I have purposely presented *the* land ethic as a

Table 2. Incidence by Word of Type 3 Definite Articles in TLE.

Term	Incidence	Word no.	Alternative(s)
The [biotic] pyramid [of life]	5	3,484; 3,491; 3,793; 4,443; 4,691	A biotic pyramid or This biotic pyramid
The average individual	1	495	An average individual
The balance of nature	1	3,447	A balance of nature
The biotic citizen	1	5,841	A biotic citizen
The biotic community	1	6,146	A biotic community
The biotic mechanism	1	909	Biotic mechanisms
The community	1	3,202	Communities
The conqueror	1	5,837	A conqueror
The conqueror role	1	778	Conqueror roles
The ecologist	1	219	Ecologists
The farmer	1	6,121	Farmers
The individual	5	316; 326; 336; 505; 534	Individuals
The land	5	577; 596; 849; 6,125, 6,542	Land
The land community	1	744	A land community
The land ethic	2	577; 6,447	A community-oriented or ecological land ethic
The land relation	1	367	Land relations
The mastodons	1	282	Mastodons
The ordinary citizen	1	883	Ordinary citizens
The trend of evolution	1	3,845	An important trend of evolution

Note. TLE = The Land Ethic.

Table 3. Incidence by Section of Type 3 Definite Articles in TLE.

Section	Usages
1. Introduction (Words 1-135)	0
2. The Ethical Sequence (Words 136-519)	8
3. The Community Concept (Words 520-1,534)	8
4. The Ecological Conscience (Words 1,535-2,335)	0
5. Substitutes for a Land Ethic (Words 2,336-3,391)	1
6. The Land Pyramid (Words 3,392-5200)	8
7. Land Health and the A–B Cleavage (Words 5,201-5,919)	2
8. The Outlook (Words 5,920-6,645)	5

Note. TLE = The Land Ethic.

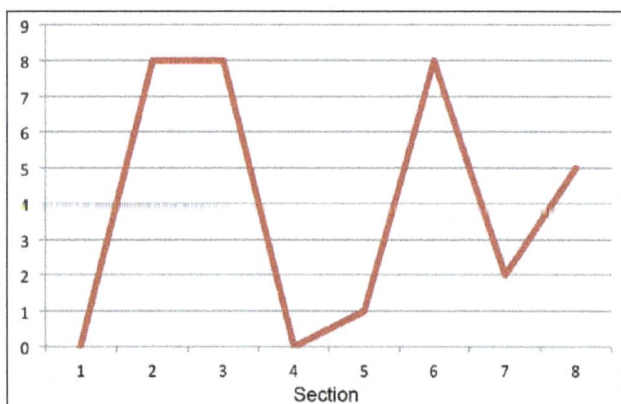

Figure 4. Frequency by section of Type 3 definite articles in TLE.
Note. TLE = The Land Ethic.

product of social evolution, because nothing so important as *an* ethic is ever 'written'" (1949/2013, p. 188, at Word 6442; emphasis added). The next two uses of *ethic* are also modified indefinitely with *an* and *any*. What Leopold did in this sentence and in the title of the essay was to promote without lengthy argument his version of a land ethic—one of admittedly two types—over all others, despite Fowler's admonition against indefensible logic. No less than *THE First Church, THE Land Ethic* stakes a claim for universality. To complete his peremptory elevation, Leopold repeats the title with its definitive article for the second time about 200 words from the end of the essay with a rhetorical technique known as *peroration* (Lanham, 1991).

Upshots

The rhetoric of *TLE* was deployed to persuade a general audience. Leopold was a prolific author who had published many scientific articles, textbooks, and tracts on land use, forestry, and wildlife. He was a recognized practitioner of scientific rhetoric in which universal statements and affectations of impartiality are valued (Bazerman, 2000). He was also a fecund writer of letters, journals, and phenological data, who left a trove for scholars after his untimely death in 1948. In that same year, he crafted *TLE* with rhetoric admittedly different from what he used for persuading professional peers. He renders judgments and calls for actions. For instance, he refers to the conservation movement as "the embryo" (1949/2103, p. 172, at Word 457) for whom (at Word 6645) "The case for a land ethic would appear hopeless but for the minority which is in obvious revolt against these 'modern' trends" (p. 188). He then appeals to the broadest possible audience by evoking two universals: evolution and religion (at Word 6464):

> Only the most superficial student of history supposes that Moses "wrote" the Decalogue; it evolved in the minds of a thinking community, and Moses wrote a tentative summary of it for a "seminar." I say tentative because evolution never stops (p. 188)

Leopold, a lover of music, here plays his prose instrument *fortissimo*. In this passage, he aligns his land ethic with the Mosaic Decalogue, further enshrining a biotic and ecological land ethic as a version of the Golden Rule of universal morality (Stace, 1937/1990).

He also connects his land ethic to evolution, which he cites in *TLE* 12 times as a noun or adjective and cites twice in the paragraph beginning, "I have purposefully presented the land ethic . . . " He is unswerving in his conviction that evolution is a universal mechanism of natural, social, and moral change. He nonetheless expressed doubt about future changes.

TLE was published in a section "The Sand County Almanac" that he had titled *The Upshot* using another Type 3 definite article. As Leopold well knew, the literal term derives from a hunting shot taken in an up-range direction. As the heading for *TLE*, it denotes uncertainty about what lies ahead; it is repeated in the section titled with another Type 3 definite article "The Outlook." He said, "Perhaps the most serious obstacle impeding the evolution of a land ethic is the fact that our educational and economic system is headed away from, rather than toward, an intense consciousness of land" (1949/2013, p. 187, at Word 5968). Weary from war, plagued by illness, and disappointed that his colleagues had not turned—as he had—away from exploiting natural resources toward their stewardship, Leopold was not optimistic.

He had been present at many changes, including forestry and ecology as disciplines of applied science; conservation policy at public and private agencies; and as a pioneer researcher and developer of wildlife and soil and water management. He was an internationally renowned scholar whose doubts set a tone for environmental communications in decades to come, fusing rhetorics of science and morality to skeptically describe present harms and warn of impending dangers (AtKisson, 1999). Like Socrates, his skepticism cohabited with his idealism.

Conclusion

TLE was published posthumously as an epilogue to the descriptive prose of *A Sand County Almanac* and as a reflection following his observations of life on a small plot of land. In this analysis, one small but important effect in *TLE* was Leopold's deft usage of definite articles to confer universality on his proposed ethic. He eschewed fallacies or hyperbole that have plagued rhetorics since Plato's time. Twain, Whitehead, and Fowler warned that bombast can sprout in the shade of a tiny word but withers under scrutiny. Rather than that, *TLE*'s survival and prosperity have contributed to evolving rhetorics, policies, and practices in science, politics, and philosophy.

After two decades of neglect, *TLE* achieved much of its author's purpose. By its fifth decade, it had influenced a host of environmental activists who have affected environmental policies and practices in deploying their own rhetoric combining scientific rigor and spiritual fervor, as Leopold did. Dyson (2008) observed,

> There is a worldwide secular religion which we may call environmentalism, holding that we are stewards of the earth, that despoiling the planet with waste products of our luxurious living is a sin, and that the path of righteousness is to live as frugally as possible. . . . The worldwide community of environmentalists - most of whom are not scientists—holds the moral high ground, and is guiding human societies toward a hopeful future. Environmentalism, as a religion of hope and respect for nature, is here to stay

Whether or not environmentalism is a secular religion, its alignment of practical wisdom and spiritual idealism does seem to be implanted in public discourse. By now, *the land ethic* is a phrase that meets all the criteria of a successful meme, along with such others as *sustainability* or *climate change*. In coining it, Leopold combined Sophistic flattery and Socratic nobility in ways that have persuaded multitudes to join his Group B among scientists, policymakers, and citizens around the world. In Twain's words, the land ethic with its "imperial first word" has indeed joined that "rare and select and exclusive little company of the THEs of deathless glory".

Acknowledgment

Thanks to Wallace Sherlock and anonymous reviewers for astute comments on early drafts.

Declaration of Conflicting Interests

The author(s) declared no potential conflicts of interest with respect to the research, authorship, and/or publication of this article.

Funding

The author(s) received no financial support for the research and/or authorship of this article.

References

Anholt, S. (2003). *Brand new justice: The upside of global branding*. Oxford, UK: Butterworth-Heinemann.

AtKisson, A. (1999). *Believing Cassandra: An optimist looks at a pessimist's world*. New York, NY: Chelsea Green.

Bazerman, C. (2000). *Shaping written knowledge: The genre and activity of the experimental article in science*. Madison: University of Wisconsin Press.

BP. (2013). *Sustainability Review*. London, UK: Author. Retrieved from http://www.bp.com/content/dam/bp/pdf/sustainability/group-reports/BP_Sustainability_Review_2013.pdf

Burke, K. (1966). *Language as symbolic action*. Berkeley: University of California Press.

Conway, E. (2008). What's in a name? Global warming vs. climate change. *NASA News & Features*. Retrieved from http://www.nasa.gov/topics/earth/features/climate_by_any_other_name.html

Cronon, W. (2013, April 15). *The riddle of sustainability: A surprisingly short history of the future*. Lecture presented at International House Auditorium, Berkeley, CA.

Dawkins, R. (1989). *The selfish gene* (2nd ed.). New York, NY: Oxford University Press.

Dyson, F. (2008, June 12). The question of global warming. *New York Review of Books*, *55*(12), 8. Retrieved from http://bit.ly/LRctjz

Fowler, W. (1983). *A dictionary of modern English usage*. New York, NY: Crown. (Original work published 1926)

Gore, A., Jr. (1992). *Earth in the balance*. Boston, MA: Houghton Mifflin.

Hugo, V. (1877). *History of a crime* (T. H. Joyce & A. Locke, Trans.). Retrieved from http://www.gutenberg.org/ebooks/10381

Lanham, R. (1991). *A handlist of rhetorical terms*. Berkeley: University of California Press.

Leopold, A. (2013). The land ethic. In C. Meine (Ed.), *A Sand County almanac & other writings on ecology and conservation* (pp. 171-190). New York, NY: Library of America. (Original work published 1949)

Master, P. (1997). The English article system: Acquisition, function, and pedagogy. *System*, *25*, 215-232.

Meine, C. (1988). *Aldo Leopold: His life and work*. Madison: University of Wisconsin Press.

Meine, C. (2004). *Correction lines* Washington, DC: Island Press.

Plato. (1871). *Gorgias* (B. Jowett, Trans.). Retrieved from http://www.gutenberg.org/ebooks/1672 (Original work published ca. 380 B.C.E.)

Stace, W. (1990). *The concept of morals*. New York, NY: Peter Smith. (Original work published 1937)

Twain, M. (1907). *Christian Science*. Retrieved from http://www.gutenberg.org/ebooks/3187.

Vickers, B. (1989). *In defence of rhetoric*. New York, NY: Oxford University Press.

Wardy, R. (1996). *The birth of rhetoric: Gorgias, Plato and their successors*. New York, NY: Routledge.

Whitehead, A. N. (1997). *An enquiry concerning the principles of natural knowledge* (2nd ed.). Cambridge, UK: Cambridge University Press. (Original work published 1925)

Willard, B. (2007). Rhetorical landscapes as epistemic: Revisiting Aldo Leopold's A Sand County Almanac. *Environmental Communication*, *1*, 218-235.

Williams, R. (1985). *Keywords: A vocabulary of culture and society* (Rev. ed.). New York, NY: Oxford University Press.

Worster, D. (1994). *The wealth of nature: Environmental history and the ecological imagination*. New York, NY: Oxford University Press.

Author Biography

Henry St. Maurice is an Emeritus Professor of Education who holds degrees from Wesleyan University (Connecticut), the University of Vermont, and the University of Wisconsin- Madison. He is also an adjunct instructor in the Edgewood College (Wisconsin) Educational Leadership Doctoral Program.

15

Online Support Groups for Depression: Benefits and Barriers

Louise Breuer[1] and Chris Barker[1]

Abstract

This mixed-methods study aimed to explore the initial process of engagement with an online support group (OSG) for depression. Fifteen British National Health Service patients experiencing depression who had not previously used an OSG for depression were offered facilitated access to an existing peer-to-peer OSG for 10 weeks. Pre- and post-measures of depression, social support, and self-stigma were taken in addition to a weekly measure of OSG usage. A follow-up qualitative interview was conducted with a subsample of nine participants. Depression and self-stigma reduced over the 10-week period, but perceived social support did not change. There was no evidence of adverse outcomes. Perceived benefits of OSG participation included connection to others, normalization of depression, and stigma reduction. However, engagement with the OSG was generally low. Barriers included concerns over causing harm to others or being harmed oneself, feeling different from others in the group, and fears of being judged by others. OSGs may potentially reduce depressive symptoms and perceived self-stigma. However, considerable barriers may hinder people with depression from engaging with OSGs. Further work is needed to determine who will benefit most from participating in OSGs for depression and how best to facilitate engagement.

Keywords

online support group, Internet support group, social support, depression, stigma

Depression is one of the most common mental health problems, with a review of European studies estimating the 1-year prevalence of major depression as 5.7% (Wittchen et al., 2011). However, treatment may be hard to obtain or may not be sought (Meltzer et al., 2000). One potential additional source of help for people with depression is online support groups (OSGs), which have many users worldwide (Griffiths, Calear, Banfield, & Tam, 2009).

Although some people use depression OSGs even when they are experiencing severe depression (Houston, Cooper, & Ford, 2002; Powell, McCarthy, & Eysenbach, 2003), some aspects of depression, particularly negative thinking styles and low motivation, may make engagement with depression OSGs difficult. Professionals wanting to refer their clients to a depression OSG may not know how best to proceed. The only empirical evidence comes from Van Voorhees et al.'s (2013) randomized trial, which found that an induction procedure emphasizing internal motivation led to greater engagement than one emphasizing external motivation.

The present mixed-methods study aimed to explore the initial process of engagement with a depression OSG by examining the experiences of British National Health Service patients encountering such a group for the first time.

Method

Participants

British National Health Service patients who experienced depression were recruited. Adults on psychological therapy waiting lists in three London boroughs were invited to participate by their therapist; in addition, patients registered with general practitioners in these boroughs were able to self-refer in response to a study website or to information in community venues. The inclusion criteria were as follows: (a) self-identified as experiencing depression, (b) Internet access (either at home or in a public venue), (c) no previous experience of OSGs for depression, and (d) above 18 and fluent in English. The study was approved by the Local Research Ethics Committee.

Thirty-five eligible people (21, 60%, men) expressed interest. Of these, 15 (43%) went on to take part (6, 40%, women; 9, 60%, men). Their mean age was 43 (range =

[1]University College London, London, UK

Corresponding Author:
Chris Barker, Research Department of Clinical, Educational and Health Psychology, University College London, Gower Street, London, WC1E 6BT, UK.
Email: c.barker@ucl.ac.uk

Table 1. Demographic Characteristics of Interviewed Participants.

Participant	Age	Gender	Ethnicity	Employment status	Previous episodes of depression	PHQ score pre-OSG	PHQ score post-OSG	Number of posts
1	40s	Male	Asian	Unemployed because of poor health	2 or more	17****	6**	2
2	20s	Male	Asian	Student	2 or more	22*****	22*****	9
3	50s	Male	White British	Unemployed because of poor health	None	24*****	21*****	2
4	50s	Male	White British	Unemployed because of poor health	None	23*****	15****	0
5	40s	Male	White British	Unemployed because of poor health	2 or more	23*****	22*****	0
6	40s	Female	White British	Looking after home/family	2 or more	17****	14***	1
7	60s	Male	Asian	In full-time employment	2 or more	11***	11***	0
8	30s	Female	White British	Student	2 or more	17****	15****	10
9	20s	Male	Asian	Student	2 or more	13***	3*	5

Note. PHQ = Patient Health Questionnaire; OSG = online support group.
*****severe depression. ****moderately severe depression. ***moderate depression. **mild depression. *no depression.

24–60). Nine (60%) were White British, 4 (27%) Asian, 1 (7%) Black British, and 1 (7%) Greek Cypriot. Ten (67%) were university graduates. Nine (60%) were self-referrals and 6 (40%) were recruited via clinical teams. Eight (56%) were currently taking anti-depressant medication. All participants were also invited to be interviewed, 9 (2, 22%, women; 7, 78% men) accepted. Their mean age was 43.5 (range = 24-60). Five (56%) of the interview participants were White British and 4 (44%) were Asian. Eight (89%) were university graduates. (See Table 1 for individual participant characteristics.)

OSG

The OSG selected for this study was PsychCentral.com. This was a U.S.-based peer-to-peer OSG which was "standalone," that is, not linked to another intervention. It was chosen because it was a high-traffic website that had a constructive atmosphere and was moderated by a team supervised by a clinical psychologist. Permission was granted by the website's owner for this research to take place.

Design and Procedure

This was a mixed-methods study, with quantitative measures taken before, during, and after the intervention, in addition to post-intervention qualitative interviews.

Participants were emailed an information sheet. If they consented to take part, they were emailed a link to the online questionnaire, on completion of which they were directed to the PsychCentral depression forum and given guidance on how to register. They were encouraged to write a post to introduce themselves to the group, and asked to engage with the site as they wished over the next 10 weeks. At the end of each week, they were emailed a link to the OSG Usage Report. At the end of the 10 weeks, they were emailed a link to the final questionnaire and were invited to participate in a

face-to-face interview. Participants were encouraged to contact the researcher by phone or email if they had difficulties using the OSG or if they wished to discuss their experiences.

Participants whose responses to the initial questionnaires suggested that they were severely depressed or a risk to themselves or others, i.e., scores >0.3 on the risk scale of the Clinical Outcomes in Routine Evaluation [CORE] or >1 on Item 9 of the *Patient Health Questionnaire* [PHQ-9] ("Thoughts that you would be better off dead, or of hurting yourself in some way") were emailed with the suggestion that they seek additional help, in line with the study risk protocol. This procedure was followed for five participants.

Measures

Pre–post measures. The following pre–post measures were used: CORE—Outcome Measure (CORE-OM; Evans et al., 2000), a 34-item questionnaire assessing general psychological distress; PHQ-9 (Kroenke, Spitzer, & Williams, 2001), a nine-item depression scale; the Medical Outcomes Study Social Support Survey (MOSSSS; Sherbourne & Stewart, 1991), a 19-item social support measure; and Depression Self-Stigma Scale (DSSS; Kanter, Rusch, & Brondino, 2008), two subscales of which were used: "General Self-Stigma" and "Treatment Stigma."

Measures taken at end only. OSG Questionnaire (OSGQ; Chang, Yeh, & Krumboltz, 2001) is a nine-item scale measuring overall satisfaction with an OSG. *Overall usage*: Postings made to the group were identified by searching for participants' usernames. Each participant's total number of posts over the 10-week study period was recorded.

Measure taken weekly throughout the 10-week period. OSG Usage Report, designed for the current study, asked how often participants had logged on to the group during the

Table 2. Outcome After 10 Weeks.

	Baseline M (SD)	10 weeks M (SD)	t(14)	p	Cohen's d
CORE-OM	62.7 (26.6)	59.3 (25.2)	0.88	.198	0.13
PHQ-9	14.6 (7.2)	12.3 (6.8)	2.22	.021*	0.33
MOSSSS	50.3 (18.6)	49.1 (20.5)	0.42	.343	0.06
DSSS	48.6 (15.9)	43.2 (17.4)	2.13	.025*	0.33

Note. CORE-OM = Clinical Outcomes in Routine Evaluation—Outcome Measure (Evans et al., 2000); PHQ = Patient Health Questionnaire (Kroenke, Spitzer, & Williams, 2001); MOSSSS = The Medical Outcomes Study Social Support Survey (Sherbourne & Stewart, 1991); DSSS = Depression Self-Stigma Scale (General Self-Stigma and Treatment subscales; Kanter et al., 2008).
*p < .05. (one-tailed)

week, how long they had used it for, and how many times they had posted. It also had an open-ended question about experiences of the group during that week.

Interview

The semi-structured interview schedule covered three main areas: (a) depression and face-to-face social support, (b) engaging with the OSG, and (c) view of, and interactions with, others in the group. Interviews lasted about an hour and were digitally recorded. Participants were offered £10 plus expenses.

The data were analyzed using thematic analysis (Braun & Clarke, 2006). For each transcript, units of meaning were identified and collated into a tentative set of themes. Themes were compared across all the transcripts and organized into general themes and sub-themes, which were checked against the original transcripts and modified accordingly. During the process of analysis, the two authors discussed the emerging thematic structure and attempted to resolve any areas of uncertainty. The interview data were also triangulated with week-by-week written feedback from the OSG usage report.

Researchers' Perspective

Qualitative research guidelines (e.g., Barker & Pistrang, 2005) recommend that researchers disclose relevant background. The first author was a female clinical psychology graduate student with a positive but limited experience of using an OSG in a personal capacity. Her initial assumptions were that OSGs can potentially be helpful, but that there might be barriers to engagement. The second author was a male clinical psychologist who had conducted research on support interventions. The researchers attempted to "bracket" their assumptions during the course of the study to be aware of how they might influence their interpretations of the data.

Results

Outcome, Usage, and Satisfaction

Eleven of the 15 participants completed all pre- and post-measures. Missing data were handled on an intent-to-treat basis using last-observation-carried-forward.

Depression and self-stigma reduced over the 10 weeks (Table 2). In terms of reliable change (Jacobson & Truax, 1991), three people improved, 12 showed no change, and no one deteriorated. There was no change in perceived social support or the total CORE-OM.

Usage was generally low. The median percentage of weeks in which participants logged in was 30% (range = 0-80%). The median number of posts over the 10 weeks was 2 (range = 0-13). Eight (55%) participants completed the OSGQ; their mean score was 27.4 (SD = 10.7, range = 11-39), indicating a moderate level of satisfaction.

Qualitative Interviews

Context. Most participants interviewed described experiencing severe recurrent depression, which had affected their ability to work and had in several cases led to suicide attempts. Only one was experiencing a first episode. Several also experienced other mental health problems such as anxiety, post-traumatic stress disorder (PTSD), and trichotillomania. None had had any experience of OSGs for depression. One described a positive experience with an OSG for trichotillomania, and others had used non-mental health online communities. About half had been hopeful about what the OSG might offer them, for instance, looking forward to sharing ideas with others and finding people with similar experiences. The rest were less optimistic, reporting that they had not expected to benefit from the OSG; they appeared to be motivated by curiosity or a desire to further psychological research.

Thematic analysis. The themes generated from the qualitative interviews and weekly feedback were grouped into two domains (see Table 3): perceived benefits and barriers to engagement.

Benefits

I'm not alone in this. Participants valued the opportunity to make contact with others with experience of depression. Five said that they felt less socially isolated as a result of using the OSG and felt an emotional connection with other users, promoted by others' openness about their experiences of depression. Three felt that the site had helped to make depression

Table 3. Themes and Subthemes.

Domain and theme (Ns)	Illustrative quotation
Benefits	
1. *I'm not alone in this*	
1.1. Feeling connected to others (5)	There were some people who I could really empathize with. I felt like they were in exactly the same situation. I think on one occasion I actually posted something saying that I empathize to that effect . . . (P2)
1.2. The site normalized depression (3)	[Using the site] has just reminded me that this is a . . . very common thing, and the people of the world who experience this . . . that kind of normalizes it a bit. (P8)
2. *The value of talking to strangers online*	
2.1. Site feels safe and non-judgmental (4)	The thing about an illness like depression is I think the criticism you get from outside is quite harsh, painful, and people who have experienced it or have some understanding of it are, are not like that, it's safe. (P7)
2.2. Anonymity makes it easier to self-disclose (3)	It was a bit easier [online] because I knew that I was anonymous. I wouldn't necessarily have to face those people again. (P8)
2.3. Others here have no preconceptions of me (2)	But usually when you're speaking to your friends, they already have an image about you . . . in the group you're being addressed on what you actually say. (P9)
3. *Supporting others and being supported*	
3.1. I have the group to fall back on (3)	There is just this surplus of people in my life to help me that give me this assurance that I'm going to be okay, and that I'm never going to go into isolation again . . . If nothing else works out, I have the group to fall back on, and as for my problems, and I know that I'll have someone to speak to me. (P9)
3.2. I wanted to help others (2)	If they have been positively affected [by my post], then I would consider it like you know, a pat on my back, like okay, so I've done a good job, I've done something good for the day that has affected somebody's life. And that's why I would want to do it. (P9)
3.3. Reading about others' experiences helped me (2)	It's quite easy to get some support from reading the other people's experiences . . . Somebody's had something then they say what happened that made it better which helps because you think "yeah, I can do that." (P7)
4. *Could be useful for others or for me at a different time*	
4.1. It could be useful for me if I were more/less depressed (4)	I remain open-minded about it. It could be that at a different stage of the illness it would have been differently useful. (P4)
4.2. It seems helpful for other people (4)	I'm sure it's a really good site for a lot of people because there's a lot of people using it and some of them have been on there like 11 years and I'm sure they get a great deal out of it. (P3)
Barriers	
5. *Fear of negative impact on others*	
5.1. I would not be able to help others (3)	I don't think [posting] contributed anything. I didn't think it did anything for me, I didn't think it did, or would do, anything for anyone else. (P4)
5.2. I might make others feel worse (5)	So I kind of worry that I'll go on and there'll be some, probably some quite young person in their teens or whatever, just coming to terms with this, and if I come along and say "Well, here I am after 20 years and it hasn't got any better" then that's not going to be very encouraging for them so I'd rather not say anything at all. (P8)
5.3. I might cause trouble (3)	People tend to say the same things. After a while a certain sort of house style starts to apply. People all say the same stuff and I thought if I do say something it's going to be radically different from what a lot of other people have said and it might cause trouble. (P5)
6. *Fear of negative impact on self*	
6.1. There's nothing helpful that anyone can say (3)	I just didn't find the things, their replies, were particularly comforting or useful so I didn't post again for a while. (P2)
6.2. It might make me feel worse (5)	"I'd been there, I hated it, I don't want to remember" was one of them; the "no no no you're going to do yourself damage if you do that" is another one; the "no you just don't understand" or at least "you don't understand where I've been." (P4)
6.3. Fear of being judged by others in the group (2)	I felt like all my sort of social anxiety that I feel about going into a physical room full of people it was also there, even in this totally anonymous online group. It still felt as though I was walking into a big party and didn't know anyone, it was strange. (P8)

(continued)

Table 3. (continued)

Domain and theme (*Ns*)	Illustrative quotation
7. *Not my kind of people*	
7.1. People were too different from me (5)	*There was almost a feeling of "If you're well enough to be on here, relating like this, writing coherently and reading stuff, then you and I are not in the same place." (P4)*
7.2. It was hard to tell whether people were genuine (3)	*It wasn't like I was reading things from real people, I was reading things from people's online versions of themselves which put me off a bit. (P5)*
7.3. Discussion was not serious enough (3)	*I know they mean well the people but it's all that patting on the back and "there, there, there." It just seemed like it was more of a social gathering than a help group. (P3)*
8. *It was hard to talk (to strangers) online*	
8.1. It was hard to share things about myself (4)	*What I wanted to offload to someone was all the specific thoughts and worries I was having, and I didn't feel I could put them down because they were . . . because I didn't know these people, they wouldn't really know what I was talking about, or it would be too personal . . . (P8)*
8.2. Having too much time to think made it difficult to write (2)	*I found it harder online because you've got to think about what to put down. You can't like . . . When I talk, things come out totally different. (P3)*
8.3. The value of people who really know me (4)	*People on the site didn't really know me or my history at all and they couldn't be expected to. But that makes a difference because people might just come up with, you know, a kind person might make a suggestion of something that was good for them but it might be totally inappropriate for me, but anyone who really knew me wouldn't have suggested that. (P8)*

seem like a common, universal experience, which they found reassuring. This was linked to the size and the international nature of the OSG.

The value of talking to strangers online. Some participants felt more comfortable talking about depression to strangers online than to their friends and family. They attributed the safe, non-judgmental nature of the OSG to the fact that users had personal experience of depression. Three felt that the anonymity enabled them to disclose more freely. Two spoke about the helpfulness of talking with people who did not have any preconceived ideas about them.

Supporting and being supported. Participants felt supported from being in contact with the group, even without any actual communication. Three said that knowing that it was there, even if they were not using it, was helpful. They appreciated being able to access the group immediately and at any time, and some contrasted this with their experience of waiting a considerable time for therapy. Two talked about wanting to help others as a result of reading their posts and the sense of reward that comes from offering help, and two others talked about the support they had got from reading about others' experiences. One felt that seeing other people going through cycles of low moods reminded her that she would once again come through her own current low patch, and another spoke about the way in which he had been inspired to try different ways of coping.

Could be useful for others or for me at a different time. Even if they had not found the site helpful, participants without

exception expressed the view that the site could potentially be a useful resource for people with depression, and commented that users appeared to be supporting each other and seemed to benefit. Four felt that they had been severely depressed at the time of the research: One spoke about finding it difficult to even motivate himself to log on. Two felt that the group may have been more useful to them if they were less depressed, although others said that they would be more likely to use the site if they were having a bad time.

Barriers

Fear of negative impact on others. Participants' concern for the other people involved with the group was striking, and some cited the potential negative consequences that their involvement could have for others as one of the reasons that they did not contribute more. In some cases, the concern that they would not be able to help others held participants back from seeking support, as they felt strongly that their relationship with the site should be reciprocal. Four participants said that they sensed that others in the group were vulnerable and worried that their contributions could make people feel worse: one said that hearing about her long history with depression might be discouraging for someone who was newly diagnosed. Three worried that they might get into conflict with others in the group and had held back from posting because they were afraid that they would express anger or criticism.

Fear of negative impact on self. Participants were concerned that spending time on the site might make their depression worse, particularly if they were not feeling especially depressed. Three felt that others on the site could not help them, and one who posted messages did not find the replies helpful. One expressed the view that when people were depressed they were too withdrawn to interact and when they were well they were afraid of interacting with other depressed people for fear of being pulled back down into depression. Two participants described experiencing social anxiety in face-to-face situations, and worried about how they would come across in the group and whether others might be critical of them.

Not my kind of people. Participants mentioned various aspects of how others in the group communicated which discouraged them from becoming involved. The site being based in the United States made it hard for some participants to relate to the other users: they mentioned differences in language, social contexts, and medication names as barriers to understanding. Three felt that it was hard to gauge whether others were being genuine, making it difficult to feel comfortable sharing thoughts and feelings. Three said that the level of discussion on the site was too shallow and people's posts could be experienced as patronizing.

It was hard to talk (to strangers) online. Participants said that although they wanted to express themselves, it was difficult to do so in the OSG, because they felt embarrassed and vulnerable. Four spoke about the difficulty in establishing rapport online (sometimes attributed to factors such as the asynchronous nature of responses and the absence of body language and eye contact) which would have made them feel comfortable enough to reveal personal information. Two said that both having time to think about what they were about to say, and not having the pressure of having to produce something for a listener in front of them, were the inhibiting factors. Four felt that people who really knew them were in a much better position to help them through their depression than strangers online. In interactions with people who knew them well, these participants valued that they did not have to explain themselves and their history, that people could tell how they were feeling, and that they could remind them what they were like when they were not depressed, which instilled a sense of hope. Comments from strangers, although supportive, were experienced by some as "meaningless" whereas real friends' support had more impact.

Discussion

This mixed-methods study introduced a small sample of 15 people with depression, recruited through the British National Health Service, to a depression OSG and tracked their engagement with it. Most had moderate to severe levels of depression at baseline. Over the 10-week period of the study there were modest reductions in depression and perceived stigma, with no change in perceived social support. Overall, there were low levels of engagement with the OSG.

The reduction in depression is consistent with the results of many other studies (Griffiths, Calear, & Banfield, 2009). However, in the absence of a control group, it is not possible to attribute a causal role to OSG participation, particularly because engagement was low.

The low levels of activity may reflect the way in which people naturally engage with OSGs, that is initially spending a long time viewing the site without posting ("lurking") before joining the discussion. Lieberman (2007) found that 74% of participants in breast cancer OSGs lurked before they posted, with half of these doing so for 2 to 8 weeks. To observe the engagement process as it naturally occurs, a longer study period would be needed.

Participants not only described several benefits of taking part in the OSG but also emphasized factors that had hindered their engagement. In contrast to previous studies that have recruited participants from among existing users of OSGs (who are likely to be at least reasonably satisfied with such groups), the present study recruited people who had never before used a depression OSG, tracking their engagement and allowing the voices of those who find such groups less helpful to be heard.

Participants showed a striking concern for the well-being of others in the group and wanted to help them, but several felt unable to do so, often because their own struggles left them without resources. Sometimes this inhibited participants from using the group, as they felt that their relationship with it should be reciprocal. This raises the question of what could be done to help users access support without feeling a pressure to reciprocate immediately.

The main barriers to participation were fears of having a negative impact on others in the group, or the group's having a negative impact on oneself. Participants also worried about their depression being made worse by their interaction with the group and reported avoiding reading posts with negative content. This may reflect a more general tendency toward emotional avoidance among participants which may have affected their low engagement with the group (Cameron et al., 2005). In addition, negative beliefs about seeking help (e.g., that it is a sign of weakness) may also have inhibited participation.

These concerns are understandable in the light of the well-documented "cognitive triad" of negative beliefs about the self, the world, and the future that characterizes depressive thinking (Beck, Rush, Shaw, & Emery, 1979). Levine and Moreland (2006) suggested that in general people try to join groups that are maximally rewarding and minimally costly. Depressive thinking may have led participants to underestimate the rewards and overestimate the costs of joining the OSG. As they did not engage, there was no opportunity for them to gain evidence to disconfirm their negative preconceptions.

Limitations

This was a small-sample, mixed-methods study. The quantitative arm had low statistical power; the measures were included to give an indication of participants' overall outcomes. The sample was a convenience one, but it did have the advantages of having a range of ages and ethnic backgrounds, and of giving people access to an OSG for the first time.

Limiting participants' choice to a single OSG was problematic, as there is evidence that when people choose a group that "fits" them best, they are more likely to engage with it (Lieberman, 2004, 2007). Moreover, the choice of a predominantly North American group was with hindsight unfortunate, as cultural differences were often mentioned as a barrier to participation. Thus, it is unclear whether the low levels of engagement and barriers to participation found in this study reflect engagement with OSGs for depression in general or with this OSG in particular. When the study started, there was no comparable U.K.-based OSG in terms of moderation, positive group climate, and high volume of traffic.

Implications

There was no evidence of negative effects from the OSG, and some participants reported benefits, which suggested that clinicians need not be concerned by their clients' involvement with OSGs. It appears that prospective members who anticipate negative effects simply do not engage with such groups. Many of the barriers to participation relate to people's fears of what might happen were they to take part, rather than to actual experiences of participation. It is unclear to what extent these fears are related to cognitive aspects of depression. Depressed clients may benefit from some therapeutic work to help them test the accuracy of their negative predictions in the initial stages of engaging with such a group.

Some of the barriers identified in this study are related to the negative thinking style, withdrawal and lack of motivation characteristic of depression. It would be valuable to design an intervention to help participants overcome these initial barriers and evaluate if this had any impact on engagement. If barriers to active participation can be minimized, people with depression may then be able to derive more benefit from online support.

Acknowledgments

Thanks to Dr. John Grohol for permission to use PsychCentral for the purposes of this study.

Declaration of Conflicting Interests

The author(s) declared no potential conflicts of interest with respect to the research, authorship, and/or publication of this article.

Funding

The author(s) received no financial support for the research and/or authorship of this article.

References

Barker, C., & Pistrang, N. (2005). Quality criteria under methodological pluralism: Implications for conducting and evaluating research. *American Journal of Community Psychology, 35* (3-4), 201-212.

Beck, A., Rush, A., Shaw, B., & Emery, G. (1979). *Cognitive therapy of depression*. New York, NY: Guilford Press.

Braun, V., & Clarke, V. (2006). Using thematic analysis in psychology. *Qualitative Research in Psychology, 3*(2), 77-101.

Cameron, L., Booth, R. J., Schlatter, M., Ziginskas, D., Harman, J. E., & Benson, S. R. C. (2005). Cognitive and affective determinants of decisions to attend a group psychosocial support program for women with breast cancer. *Psychosomatic Medicine, 67*(4), 584-589.

Chang, T., Yeh, C. J., & Krumboltz, J. D. (2001). Process and outcome evaluation of an on-line support group for Asian American male college students. *Journal of Counseling Psychology, 48*(3), 319-329.

Evans, C., Mellor-Clark, J., Margison, F., Barkham, M., Audin, K., Connell, J., & McGrath, G. (2000). CORE: Clinical outcomes in routine evaluation. *Journal of Mental Health, 9*(3), 247-255.

Griffiths, K. M., Calear, A. L., & Banfield, M. A. (2009). Systematic review on Internet support groups (ISGs) and depression (1): Do ISGs reduce depressive symptoms? *Journal of Medical Internet Research, 11*(3), e40.

Griffiths, K. M., Calear, A. L., Banfield, M. A., & Tam, A. (2009). Systematic review on Internet support groups (ISGs) and depression (2): What is known about depression ISGs? *Journal of Medical Internet Research, 11*(3), e41.

Houston, T., Cooper, L., & Ford, D. (2002). Internet support groups for depression: A 1-year prospective cohort study. *American Journal of Psychiatry, 159*(12), 2062-2068.

Jacobson, N., & Truax, P. (1991). Clinical significance: A statistical approach to defining meaningful change in psychotherapy research. *Journal of Consulting and Clinical Psychology, 59*(1), 12-19.

Kanter, J., Rusch, L., & Brondino, M. (2008). Depression self-stigma: A new measure and preliminary findings. *Journal of Nervous and Mental Disease, 196*(9), 663-670.

Kroenke, K., Spitzer, R. L., & Williams, J. B. (2001). The PHQ-9: Validity of a brief depression severity measure. *Journal of General Internal Medicine, 16*(9), 606-613.

Levine, J. M., & Moreland, R. L. (Eds.). (2006). *Small groups, key readings*. New York, NY: Psychology Press.

Lieberman, M. (2004). Self management in online self help groups for breast cancer patients: Finding the right group, a speculative hypothesis. *International Journal of Self Help and Self-Care, 2*(4), 313-328.

Lieberman, M. (2007). The role of insightful disclosure in outcomes for women in peer-directed breast cancer groups: A replication study. *Psycho-Oncology, 16*(10), 961-964.

Meltzer, H., Bebbington, P., Brugha, T., Farrell, M., Jenkins, R., & Lewis, G. (2000). The reluctance to seek treatment for neurotic disorders. *Journal of Mental Health, 9*(3), 319-327.

Powell, J., McCarthy, N., & Eysenbach, G. (2003). Cross-sectional survey of users of Internet depression communities. *BMC Psychiatry, 3*(1), 19. doi:10.1186/1471-244X-3-19

Sherbourne, C., & Stewart, A. (1991). The MOS social support survey. *Social Science and Medicine, 32*(6), 705-714.

Van Voorhees, B. W., Hsiung, R. C., Marko-Holguin, M., Houston, T. K., Fogel, J., Lee, R., & Ford, D. E. (2013). Internal versus external motivation in referral of primary care patients with depression to an internet support group: Randomized clinical trial. *Journal of Medical Internet Research, 15*(3), e42.

Wittchen, H.-U., Jacobi, F., Rehm, J., Gustavsson, A., Svensson, M., Jönsson, B., . . . Steinhausen, H.-C. (2011). The size and burden of mental disorders and other disorders of the brain in Europe 2010. *European Neuropsychopharmacology, 21*(9), 655-679.

Author Biographies

Louise Breuer, DClinPsy, is a recently qualified clinical psychologist working in the British National Health Service. She is currently employed in a sexual health service, affiliated with University College Hospital, London. She has a research and a clinical interest in online support groups.

Chris Barker, PhD, is a professor of clinical psychology at University College London. He has had a longstanding research interest in the process and outcome of psychological help and support, across a range of populations and settings, particularly mutual help and peer support. He has also published on research methodology in clinical and community psychology.

16

Computer Users Do Gender: The Co-Production of Gender and Communications Technology

Lori Leach[1] and Steven Turner[2]

Abstract

The so-called "digital gender divide" has encouraged studies attempting to demonstrate the co-production of gender and information technology. Vivian Lagesen has criticized many of these attempts for failing to provide fully symmetrical accounts. Here we describe and analyze beliefs and practices concerning computers, gender, and technology evinced by managers in a network of public sites (Community Access Centers) created to provide community access to digital technology in the Canadian province of New Brunswick. From those results, we argue, among other conclusions, that distinguishing more carefully between the gendered uses of new technologies and the gendered forms of attraction associated with them produces a more fully realized and more perfectly symmetric understanding of how gender and communications technologies are co-produced. We show that the concepts of actor-network theory facilitate that analysis, and so interpret the study as supporting and extending Lagesen's program.

Keywords

technology, computer, gender, actor-network theory, co-production, New Brunswick, Lagesen

In the 1990s, the Canadian province of New Brunswick moved to create a network of Community Access Centers (CACs), which would give rural and small-town citizens access to computers and Internet services and training. The Community Access Program (CAP) was to address the so-called "digital divide" and democratize access to the new Information and Communications Technology (ICT; Rideout, 2000, 2001). The CAP was not intended to address gender questions, but women quickly came to predominate among the managers and clients of the centers. The centers therefore emerged as ideal sites for positing feminist questions about technology, society, and gender, and for addressing the issue of women's access to and adoption of ICT: the so-called "ICT gender gap."

This article reports the empirical findings of a study conducted with the managers of New Brunswick CAP Centers by one of us (Lori). It begins by discussing current issues and literature on the "gender digital divide" as well as certain theoretical and methodological problems related to studies of technology, society, and gender. It then describes the CAP initiative and the study, and goes on to outline the results and their significance.

Introduction: ICT, Gender Gaps, and Co-Construction

Studies of the relationships among technological change and adaptation on the one hand, and gender stereotyping and women's experiences on the other, have multiplied rapidly in the diverse literatures on science and technology, media, and gender (Wajcman, 1995, 2010). Some of that growth has been fueled by worry over the so-called "ICT gender gap": the concern that girls and women still lag behind their male counterparts in the adoption of and use of digital information and communications technology (ICT). The "gap" may refer to the use of ICT in daily living, or to the claim that women's participation in ICT as university students and industry-professionals has decreased proportionally from highs in the 1980s. By the 1990s, earlier concerns over a socioeconomic and rural-urban "digital divide" had given way to concern over a digital divide across gender lines in many countries. That concern has prompted efforts of many kinds to make ICT more attractive to women, as well as debates over the success of such programs.

Whether a digital gender gap persists, or is disappearing with the maturation and diffusion of ICT, has prompted debate and uncertainty. Canadian data, for example, show that male adults use the Internet slightly more than women (81%-79% in 2009), that the ratio has changed little since

[1]Government of New Brunswick, Fredericton, Canada
[2]University of New Brunswick, Fredericton, Canada

Corresponding Author:
Lori Leach, Department of Post-Secondary Education, Training and Labour, Government of New Brunswick, Fredericton, Canada, E3B 5H1.
Email: lori.leach@gnb.ca

2005, and that age and socioeconomic status are far more important determinants than gender of ICT access and use (Statistics Canada, 2014). Thomaz Drabowicz (2014) draws on the 2006 PISA data (Program for International Student Assessment) for a comparative study of self-reported computer and Internet use among adolescents in 39 countries. He finds higher rates of access and use among boys in every country, but, unexpectedly, also that the size of the digital gender gap is not correlated with other social and cultural measures of gender equality. In a major study, Knut H. Sørensen et al. acknowledge that ICT gender differences in daily life in Europe and the United States have narrowed on some fronts, such as Internet use and ownership of mobile phones, and that women's use of social media outweighs that of men. But they conclude that "available statistical information does not support claims of a disappearing digital gender gap" (p. 24), whether in the proportion of women and men accessing and using the Internet, in time spent on the net, or in Internet and computer skills.

The under-representation of women in higher computing education and among professional ICT specialists remains "very substantial and apparently stable in most (Western) industrialized countries" (Sørensen, Faulkner, & Rommes, 2011, p. 24). For the United States, Thomas J. Misa (2010) has shown that the proportion of women studying computer science has fallen from 37% in the mid-1980s to 15% around 2007, and from 39% to 29% of the American white-collar computing workforce over roughly the same period (pp. 5-6). But Mellström (2009) has used the example of Malaysia (where women are highly represented among computer science students and computing professionals) to argue that "ethnic and class inequalities often are as important as gender differentials" in determining career access (p. 891).

Efforts to explain and diagnose the digital gender gap have varied widely. Sørensen et al. (2011) claim to detect three distinct but overlapping "narratives" in the existing literature which perform that function. The earliest narrative, linked to feminist technology-and-society theorists such as Judy Wajcman and Wendy Faulkner, focuses on the strong association of technology with masculine culture, and with the corresponding exclusion of women (Faulkner, 2007). That exclusionary association, reinforced by social and institutional pressures, leads to the voluntary abandonment by women of many technologies (including ICT) that are perceived as "gender inauthentic." Sørensen et al. detect a more recent variant of this account in what they call "chilly climate" narratives. These focus less on the gender-exclusionary quality of technology in general, and more on the particular history and unique features of ICT. In particular, they portray computing-culture as originally gender neutral or even woman-friendly, but as increasingly subject to "hacker" or "geek" stereotypes since the 1980s. These have subsequently discouraged women's participation in computing and led to "leaky pipeline" effects, as women enter, and subsequently leave, the field.

A third contemporary narrative, one allegedly more conducive to women's participation in ICT than the first two, Sørensen et al. call "the woman communicator." They identify this narrative with "cyberfeminist" approaches. Scholars writing in this tradition may be highly critical of ICT and its culture under current patriarchal structures, but they are typically optimistic about women's engagement with the Internet and net-based communications. Sometimes they invoke women's talent for communication and networking to predict a coming, utopian convergence of ICT, female participation in and transformation of ICT, and women's culture. Although this approach has been criticized as essentialist, it more commonly seeks to transcend discriminatory gender differences in ICT use and access. Sonja Bernhardt (2014), for example, has argued that new forms of consciousness promoted by the digital social media and global communications will kick-start a new, radical individualism among the young, and make it possible for them (both men and women) to transcend the cultural pressures of the past that have blinded them to the potential of ICT to serve their life-interests.

A commonly used theoretical framework for analyzing the digital gender gap today, and one applicable to all the so-called narratives, is rooted in the "social-shaping" tradition within feminist studies of technology. Social shaping sought to demonstrate how new technologies often owe their evolving form and acquired social meaning much more to their users than to their ostensible producers, who are typically unable to dictate or foresee the meaning of their creations. Those users are very often women, especially for communications and household technologies. The analysis, therefore, blurred the line between "making" and "using" a technology in ways that commonly emphasize women's agency (Berg, 1994; Cockburn & Ormrod, 1993; Lie, 1995; Lohan, 2001; Martin, 1991; Oudshoorn, Rommes, & Stienstra, 2004; van Zoonen, 1992; Wajcman, 1991, 2002).

Social shaping was directly extended to the analysis of technology and gender, often under the label, "co-construction." It was further popularized within science and technology studies by Sheila Jasanoff in 2004, as "co-production." In this framework, analysts seek to show, first, how the social meaning of an emerging technology is shaped by prevailing gender conceptions, and, second, how that evolving technology in turn shapes and reshapes cultural understandings of gender and individual forms of gender consciousness. Especially in its co-production form, this approach invokes a performative understanding of gender as "something done" and continuously re-created by social action. It seeks a symmetrical analysis by imposing the same understanding on new technologies, treating them as sociotechnical assemblages, the meaning of which is shaped extensively by use.[1]

Demonstrating co-production via empirical studies, however, has proved a challenge. In 2012, Norwegian scholar Vivian Lagesen charged that feminist studies have typically offered asymmetrical treatments of gender and technology, including information technology. Often, she claims, established gender roles are taken as the social "given," and the analysis proceeds by showing how technology is interpreted to comply, or how it "becomes gendered." Only rarely, she says, does the analysis proceed in the other direction, to show how technology helps to re-shape or co-construct gender itself. As an example, she cites a famous analysis of the microwave oven, which was originally marketed as a "brown" technology (like televisions and stereo sets) mainly for men, and then "re-invented" (this time successfully) as a "white good" (like refrigerators and kitchen ranges) to be sold to women. The analysis cleverly demonstrates how gender-based user-roles shape the technology and its meaning, but it declines to discuss how the technology might have impacted gender roles themselves. Similar criticisms have been offered by Catharina Landström (2007) and, in a different context, by Julia Nentwich and Kelan (2014). Nentwich and Kelan note that the concept of "doing gender" is common in their field (labor studies), but that in almost three quarters of the empirical studies, the concept is appropriated only "as a way of grounding, legitimating or validating [authors'] research findings," and plays no substantive role in the analysis. Performative notions of gender underlie co-production analysis, and according to Lagesen's critique, are equally difficult to demonstrate persuasively in gender and technology studies.

Lagesen offered two recommendations for escaping this practical impasse. The first is that feminist technology studies should embrace so-called actor-network theory (ANT). Following the action-theoretical schema pioneered by Bruno Latour (1999, 2005), analysts must first observe how new elements (in this case, new technologies) introduce uncertainties, tensions, or controversies (often about gender roles and capabilities), and then follow closely "how involved actors work to overcome the uncertainties, to stabilize the controversies," usually by recourse to new standards and discourses, or the building of new networks of interest and compatibility (Lagesen, 2012, p. 446; Quinlan, 2012). Her second recommendation is that these efforts should be embodied in empirical case studies of encounters with new technologies which would provide symmetrical analyses of how technology and gender is co-produced. Lagesen concludes her analysis by inviting readers to join the effort and produce empirical results on the "reassembling" and "stabilizing" of gender as part of the technology and gender co-production.

We concur with Lagesen that analysis in terms of co-production, as well as the ANT perspective itself, has much to offer studies of technology and gender, especially issues of ICT, the digital divide, and the contemporary encounter of women with computers and information technology. We also share Lagesen's perspective that attempts to distinguish between technology and the social meaning of that technology is generally unproductive for gender analysis and insight into the process of co-production. In what follows we respond to Lagesen's call and offer an empirical,

gender-focused examination of an encounter between human actors and a new technology (the personal computer) in one specific context and ask how traditional gender understandings were de-stabilized and re-formulated for those actors as a result of the encounter.

The Setting and the Study: New Brunswick's CAP and Its Actors

In the early 1990s, the federal government of Canada began laying the groundwork for what was popularly called the "information highway." That highway, it was idealistically hoped, was to carry the nation toward a knowledge- and information-based economy that would in turn create a new society that would address the needs of all individuals economically, socially, and culturally (Information Highway Advisory Council, 1997). Geographic obstacles would be eliminated, enabling the most remote community to mobilize and reorganize itself to benefit from the economic and social opportunities of computer technologies. The federal government's "Community Access Program"[2] was intended to democratize access to the information highway by bridging the socioeconomic gap between the digital "haves" and the "have nots," that is, the gap between those who had the means to access computer technologies and those for whom computer technology was not accessible—the disenfranchised, low-income minorities (Dickinson & Sciadas, 1996, 1997, 1999).

Most of Canada's provinces and their governments readily bought into the federal initiative and none more enthusiastically than the province of New Brunswick in eastern Canada, under then-premier Frank McKenna. New Brunswick was economically depressed, mostly rural, and isolated from large urban markets. The information highway promised a way to promote the province's human capital, foster digital modernization, and encourage economic opportunity. But ensuring equality of access was seen as essential. Computer technology was the "new toy" and up to this point in time, tinkering with it was reserved for those with vast computer technical knowledge and special expertise. Under the CAP, access to computers was to become available to the public and affordable to all.

New Brunswick CAP sites, funded jointly with federal and provincial money, were established first in rural communities where need was perceived as the greatest, and housed primarily in free spaces such as schools, libraries, and community centers. A CAP Center offered anywhere between 3 and 20 computer terminals for public use. Moll and Fritz (2007) report that at its peak, CAP had opened 8,800 sites across Canada, while in New Brunswick, the number of active sites peaked at 250 in 2001 (Connect NB Branché database, 1992-2012).

In Canada as elsewhere, political enthusiasm for the digital revolution proved short-lived, and challenges mounted for the provincial CAP networks. The speed with which computer technologies were growing and changing during this time (and still today) became problematic. The digital gap was not eliminated, as was initially anticipated, and the continual decrease in federal funding forced provinces and local communities to assume the burden of maintaining the CAP network. By 2005, only 3,786 sites remained open in Canada and only 109 in the province of New Brunswick. Well beyond its original 4-year mandate, CAP survived until 2012 when the federal government officially disbanded the program. Today in New Brunswick, remnants of the former CAP centers exist in small numbers as "Community Adult Learning Centers" and focus primarily on adult literacy, including "digital literacy."

Gender equality and women's access was not an explicit motivation for the federal government in the development of the CAP. In New Brunswick, when CAP sites became operational in the mid 1990s, they were initially operated by employee-managers who were primarily men.[3] Managers' main function was to provide technical expertise in local communities to those interested in learning how to operate this new technology. As the network evolved, however, the position of Center manager shifted from one of pure technical expertise to one focused on social networking, promotion, and teaching basic computer skills. Gradually, women began to assume these positions, until CAC managers came to constitute a mostly female and decidedly low-wage ($11.00/hr) workforce.[4] The evolving role of the network placed new demands on the CAP manager. She or he was increasingly required to be a multitasking, creative, entrepreneurial-type individual who had excellent communication skills and the ability to teach others. The need for technical skills was not abandoned, but increasingly CAC managers, especially women, were recruited without special technical backgrounds and were expected to learn those skills on the job in the process of using and teaching the technology themselves.[5]

The gendered nature of the manager-position made the CACs valuable sites for formal investigation of gender-issues in women's encounter with ICT. One of us (Lori) had more than a decade of experience of working with the provincial government in managing New Brunswick's CAP, and would draw on her familiarity with the CAP to launch a program of research focused on the CAC network. Amid whispers of the federal government "sunsetting" the CAP program in the early 2000s, the study titled "Computer Users Do Gender: The Co-Production of Gender and Communications Technology" took form (Leach, 2011). The study was to explore the subjective experience of men and women who worked with computer technology daily and who had shared pleasures in, and identification with, technology. It drew heavily on the literature of gender and technology (Berg, 1994, 1997; Berg & Lie, 1995; Broos, 2005; Broos & Roe, 2005; Cockburn, 1983, 1985a, 1985b, 1992; Cockburn & Fürst-Dilic, 1994; Cockburn & Ormrod, 1993; Faulkner, 2000, 2001, 2004; Hacker, 1981, 1989, 1990; Lagesen, 2007;

Moll & Shade, 2001, 2004; van Zoonen, 2002; Wajcman, 1991, 1999, 2000, 2004, 2006; Wajcman & MacKenzie, 1999), as well as on recent studies on the digital gender gap and inclusion initiatives, such as that associated with the European SIGIS project (Lie & Sørensen, 2003; MacKeogh & Preston, 2003; Oudshoorn, Rommes, & van Slooten, 2004). The results support the premise of two-way (technology *and* gender) mutual shaping and, in doing so, rise to the call from Lagesen (2012) for more research on "explor[ing] what different ways of relating to technologies may mean for the doing of gender" (p. 447).

The study used a sequential, mixed-methods design; data collection was two-phased. It began with a web-based survey directed at all New Brunswick CAC managers. At the time of this study, of the 109 CAC managers of Access Centers in New Brunswick, provincial government data determined that 92 (84%) of the centers employed a female manager; 76 (70%) were rurally located and all managers were adults, between the ages of 18 and 65 years old (*Connect NB Branché Indicator Survey*, 2008). Fifty-nine of the 109 CAC managers in the province answered the questionnaire. Forty-seven of the 59 survey respondents were female and 12 were male. The survey instrument was self-administered. It was "live" on the Internet and available to participants for a 1-month period. It used both structured open-ended and closed questions. There were 23 survey questions. Questions were presented as multiple-choice (8 questions), yes or no (11 questions), and Likert-type rating scale (4 questions). In addition, empty text boxes were provided when the survey asked an open-ended question, prompting the user to describe or explain their thoughts/opinions/experiences. In December, 2008, one of us (Lori) conducted face-to-face interviews with 11 CAC managers, 7 of whom were female and 4 of whom were male. Open-ended questions in a semistructured interview format were utilized. All interviews were audio-taped and transcribed, and on average interviews lasted approximately 1 hr.

In both phases of the empirical research, the managers were examined in their dual roles both as users and teachers of computer technologies. As CAC managers, they spent their workday observing and assisting men and women using computer technology at CACs. This experience made their observations of and reports about CAC users important in terms of providing information on gendered computer technology use patterns and behaviors. At the same time, their unique personal and professional experience with computers made them invaluable direct sources of attitudes and information about gender and ICT. Therefore, this research focused on CAC managers' experiences with technology as representative types (women and men), and as privileged informants (their observations of female and male CAC users' experiences with computer technologies). The study sought to examine computer use practices, attraction to the technology, and the local cultures that were created in New Brunswick CACs.

Managers on the Gendered Nature of Technology

The CAC managers were uniquely positioned, first as observers of gendered behavior and attitudes toward technology among their Center clients, and second as user/agents/shapers of computer technology in the social context of the CAP Center. Unlike the subjects of most previous studies of computer users, most female managers in the study were older (41-55 years), and came mostly from the more rural areas of a rural Canadian province. Few had formal computer training of any kind before taking the CAC positions, and while most reported 10 to 15 years of experience with computers (counting their CAC service), nearly 20 reported significantly less experience; not all had computers at home, and not all of those had Internet service. In this, the female managers differed significantly from the male managers. These men, while a minority among managers, were younger, with longer experience with computers than the women. The female managers in the study, bringing little expertise or training to their positions, mostly learned on the job, in the process of teaching computers skills, and helping others on the wrong side of the digital divide to use the new communications devices. In this regard, the CAC managers may more closely resemble the general population, who self-learn computer skills via trial and error to cope with the plethora of new applications on desktop, laptop, and handheld devices. They certainly resemble the classical Latourian actor, challenged by new social demands and expectations, all mediated by a technology whose meaning and possibilities are still undefined and in flux.

Of course, incorporating gender stereotypes into the analytical categories of the study ran the risk of acting as a self-fulfilling prophecy and mitigating against more nuanced understandings. Nevertheless, it was imperative to examine how the managers, as key informants, shared common stereotypes about women and technology, and how they observed them to be shared by others. Starting with such gender dualisms, we believed, was essential to our aim to later problematize them.[6] Questions both on the survey and in the interviews prompted them to classify common technologies as "used mostly by men" or "used mostly by women." Predictably, diesel motors and air pumps scored heavily as "used by men" and domestic technologies such as irons, vacuum cleaners, and baby monitors, and that classical feminized business machine, the typewriter, were seen overwhelmingly as "used by women." Interestingly, electronic and communications technologies (telephones, fax machines, computers, scanners, DVD players) were scored predominantly as "used by both." But only 9 out of 59 respondents to the survey were prepared to associate the computer more with one gender than another. We take this as a sign that the "social meaning" of the personal computer is still in flux and in formation, and has not yet undergone the assimilation to gender stereotypes that most or all technologies of the past seem to inevitably experience.

That is not a characteristic of all digital technologies, however; the survey showed that the X Box, the classic gaming device, is viewed overwhelmingly as used by men. The survey also showed that male managers were somewhat less willing to classify technologies and devices as strongly gendered by use as were female managers. This may, of course, reflect only reluctance among the male participants to give voice to derogatory gender stereotypes. However, we speculate that growing ubiquity of computers and digital communications devices in the modern workplace (both still relatively "gender neutral" and both used extensively by men and women in that context) may be carrying over into gradual reshaping of attitudes about gendered technology.

Managers were also asked how they and their CAC clients used computer technologies for personal ends, and how those uses differed by gender. The survey and interviews confirmed the now-ubiquitous research finding that for men, gaming is the predominant non-work-related use of the technology; 61% of the managers listed it among the uses they observed among their male clients, but only two of them confessed personally to gaming. One young male manager commented as follows:

> In my experience at the CAC the war type gaming there's a lot of guys of all ages in there . . . I guess you'd call them hard-core online gamers . . . [T]here isn't any women that get involved. . . . They don't want to waste their time doing it. They typically don't need the whole male ego shoot them up, kill them type [of activity]. (Leach, 2011, p. 209)

The female managers in the study reported rather different patterns of personal use from the men, and those patterns applied both to their clients and themselves. Email communication and online chatting replaced gaming by men among the most frequently reported uses (50% of all the uses reported for women, only 11% of those reported for men), and general information searching on the web was reported twice as frequently as applying to women than to men (46% vs. 21%). The informants were fully aware that these use-patterns reflected gender-stereotypical behavior. One younger female manager wrote,

> [W]omen use [computers] for things like support groups, things like connecting with other people, information, connecting with programs, things they may not have had access to fifteen years ago . . . A stay at home mother today has access to everything that women did not have access to then. (Leach, 2011, p. 209)

Two other managers confirmed:

> Men don't communicate as much as women. You know, with grandkids far away and things. Here, men don't seem to be the main users; it is women for genealogy or patterns online or email. (Leach, 2011, p. 210)

I guess women tend to do more Internet searching. Theirs is more communication-based Mail and Internet. (Leach, 2011, p. 211)

Not all computer uses showed different gender patterns, of course. Both men and women were reported as using computers for hobbies and other personal interests (29% and 36%), and for online services like banking (7% and 11%). Men were reported as using the computer to search for employment more than women. Perhaps the most interesting difference between managers' personal use of computers and those uses they reported observing in clients was their higher use of online services, an outcome perhaps reflecting their greater confidence in the technology. In general, we conclude that the female managers in the study showed markedly greater use of computers for emotionally connected activities, including means of coping with isolation, means of coping with routines of daily life, and online support. The women in the study demonstrated a flexibility to be connected to both stereotypically masculine practices and stereotypically female practices of connecting with others and expressing emotional needs (Leach, 2011).

Having established gender differences among the predominant uses that the managers (and their clients) made of personal computers, the study went on to ask whether the managers saw gender differences in the approaches to computer use among their clients: whether men and women differed in how they went about learning to use computers, solving problems, and dealing with frustrations. Managers mostly agreed that their female clients initially lacked confidence and were hesitant about their learning tasks in comparison with male clients, but several reported that with experience, their female clients usually attained a confidence and competence equal or superior to men. They reported their female clients as more willing to ask questions, more flexible, more patient with the technology and its outcomes, less concerned than men with deep control and knowledge about the hardware. One manager reported that the

women I teach in general seem to be less worried about how it happens if they know what the result or outcome was, even if they don't understand double-clicking could get to there or get to there . . . If they can do that [one way] that's good enough . . . [T]he majority of men will [want me to] really break [the double-click action] down. They want to know how they might have gotten [to the same result in another way] and want to try it again. (Leach, 2011, pp. 233-234)

Another claimed that

women seem to maybe, be able to adapt easier to a non-linear style of experimentation. . . . Women are a bit more open to exploring things as they go than a man. (Leach, 2011, p. 237)

A male manager contrasted his own operating style with those of his female clients:

Myself, when I attack computer problems, or program, or anything, I want to know the "nuts and bolts" of how it works, first. Then I'll use the darn thing as best I can. I know quite a few women that that's not the approach they would take to that problem. They would learn how to use the program, and not really care about what was going on behind the screen. (Leach, 2011, p. 239)

Co-production analysis demonstrates the insufficiency of treating technological innovation exclusively as a supply-driven phenomenon dependent upon scientific knowledge, market forces, or engineering rationality. New technologies are rather shaped by social interpretations and user needs and interests, which only gradually stabilize new technologies and fix their social meaning. So far-reaching is this shaping that distinctions between a technology and the social matrix of its mobilization lose analytic interest. The contribution of feminist studies of gender and technology has been to show that prevailing gender concepts, reified through the active participation of women users, is one key factor in the mutual shaping of a new technology. This is particularly true for the computer, the most protean of modern technological innovations.

The study of CAC managers confirms the many others that document the distinct use-patterns of the computer by men and women, and the gender-specific styles of use, learning, and operation. In the spirit of co-production analysis, it indicates how profoundly gender shapes the technology itself. The fact that the personal computer despite its long history is still today viewed as a "flexible technology" associated with neither men nor women is, we contend, due to the fact that men and women respectively have carved out overlapping but still separate realms of use and operating style that the CAP managers, hard at work in the trenches of the digital revolution, observe each day.

CAC Managers as Latourian Actors

Literature on the digital gender divide has focused on two, somewhat different problems of inclusion. One is the comparative extent to which women have adopted ICT for personal and job-related use. The other is women's inclusion in advanced computing study and professional roles in ICT. The CAP managers occupy a fascinating middle-ground. As a group, they lacked technical training and professional status, and so encountered the computer as a tool for personal use, much as their clients did. But as self-taught local-experts and trainers, they cultivated a job-related identity defined around ICT expertise and skills, much as ICT professionals do. The study probed questions of gender authenticity and self-concept that managers would evince in light of that expertise and their role as advisors and trainers. Those questions were central to analysis in terms of co-production. As Lagesen insists, new technologies produce uncertainties and ambiguity for actors, and it is in the resolution of those tensions, the stabilization of the resulting sociotechnical assemblages—if it occurs—that the co-production of technology and gender can be demonstrated symmetrically.

The empirical results richly demonstrated the uncertainties and ambivalence among the female CAP managers resulting from their expertise and interests. Not surprisingly, these jobs attracted women who had already evinced a standing interest in technology, especially electronic technologies, often from an early age. Nearly half (47%) had played with toy cars and electronic gadgets when they were children, a proportion transgressional of the traditional stereotypes of "girl's toys" and "boy's toys." "It's my nature to take things apart and find out how it works," one rural female manager noted (Leach, 2011, p. 221). Another commented on patterns of male socialization, noting that boys "like to take things apart . . . because it is a mechanical thing in the male. It comes back to . . . playing with cars with Dad, and looking under the hoods of vehicles." Asked whether she too had played with cars with her father, she replied with a laugh, "No, I used to take radios apart," adding "My mother was not impressed" (Leach, 2011, pp. 220-221).

The "chilly climate" narrative on women's recruitment into ICT, cited above, has focused heavily on the rapid emergence of the "geek" stereotype in the 1980s, as rendering professional and intellectual engagement with computers as gender inauthentic for women. The effect of the geek stereotype was sufficiently powerful to slow or drastically reverse recruitment patterns into computer science and ICT careers (Dunbar-Hester, 2008; Lagesen, 2007; Peddle, Powell, & Shade, 2008; Wajcman, 2004; Webster, 2010). Adopting computers for personal use and job-related use alone might not create gender ambiguities, but constructing a job-related identity around ICT skills might cross that dangerous boundary. The study sought to discover whether the managers feared that their unexpected new roles as community computer experts threatened their femininity, or social "normality" in the eyes of others. It asked the managers whether they thought others might consider them a "geek" in light of their employment and their skills. A substantial majority (77%) of the female CAC managers answered "no," while a smaller proportion of male managers (50%) did so. The result suggests that the geek stereotype may be a greater threat to men in the field than to women, or it may suggest that women mangers more willingly embrace elements of the stereotype (skill, commitment), at least within the protected setting of the CAP Centre. Many of the managers coupled their responses with an attempt to redefine the geek stereotype in ways that avoided or minimized its common, derogatory connotations. More than half the managers defined the "computer geek" for us mainly as someone possessing mastery, either technical or general in nature, over the technology; slightly less than half invoked the derogatory image of the geek as "obsessed" with the technology, as someone for whom "all they can talk about is computers" and "attached to a computer morning, noon, and night" (Leach, 2011, p. 153).

Interestingly, female managers were more prone to think of the "geek" as someone "obsessed" than were men. We conclude that some elements of the geek stereotype trouble the managers. For women, those elements did not suffice to render their roles as managers "gender inauthentic," but they did reflect an underlying tension between technical expertise and traditional notions of gender appropriateness.

For the female managers, therefore, their personal encounter with the computer as a technology offered career opportunity, fulfillment, and pleasure, combined with the threat of de-stabilizing traditional gender expectations and social conformity. Whatever ambivalence or tension that encounter may have entailed, it did not prevent the CAP managers from experiencing their work in the CACs as overwhelmingly positive. Women reported working in a technology environment had increased their self-confidence (83%), self-esteem (72%), ability to work with technology (94%), logical thinking (55%), and capacity to understand ICT (91%). Male managers also reported overall positive benefits, but to a much lesser degree than did women managers (10-15 percentage points lower on most key measures). The significantly greater impact of CAP jobs on women than on men reflects, we think, lack of other comparable opportunities for technology involvement for the women attracted to CAP positions, and their own success in creating a feminized environment uniquely nourishing to themselves and their clients—the subject of a later discussion.

The theme of acquiring computer technology as liberating and empowering was echoed in many of the interviews with female managers, some of whom spoke directly to their satisfaction in transgressing traditional gender expectations:

> I learned about computers and I [became] the go-to person. I'd do all the wiring for speakers like for surround sound . . . DVD player programming . . . You know the cable guy, the TV repair guy . . . Anything to do with electronics or to do with televisions or, uh, equipment hooking up speakers or anything like that . . . Now I do all that! . . . It's more of an ego boost because it's nice to have someone, especially in the male department, ask me [for help] . . . A man wants me to hook up his speakers! . . . And I have to say there are times when I do smirk. 'Cause it is such a, like a contrast. Because I was raised . . . like mother in the kitchen and father is the breadwinner. (Leach, 2011, p. 222)

Another sounded the same theme:

> Well, yes, I guess I think so. I do get people who approach [me] and they ask, "Can you come to my house and look at the mess I made [of my computer]." And, you know, you feel better when you are able to fix it for them. (Leach, 2011, p. 225)

It is important to the analysis that the position of CAP manager is not only one of technical expert and resource person but also one that involves teaching, helping, and social networking with the diverse, largely unskilled, and generally non-affluent clients who visit the Center. Many of the female

CAP managers sought a resolution of the tension between their status as technical experts and traditional gender expectations by emphasizing how the skills required in their positions proved to be inseparable from the "soft" attraction of helping others. One wrote that her most memorable experience of working at the CAC was "watching the face of a 78-year-old woman seeking a copy of her mother's picture scanned and come up so clear on the computer; an amazing gift" (Leach, 2011, p. 226). Another female manager, one who prided herself on her technical skills, expressed surprise at the unexpected satisfaction she derived from the softer side of her CAC job: "I have surprised myself at some point when it really became obvious to me that I love teaching. And, it's like, where did that come from?" (Leach, 2011, p. 226). For this manager, as for others, the CAP Centre proved to be a network site at which tensions arising from involvement with a new technology could be negotiated and resolved. Managers' personal envisioning of their relationship with digital technology both confirmed and simultaneously problematized the gender dualisms at work in the relationship.

The CAC Network as a Feminized Space

The categories of actor-network analysis serve well to explicate the role of New Brunswick's CACs. They functioned as a support network—the site for a web of associations and interrelationships that redefined and reconciled interests around the disruptive effects of a new technology. Clients from the community, older citizens, and people disadvantaged by lack of access to computers and frequently intimidated by the challenge computer use posed, found in the CAC personal assistance and local support for their first ventures with the new technology. CAC managers found outlets for technical interests and proclivities that previously, especially for the women managers, had had no satisfying and non-threatening ways of expression. The job-expectations of the manager's role integrated the "hard" requirement of technical skills with the "soft" requirements of caring, teaching, and nurturing, and in so doing reconciled self-image and social expectations for women managers in ways that few other work-spaces would have done. None of this was intended. The interests of the governments that initiated the program lay in rural modernization, workforce enhancement, and the development of social capital. The evolution of the CAP toward a network populated by women site-managers and women clients was an unplanned and unintended consequence. But like most classic Latourian networks, it nevertheless evolved to accommodate actors and interests very different from those involved in the original negotiation.

To this role, gender was essential. The success of the CACs lay in their realization of a "feminized space," where clients' encounter with the technology was explicitly and instinctively tailored by manager-instructors to recognize

and respond to gender-based needs. In that sense, the encounter with a new technology offered by a CAC was very different from that offered by any formal training course in a job-training center, community college, or university. One female manager, explaining why in her opinion women made the best CAC trainers, offered the following:

> Women are just a little more tolerant, patient, and understanding than men seem to be . . . And women are only now finding out about [technology] . . . so I think that women . . . when it comes to dealing with other people, you know, you can attract more bees . . . you know more flies with honey . . . Women are not as intimidating as men are. So it would be easier for women to come back because they would feel comfortable with another woman. And men would come back because men would feel that they weren't competing . . . It's bad enough that [men] have to swallow [their] pride and come in and have someone explain things. But to have to come in and have you know Mr. Sharpe . . . sit down beside you. (Leach, 2011, p. 245)

One young male manager agreed:

> [Women] provide more of a welcoming open environment. Maybe it's less threatening because women are not perceived in those technology savvy roles. (Leach, 2011, p. 244)

Most of the female managers interviewed admitted to using very different approaches to introducing male and female clients to the personal computer, and to doing so on ways that capitalized upon gender stereotypes in regard to technology attraction:

> When I [teach a] computer basics [course], first I'll have a computer that is torn apart so that they can actually see inside. Because people are under such misconceptions about what's inside the computer, and how it works. . . . The men that take the course, I can always catch their attention at that time. (Leach, 2011, p. 246)

Another said,

> I know when teaching the Internet to male and female clients that going to sites that would interest them is a boost. For example, for women I take them to the Sears outlet site and men I will ask what their interests are and often end up on hunting sites or car dealer sites. I notice in the body language that the client becomes more excited and engaged in what they are learning and will often then forget about their nervousness when preoccupied in exploring something that interests them. I also find with men if you use the words "if you want" or "if you would like to do this next" before getting them on to the next step, they listen better. Where with the women, there is no need to do that extra coaxing. Praise works well on both sexes. (Leach, 2011, pp. 248-249)

Not all female managers admitted to gender-based manipulation in the interests of better technology training. The results showed, in fact, that the more highly a manager self-rated her technical skills, the less likely she was to admit to gender-tailored teaching styles. "My teaching style changes with each student and depends solely on their need and expectations," wrote one such manager. "Is there a difference in style regarding gender of the student? I would not say there is a marked difference" (Leach, 2011, p. 250). But the insistence on training styles tailored to individual needs, gendered or not, was enough to confirm the status of the CAC as a feminized space.

Sørensen et al. (2011) have discussed the creation of "women-centered ICT spaces" as "inclusions strategies" (pp. 191-214). They conclude that initiatives such as women-only training programs, online networks for women ICT professionals, and online interactions for websites and magazines that specifically target women have been very successful. They are cautious, however, in their assessment of whether women-centered spaces can effectively influence the gender-coding of ICT. They associate women-centered spaces with essentialist narratives that presuppose women's essential skills and predilections for networking and communication. The CAP study, however, suggests that discussing "feminized spaces" may start from a gender-dualist perspective, but need not result in a dichotomous categorization. Feminized spaces (not to be confused with "women-only" spaces) are capable of promoting women's adoption of computers and influence the perceived gendering of ICT without confirming essentialist presuppositions.[7] The results do not suggest that women in CACs made good technology teachers because they, somehow more than men, tailored their teaching to gendered needs and expectations (though they may have done so). Rather, in the space of the CAC, clients experienced computer technology predominantly as mastered and mobilized by women. They encountered it as fully integrated into the interests, pleasures, abilities, and lives of women. As a consequence, they were able to see the technology itself, if not as "gendered feminine," or "gendered neutral," at least as a technology relevant and accessible to men and women alike, both as a source of personal fulfillment and career potential. CAP centers offered a site for gender mainstreaming with the potential to reduce if not eliminate the gender-loading of the technology.

Vivian Lagesen reached similar conclusions in 2007, in her critical analysis of strategies intended to enhance recruitment of women into university faculties of computer science and to ensure their self-identification with the field. Lagesen's analysis did not support the assumption that the computer science discipline had to change in a fundamental way to recruit female students. The key to success lay in increasing the number of female students already in the field and in increasing the number of female teachers, as well as efforts to make female students feel welcomed. When the number of women involved in the program increased, the symbolic image of the field changed from masculine to more neutral or gender diverse (Lagesen, 2007). Women entering the field did not report doing so to be "masculine," suggesting that the gender symbolism of computer science is dynamic. Lagesen

concludes optimistically that "gender symbolism in this field is much more dynamic and fluid than held in the literature on computers and masculinity" (pp. 87-88). The Canadian case study presented here confirms that optimism, and suggests that altering the gender symbolism surrounding computer technology may be easier in the "low-tech" and informal community context of the CAC network than in the formal, quasi-professional structures of a university faculty.

The Construction of Gender in a CAP Center

Lagesen claims that the greatest challenge of a case analysis of the co-construction of technology and gender lies not in demonstrating how existing gender expectations shape the reception, use, and social meaning of the technology, but in demonstrating how the technology shapes and re-shapes gender itself. The CAP study responds to that challenge by demonstrating that the encounter with the personal computer, as experienced by female clients and managers within the feminized space of the local CAP Center, enhanced empowerment, self-fulfillment, and the capacity to integrate technology with a personal, gendered life. We conclude by arguing for the technological shaping of gender in a stronger sense, by drawing on an unexpected result that emerged from surveys and interviews.

Human beings are attracted to technologies partly for their utilitarian benefit, that is, the many uses to which technologies can be put to further our particular life goals and interests. But most technologies, including ICT, also provide affective rewards to users in their own right—pleasure, fun, satisfaction, stimulation, compensation—beyond all considerations of utilitarian application. And in ICT, the capacity for pleasure has long been considered one of the most important markers of how men and women's engagement with the computer differ. Men, so the argument goes, find affective reward in the possession of "technical knowledge" and "technical mastery," while women's attraction to the technology is more exclusively mediated by "use-knowledge" and skill at practical application. This gendered difference is thought to be expressed most clearly in the enthusiasm men and boys demonstrate for computer gaming. Gaming is often claimed to combine enthusiasm and technical skills, and so to instill in boys what Gansmo (2011) has called "boys room competence." This competence carries over into personal adoption of ICT, comfort with ICT in the workplace, and choices about ICT careers. As such, it provides a partial explanation for the digital gender divide. Girls and women, on the other hand, treat computers as tools rather than toys; they value ICT for its ability to enhance other life goals rather than as a direct source of pleasure and reward. And while this more pragmatic and utilitarian embrace of the technology has often been considered "healthier," it is thought to be one of the exclusionary factors underlying the digital gender divide, in accordance with Sørensen et al.'s first explanatory narrative, which emphasizes the intrinsic link between technology and masculinity.

Helen Gansmo (2011) provides a trenchant review of the literature regarding these claims and its development. It is, she noted, founded on the venerable contrast between "hard and soft mastery" in programming and computer use (associated respectively with boys and girls) which Sherry Turkle (1984) developed in her classic work *The Second Self: Computers and the Human Spirit*. This contrast was deepened in the 1980s by Turkle and others, who claimed to analyze the affective roots of hacker culture and personality, which was regarded as largely alien to women (Gansmo, 2011). More recently, Gansmo argues, this alleged gender dichotomy has been challenged by more nuanced studies, showing that women may be motivated by fun and pleasure as much as men, although what women find pleasurable in ICT may be different. But newer studies have not displaced, in her judgment, the "gender dualised toy/tool dichotomy." And that dichotomy, and its alleged significance for the digital divide, has stimulated inclusionary initiatives to persuade girls that "having fun with computers" is not gender inauthentic. Those initiatives have included computer parties, participatory sites aimed at girls, commercial development of computer games designed to cater to the imagined interests and desires of girls, and a large literature on girls and gaming.

It was to be expected that CAC managers' attitudes toward the computer would have demonstrated those dualistic, stereotypical preferences: toy/tool, pleasure/use, technical-mastery/use-mastery. To some extent, they did. The survey and interviews inquired after the *uses* which managers and their clients made of personal computers. As explained above, managers reported a range of uses sharply delineated by gender and largely conforming to stereotypical expectations of the different uses to which men and women put personal computers in their daily lives. Then, to tap the affective dimension of computer use, the study also asked the managers cooperating with the survey to rate the level of *enjoyment* that they obtained from working with computer technologies. They were presented with the opportunity to explain what gave them enjoyment and why. In general, male managers reported higher levels of enjoyment: Three-fourths reported the highest level of enjoyment (5 on our 5-point scale), while only half of the female managers reported enjoyment at that level. Moreover, the male managers with few or no exceptions described the sources of the pleasure that they drew from computer use in terms readily associated with technical knowledge or hard mastery: technological problem solving, learning new things about computers and their operation, creation of new computer applications, fascination with technology in general, and discovering increasingly easy technical applications.

The female managers in the survey all reported high levels of enjoyment from working with computers (although not, as noted, levels as high as those reported by male managers). But their reports of the kinds of things that gave them pleasure differed from those of the male managers, and also differed from expectations, in several respects. First, almost half of the write-in responses offered by women managers were readily codeable into the areas stressing the pleasures of technical knowledge and hard mastery, including a general fascination with technology that had dominated the responses of male managers. "I like challenges and finding answers to problems," wrote one, "There's great satisfaction in solving problems on your own." "I think it is mostly the type of person I am," wrote another. "I love working with electronics and the like" (Leach, 2011, p. 174). The results of the survey problematized the dualistic and stereotyped expectation that technical knowledge and hard mastery would have no affective appeal to women. Of course, the work of a CAC manager may have produced a self-selection for the capacity to take pleasure in technical knowledge, but that self-selection would have acted upon managers of both genders.

Second, the female managers in the survey reported enjoyment that was associated with the skills they possessed in using the computer for particular applications and purposes. And they did so in significant numbers—about half of all the responses—whereas male managers almost never produced responses so codeable. Ease and speed of access to information, the pleasure of skill acquisition, skill in teaching others, and skill in communicating widely and quickly—these were among the codeable responses that female managers produced which the men of the sample did not. For the female managers, pleasure-in-use and pleasure from skill and knowledge in use blended together in ways that seemed to find no direct counterpart among the male managers in our study, even though both genders were prompted with the same questions and the same opportunities to respond. This outcome roughly confirms the stereotypical expectation that women are attracted to use-knowledge, with the added finding that pleasure is associated with the possession and exercise of use-knowledge.

Both these findings together, however, indicate that what we call the "spectrum of attraction" to computers (from hard to soft knowledge and from hard to soft mastery) is gendered, but it is gendered in ways that differ from, and that deeply problematize, the stereotypical expectations of widely accepted gender dualisms. Asked why and how they took pleasure in working with computers, the women of the study produced many more reasons than men, and they produced responses that ranged over the whole spectrum of attraction, whereas male responses were focused and limited at the hard-mastery end of that spectrum. On the basis of that evidence, we suggest that women's responses to computer technologies are more flexible, more multifaceted, and (at least at the level of affective response) more open than that of men are likely to be. This is novel. Traditionally biased expectation that gendered response to technology should be polarized around male and female norms leaves us unprepared for patterns of gendered response which do not reflect a simple polarization by gender, but which do not also suggest that gender is secondary or irrelevant. They invite deep re-thinking of what gender means in relation to technology.

Interviews with CAC managers of both genders confirmed that tentative survey finding. Male managers proved much more inclined to boast about their technical understanding of the machines they used daily than did female managers. Prompted to explain the pleasures they took in working with computers, they readily resorted to the language of hard mastery, including (on occasion) that of "conquering" the machine and being driven by the need to master it. "I've always been a tinkerer, you know," wrote one male manager,

I'll rip things apart to see how they work. . . . Learning about [computers] has definitely freed me up to not fear anything—any problem, it's a matter of time before it can be figured out. (Leach, 2011, p. 219)

"I've always been a tech junkie," another told us.

My wife banned me from the Future Shop, the whole nine yards. It's just I have always been someone who has followed technology and wants all the latest toys. . . . My first computer hooked up to the television set in the eighties, and my brother and I sat there taking it apart and writing programs so we could play games. . . . I have always looked at mechanical stuff like a clock you know, you've got the gears and the grinding and there's an actual physical aspect to it. (Leach, 2011, p. 219)

As expected, significant numbers of female managers also reported pleasure and satisfaction in the technical knowledge and hard mastery they possessed, but they did not do so as readily as the men. Drawn out by conversation, one manager admitted, "Yep, you give me the parts and I can build 'er!" Asked how she discovered her satisfaction in technology knowledge, she went on,

The first programming courses I took were probably in the '70s . . . and I loved that! But I have kind of a mind that likes that kind of thing. And then, really getting into it and I can remember [writing] a software program . . . based on a mathematical game. (Leach, 2011, p. 221)

Another commented on the pleasure she took from acquiring technical understanding: "I'm not intimidated at all [with electronics] now . . . I understand now more of two things work together . . . whereas before the stuff [in] there . . . made no sense to me" (Leach, 2011, p. 222).

As in the survey data, the female managers interviewed expressed a broader spectrum of attraction to computer use when queried them about what gave them satisfaction and pleasure. These included not only technical knowledge and hard mastery, but "soft" skills, such as facility in communicating and searching for knowledge over the Internet and their understanding of how to assist other users and learners.

The helping theme came up frequently during interviews with female managers. "Computers have made me who I am," commented one manager, "and I just love to help to show this to others, you know." Helping adults navigate the 'new frontier' and gain confidence with knowledge" (Leach, 2011, p. 225). Female managers described the pleasures they took from their interaction with the computer as lying along a broader spectrum of attraction than did male managers, a spectrum that ranged from the extremes of hard mastery to the "soft" pleasures of social uses and effective teaching.

Some of the interviewees seemed quite aware of the greater openness to computers suggested by this spectrum of responses and the attitudes behind it. One woman commented with asperity,

In my world, it is the women who have all the technology skills and it is the men who know very little, you know, about technology. They might know how to put [a computer system] together, but to actually use one for a purpose other than porn or games . . . You know, they just don't have a clue. So we [women] are more productive, we use it in more productive ways. (Leach, 2011, p. 229)

Many speculated that the emergence of information technology was undermining traditional stereotypes about men, women, and technology, opening up technology-related opportunities for women, and creating opportunities for women to apply their particular styles of learning and use. "Women seem to maybe be able to adapt easier to a non-linear style of experimentation [with computers]," wrote one male manager. "Women are a bit more open to exploring things as they go than a man" (Leach, 2011, p. 237). Observations of that kind reflect a tacit recognition of the differing spectrum of attraction evinced by men and women, a spectrum on which women may see the benefits of use-knowledge differently than men, and so report a distinctly different pattern of affective response. They also represent a subtle but very important shift in the subjectivity of gender itself, here a product of technological change, and another episode in the ongoing co-construction of gender and technology that Vivian Lagesen has challenged observers to document.

Conclusion

The empirical study discussed in this article has focused on the managers of CACs created in the Canadian province of New Brunswick in response to the problem of the so-called "digital divide"—initially understood as a socioeconomic and rural/urban divide. We have addressed the "other" digital divide—the gender divide—and tried to demonstrate the co-production of technology and gender from the experiences and beliefs of the CAC managers, as well as the efficacy of actor-network perspectives in understanding their encounter with personal computers within the CAC network. Particularly for the female managers, the encounter was seen to be an ambivalent one, experienced simultaneously as rewarding and empowering on the one hand and threatening to traditional gender expectations on the other. That ambiguity was resolved through the active creation of a feminized space within the CAP network, which in turn facilitated actors' reconceptualizing the way in which computers, as a technology, are thought of as gendered objects and systems. The context in which CAC clients encounter the computer is one in which the technology and its human agents that use, teach, and communicate its social meaning, blur together into a state of mutual interdependence.

The case study also addresses the co-construction of technology and gender. For managers, the social meaning of the computer is actively shaped by gender-based attitudes and use-preferences, and we have documented these dualisms at work in the attitudes and performances of the participants. But the other side of the equation has also been addressed. By teasing apart the experiences of use and the experiences of pleasure in a new technology, the study has moved toward a less dichotomous characterization of the findings and opened interpretive space beyond the dualisms. The results have suggested that women's gendered relationship to technology is different from what has been taken for granted, and is being reshaped by the encounter with ICT. The analysis therefore satisfies Lagesen's call for symmetrical accounts of the co-production process, and may provide useful perspectives on inclusionary projects for women in ICT.

Declaration of Conflicting Interests

The author(s) declared no potential conflicts of interest with respect to the research, authorship, and/or publication of this article.

Funding

The author(s) received no financial support for the research, authorship, and/or publication of this article.

Notes

1. This theoretical perspective is compatible with the feminist post-humanism advocated by Karen Barad, although her "agential realism" carries skepticism about the technology–society dualism further than do traditional formulations of co-production (Barad, 2003).
2. The Connecting Canadians program included other programs such as SchoolNet, VolNet, LibraryNet, and Smart Communities programs. "Altogether, several hundred million dollars were spent through these programs in support of roughly 10,000 community-based ICT initiatives ranging from community web portals, public Internet access sites, and community technology centers to computing hardware for schools and network infrastructure for rural and remote communities" (Clement et al., 2004, p. 10).
3. That said, according to Shade (2006), Canada was one of the few countries to consider gender equity in public policy deliberations including concern specifically for gender and information

technology. Early Canadian policy formulations on the promotion of universal access to computer technology and the Internet recommended that initiatives consider gender as an important category to include in universal access definitions (Shade, 2006). Unfortunately, gender mainstreaming of Canadian access-to-internet programs never did materialize (Shade, 2006). The Federal government shifted the focus to technical broadband issues and an e-commerce strategy to address the growing trend for a "knowledge-based" economy, and gender considerations became subsumed by Industry Canada's Connecting Canadians agenda and other international projects (Rideout, 2003).
4. In this respect, as in others, the New Brunswick Community Access Centers (CACs) program resembled Scotland's Ardmore Network, studied by Tine Kleif and Faulkner (2003), as one of 48 European case studies constituting the recent SIGIS project.
5. Given the pay of $11/hr, it would be unrealistic to attempt to attain advanced technically trained employees.
6. We sometimes refer to managers and clients as "male" or "female" to distinguish "gender." For an excellent review of the non-consensus use of the terms *gender* and *sex* in academic literature, see Muehlenhard and Peterson (2011). They conclude that distinguishing between sex and gender was a valuable contribution at one time. For feminists especially, it provided a way to reject biological determinism that linked biology with rigid sex roles and expectations and, more recently, provided a way to understand trans-sexualism where the biological sex of an individual did not match their gender identity. However, they predict that "as researchers learn more, the distinction between sex and gender may become less important or meaningful" (Muehlenhard & Peterson, 2011, p. 801).
7. We recognize that male managers can be actors in the "feminized space" as well, and that men can have varying arrays of "feminine" attributes. This thinking takes "gender" beyond "sex" and "feminine" qualities into the realm of technology work.

References

Barad, K. (2003). Posthumanist performativity: Toward an understanding how matter comes to matter. *Signs: Journal of Women in Culture and Society*, *28*, 801-831.

Berg, A.-J. (1994). Technological flexibility: Bringing gender into technology (Or was it the other way around? In C. Cockburn & R. Fürst-Dilic (Eds.), *Bringing technology home: Gender and technology in a changing Europe* (pp. 26-52). Philadelphia, PA: Open University Press.

Berg, A.-J. (1997). *Digital feminism* (Rapport Nr. 28). Dragvoll: Senter for Teknologi of Samfum, Norwegian University of Science and Technology.

Berg, A.-J., & Lie, M. (1995). Feminism and constructivism: Do artifacts have gender? *Science, Technology & Human Values*, *20*, 332-351.

Bernhardt, S. (2014). *Women in IT in the new social era: A critical evidence-based review of gender inequality and the potential for change*. Hershey, PA: Business Science References.

Broos, A. (2005). Gender and information and communications technologies (ICT): Anxiety, male self-assurance and female hesitation. *Cyberpsychology & Behavior*, *8*, 21-31.

Broos, A., & Roe, K. (2005). Marginality in the information age: Is the gender gap really diminishing? *Communications*, *30*, 251-260.

Clement, A., Gurstein, M., Longford, G., Luke, R., Moll, M., Shade, L. R., & DeChief, D. (2004). The Canadian Research Alliance for Community Innovation and Networking (CRACIN): A research partnership and agenda for community networking in Canada. *The Journal of Community Informatics*, *1*(1). Retrieved from http://www.ci-journal.net/index.php/ciej/article/viewArticle/207

Cockburn, C. (1983). *Brothers: Male dominance and technological change*. London, England: SAGE.

Cockburn, C. (1985a). Caught in the wheels: The high cost of being a female cog in the male machinery of engineering. In D. MacKenzie & J. Wajcman (Eds.), *The social shaping of technology* (pp. 126-133). Milton Keynes, UK: Open University Press.

Cockburn, C. (1985b). *Machinery of dominance: Men, women and technical know-how*. London, England: Pluto.

Cockburn, C. (1992). The circuit of technology: Gender, identity and power. In R. Silverstone & E. Hirsch (Eds.), *Consuming technologies: Media and information in domestic spaces* (pp. 18-25). London, England: Routledge.

Cockburn, C., & Fürst-Dilic, R. (1994). *Bringing technology home: Gender and technology in a changing Europe*. Philadelphia, PA: Open University Press.

Cockburn, C., & Ormrod, S. (1993). *Gender and technology in the making*. London, England: SAGE.

Connect NB Branché database. (1992-2012). *Department of post-secondary, education, training and labour, province of New Brunswick* (Unpublished report). New Brunswick, Canada: Government of New Brunswick.

Connect NB Branché Indicator Survey. (2008). Department of post-secondary, education, training and labour, province of New Brunswick (Unpublished report). New Brunswick, Canada: Government of New Brunswick.

Dickinson, P., & Sciadas, G. (1996, December). *Access to the information highway* (Canadian Economic Observer, Catalogue No. 11-010). Ottawa, Ontario: Statistics Canada.

Dickinson, P., & Sciadas, G. (1997, June). *Access to the information highway: The sequel* (Services Indicators, Catalogue No. 63-016). Ottawa, Ontario: Statistics Canada.

Dickinson, P., & Sciadas, G. (1999, February). *Canadians connected* (Canadian Economic Observer, Catalogue No. 11-010). Ottawa, Ontario: Statistics Canada.

Drabowicz, T. (2014). Gender and digital usage inequality among adolescents: A comparative study of 39 countries. *Computers & Education*, *74*, 98-111.

Dunbar-Hester, C. (2008). Geeks, meta-geeks, and gender trouble. *Social Studies of Science*, *38*, 201-232.

Faulkner, W. (2000). The power and the pleasure? A research agenda for 'making gender stick' to engineers. *Science, Technology & Human Values*, *25*, 87-119.

Faulkner, W. (2001). The technology question in feminism: A view from feminist technology studies. *Women's Studies International Forum*, *24*, 79-95.

Faulkner, W. (2004). *Strategies of inclusion: Gender and the information society (SIGIS)*. Edinburgh, Scotland: University of Edinburgh, European Commission IST Programme.

Faulkner, W. (2007). Nuts and bolts and people: Gender-troubled engineering identities. *Social Studies of Science*, *37*, 331-356.

Gansmo, H. J. (2011). Fun and play in digital inclusion. In K. Sørensen, W. Faulkner, & E. Rommes (Eds.), *Technologies of inclusion: Gender in the information society* (pp. 109-113). Trondheim, Norway: Tapir Academic Press.

Hacker, S. (1981). The culture of engineering: Women, workplace and machine. *Women's Studies International Quarterly*, *4*, 341-353.

Hacker, S. (1989). *Pleasure, power and technology*. Boston, MA: Unwin Hyman.

Hacker, S. (1990). *Doing it the hard way: Investigations of gender and technology*. Boston, MA: Unwin Hyman.

Information Highway Advisory Council. (1997). *Preparing Canada for a digital world*. Ottawa, Ontario, Canada: Author.

Jasanoff, S. (Ed.). (2004). *States of knowledge: The co-production of science and the social order*. New York, NY: Routledge.

Kleif, T., & Faulkner, W. (2003). Rural community resource centres: A case of "de facto" women's inclusion in the information society. In M. Lie & K. H. Sørensen (Eds.), *Strategies of inclusion: Gender the information society. Vol. I Experiences from public sector initiatives* (Report 2003-63) (pp. 237-272). Trondheim, Norway: Norwegian University of Science and Technology, Centre for Technology and Society.

Lagesen, V. A. (2007). The strength of numbers: Strategies to include women into computer science. *Social Studies of Science*, *37*, 67-92.

Lagesen, V. A. (2012). Reassembling gender: Actor-Network Theory (ANT) and the making of the technology in gender. *Social Studies of Science*, *42*, 442-448.

Landström, C. (2007). Queering feminist technology studies. *Feminist Theory*, *8*, 7-26.

Latour, B. (1999). On recalling ANT. In L. Law & J. Hassard (Eds.), *Actor network theory and after* (pp. 15-25). Oxford, UK: Blackwell.

Latour, B. (2005). *Reassembling the social: An introduction to the actor-network-theory*. Oxford, UK: Clarendon Press.

Leach. (2011). *Computer users do gender: The co-production of gender and communications technology* (PhD thesis). University of New Brunswick, Canada.

Lie, M. (1995). Technology and masculinity: The case of the computer. *The European Journal of Women's Studies*, *2*, 379-394.

Lie, M., & Sørensen, M. H. (Eds.). (2003). *Strategies of inclusion: Gender the information society. Vol. I Experiences from public sector initiatives* (Report 2003-63). Trondheim, Norway: Norwegian University of Science and Technology, Centre for Technology and Society.

Lohan, M. (2001). Men, masculinities and "mundane" technologies: The domestic telephone. In E. Green & A. Adams. (Eds.), *Virtual gender: Technology, consumption and identity* (pp. 149-162). New York, NY: Routledge.

MacKeogh, C., & Preston, P. (Eds.). (2003). *Strategies of inclusion: Gender in the information society. Vol. II Experiences from private and voluntary sector initiatives* (Report 2003-65). Trondheim, Norway: Norwegian University of Science and Technology, Centre for Technology and Society.

Martin, M. (1991). *Hello central? Gender, technology, and culture in the formation of telephone systems*. Montreal, Quebec, Canada: McGill-Queen's University Press.

Mellström, U. (2009). The intersection of gender, race and cultural boundaries, or why is computer science in Malaysia dominated by women? *Social Studies of Science*, *39*, 885-907.

Misa, T. J. (Ed.). (2010). *Gender codes: Why women are leaving computing* [IEEE Computer Society]. Hoboken, NJ: John Wiley.

Moll, M., & Fritz, M. (2007). *Community networks and Canadian public policy: Preliminary report on the CRACIN survey of community networks.* Toronto, Canada: Canadian Research Alliance for Community Innovation and Networking.

Moll, M., & Shade, L. (2001). *E-Commerce vs E-Commons: Communications in the public interest.* Ottawa, Ontario: The Canadian Centre for Policy Alternatives.

Moll, M., & Shade, L. (Eds.). (2004). *Seeking convergence in policy and practice: Communications in the public's interest* (Vol. 2). Ottawa, Ontario: Canadian Centre for Policy Alternatives.

Muehlenhard, C. L., & Peterson, C. D. (2011). Distinguishing between sex and gender: History, current conceptualizations, and implications. *Sex Roles, 64,* 791-803.

Nentwich, J. C., & Kelan, E. K. (2014). Toward a topology of "doing gender": An analysis of empirical research and its challenges. *Gender, Work, & Organization, 21,* 121-134.

Oudshoorn, N., Rommes, E., & Stienstra, M. (2004). Configuring the user as everybody: Gender and design cultures in information and communication technologies. *Science and Human Values, 29*(1), 30-63.

Oudshoorn, N., Rommes, E., & van Slooten, I. (Eds.). (2004). *Strategies of inclusion: Gender in the information society. Vol. 3: Surveys of women's user experience* (Report 2004-66). Trondheim, Norway: Norwegian University of Science and Technology, Centre for Technology and Society.

Peddle, K., Powell, A., & Shade, L. (2008). Bringing feminist perspectives into community informatics. *Atlantis, 32*(2), 33-44.

Quinlan, A. (2012). Imagining a feminist actor-network theory. *International Journal of Actor-Network Theory and Technological Innovation, 4*(2), 1-9.

Rideout, V. (2000). Public access to the internet and the Canadian digital divide. *The Canadian Journal of Information and Library Science, 25,* 2-3.

Rideout, V. (2001). *Bridging the digital divide in Atlantic Canada communities.* Ontario, torrent: Human Resources Development Canada Office of Learning Technologies.

Rideout, V. (2003). Canadians connected and unplugged: Public access to the internet and the digital divide. In M. McCauley, E. Peterson, B. Artz, & D. Halleck (Eds.), *Public broadcasting and the public interest* (pp. 192-203). Waltham, MA: Focal Press.

Shade, L. R. (2006, February). *Stirring up the pot? Integrating gender into OCT policy, practice, and evaluation* (CRACIN Working paper No. 13). Available from www.cracin.ca

Sørensen, K. H., Faulkner, W., & Rommes, E. (2011). *Technologies of inclusion: Gender in the information society.* Trondheim, Norway: Tapir Academic Press.

Statistics Canada. (2014). *Summary table. Internet use by individuals in Canada.* Retrieved from www.statcan.gc.ca/tables-tableaux/sum-som/L01/cst01/comm35a-eng.htm

Turkle, S. (1984). *The second self: Computers and the human spirit.* New York, NY: Simon & Schuster.

van Zoonen, L. (1992). Feminist theory and information technology. *Media, Culture & Society, 14,* 9-29.

van Zoonen, L. (2002). Gendering the internet: Claims, controversies and cultures. *European Journal of Communications, 17,* 5-23.

Wajcman, J. (1991). *Feminism confronts technology.* Cambridge, UK: Polity Press.

Wajcman, J. (1995). Feminist theories of technology. In S. Jasanoff, G. E. Markle, J. C. Petersen, & T. Pinch (Eds.), *Handbook of science and technology studies* (pp. 189-204). Thousand Oaks, CA: SAGE.

Wajcman, J. (2000). Reflections on gender and technology studies: In what state is the art? *Social Studies of Science, 30,* 447-464.

Wajcman, J. (2004). Addressing technological change: The challenge to social theory. *Current Sociology, 50,* 347-363.

Wajcman, J. (2004). *TechnoFeminism.* Cambridge, UK: Polity Press.

Wajcman, J. (2006). The feminization of work in the information age. In M. Fox, D. Johnson, & S. Rosser (Eds.), *Women, gender, and technology* (pp. 80-97). Chicago: University of Illinois Press.

Wajcman, J. (2010). Feminist theories of technology. *Cambridge Journal of Economics, 34,* 143-153.

Wajcman, J. & MacKenzie, D. (Eds.). (1999). *The social shaping of technology* (2nd ed.). Buckingham, UK: Open University Press.

Webster, J. (2010). [Review of the book *Gender codes: Why women are leaving computing,* by T. J. Misa].

Author Biographies

Lori Leach is a researcher with the Provincial Government in the province of New Brunswick, Canada, and, an instructor at the University of New Brunswick.

Steven Turner is retired from the University of New Brunswick as professor in the history and social studies of science.

17

Comparative Study of the Availability and Use of Information Technology in the Subject of Education in Public and Private Universities of Islamabad and Rawalpindi

Saima Yasmeen[1], Muhammad Tayyab Alam[1],
Muhammad Mushtaq[1], and Maqsud Alam Bukhari[1]

Abstract

The study was designed to compare availability and use of information technology in the subject of education in public and private universities of Islamabad and Rawalpindi. The objectives of the study were, first, to highlight the status of information technology in public and private universities of Islamabad and Rawalpindi; second, to compare the availability and utilization of resources for information technology in public and private universities of Islamabad and Rawalpindi; third, to find out the problems in the use of information technology in public and private universities of Islamabad and Rawalpindi and to recommend strategies for better use of information technology in public and private universities of Islamabad and Rawalpindi. The design of the study was descriptive, and it was a survey study. Two questionnaires were used for data collection: one for teachers and one for students. A stratified random sampling technique was used. Two groups of teachers were selected from public and private universities, and 50% teacher and 10% students were selected from the population. The data were analyzed in terms of percentage, and t test was also applied. A significant difference was found between the availability and usage of equipment in education departments of public and private universities of Islamabad and Rawalpindi. Difference between the students learning and teachers training skills was not significant.

Keywords

information, communication technology, education, Pakistani universities

Introduction

North Dakota Century Code (NDCC) 54.59.01 defines information and communication technology (ICT) as the use of both hardware and software services by persons or groups to perform various functions. It is not a product but a process to support infrastructure and manage it for delivering information by using voice, visuals, and video. The use of technology at large scale is changing the pattern of life that includes working, learning, communicating with each other, and other daily activities.

In all types of education, the gadgets of technology had long-lasting impacts on the process of teaching and learning. This dual-way process can be speeded and supported by using information technology for both faculty and students. The easy access to new dimensions of technologies have placed the institutions and individuals on an advantaged place to follow rapid changes

Udvari-solner and Thousand (1996) described that, in the past, technology was in the hand of "few alerts," the lone computer lab teacher, or the specialists who knew how to program the "mysterious" augmentative communication device used by students with communication limitation. In the new vision of university education in Pakistan, information technology has become an integral part, because this century belongs to a high tech era where both teachers and students need to be ready for new changes and challenges. This era has special needs and requirements of particular level of knowledge and skills as compared with the previous industrial era. There are different and changing learning environments that require special kinds of flexibility to be more effective. Media, particularly electronic media, has more strength and scope to develop conceptual understanding and analytical

[1]Foundation University, Islamabad, Rawalpindi, Pakistan

Corresponding Author:
Saima Yasmeen, Department of Education, Foundation University Islamabad Pakistan, Rawalpindi, Pakistan.
Email: saimayasmeen134@yahoo.com

ability among teachers and students, and it is imperative for teachers and students to use this medium for making teaching–learning more result oriented. In future, teacher's success in the class would be judged through media, their competency, and performance by using modern gadgets of technology. Hence, there is an urgent need for incorporation of advanced media in the teaching–learning process. Its availability and skills to use it would enhance teachers teaching skills and quality.

There is no doubt that technology-enhanced education is more effective as proven by many researches; computer-based chat rooms, different level course-based websites, and emails are some of the glimpses of technology-enabled resources. This type of learning develops interaction, team work, collaborative learning, and problem-solving skills among learners and teachers, apart from easing the tasks of collaborative designs, peer writing groups, and geographical barriers. It has removed distances as well as cultural barriers from one continent to another continent. There is a big challenge for universities and school administrations to train faculty and staff for maximum utilization of technology that seems still unaccomplished. It is obvious that availability of physical facilities of technology is not adequate because of the need for development of human resources to accomplish the desired and demanded fruits of technology.

There is no doubt that technology in the classroom has a large number of benefits but is not free of drawbacks. These disadvantages are related to the training of teachers and other staff, access to technology gadgets, and time required to implement its proper use. The above mentioned are a few of the reasons that prevent extensive use of technology in classrooms. Training is vital whenever we take up a new endeavor, and this is so with technology in education. The educationists do not consider it an end goal but a means to accomplish the goals. The expertise and skills in its use will enhance and improve the teaching–learning process; otherwise, it can be a hindrance and obstacle in achieving goals.

The inappropriate allocation of resources for classroom like fixing up computers or providing laptops in classroom may create fuss for teaching–learning process. Technology is changing rapidly, and hence, meeting new challenges and innovation is costly and may not be feasible for institutions to accommodate themselves with these resources. Quality material is available, but finding the appropriate one with relevant subjects is difficult, and hence, it is imperative to design and develop course contents as per requirement of the learner and teacher. This study was helpful in finding problems in the use of information technology in the education departments of public and private universities and gave a long-term strategy for better use of information technology.

Statement of the Problem

The study was designed to compare the availability and the use of information technology in the subject of education in public and private universities of Islamabad and Rawalpindi

Objectives of the Study

The objectives of the study were

1. to highlight the status of information technology in public and private universities of Islamabad and Rawalpindi,
2. to compare the availability and utilization of resources for information technology in these universities, and
3. to find the problems in the use of information technology in these universities.

Research Questions of the Study

Research Question 1: Is there any difference in availability of equipments among public and private universities?
Research Question 2: Is there a difference in use of ICT by public and private university students?
Research Question 3: Is there any difference in learning among public and private universities?
Research Question 4: Is there any difference in teachers training and skills of ICT among public and private universities?
Research Question 5: Is there any difference in use of ICT by teachers of public and private universities?
Research Question 6: Is there any difference in teachers learning and skills among public and private universities?
Research Question 7: What are the problems in the use of of information technology?

Education

The United Nations Educational, Scientific and Cultural Organization (Leu & Price-Rom, 2006) defined education as "an organized and sustained instruction designed to communicate a combination of knowledge, skills and understanding valuable for all the activities." Education is transformation of knowledge skills and values that are required to make citizens well adjusted in the society.

Information Technology

The information technology foundation defined it as "the technology used for the study, understanding, planning, designing, testing, distributing, supporting and operating of software, computers and computer related systems that exist for the purpose of data, information and knowledge processing" (Gorgone, Davis, Valacich, Topi, Feinstein, & Longenecker, 2003, p. 1).

Basic Computer Training Skills

Learners, particularly those knowing fundamentals of computers, feel comfortable using computers in the classrooms.

But the interaction with computers can make these newly inducted learners' concepts about technology vocabulary, including different operating systems, and their file and folder management techniques, more clear. By exploring and interacting with technology gadgets, one can also understand the functioning of the following heads:

- Folders and files organization
- Drag and drop skills
- Keyboard use
- Recycle bin
- Using different browsers
- Developing and saving different types of files and folders

Effects of Educational Technology

Krotoski (2004) found "Educational technology has a significant positive impact on achievement in all subject areas, across all levels of education, and in regular classrooms as well as those for special-needs students."

Information Superhighway

Sacramento School District in California (1987-1992) reported that "students using multimedia and telecommunications showed improved attitudes toward reading, social studies, and science, and became more active and independent in learning. Some also showed improved reading scores" (Sivin-Kachala & Bialo, 1994). A survey of 550 teachers who use telecommunications technology in the classroom reported that

Inquiry-based analytical skills—like critical thinking, data analysis, problem solving, and independent thinking—develop when students use a technology that supports research, communication, and analysis. However, telecommunication does not directly help their performances on state- or city- mandated tests.

Technology's Impact on Learning

There is a strong relationship between technology-based instruction and academic test scores of the students, which advocated that the institutions need to focus on development of technology skills to enhance their learning and improve sources of open access learning to become successful in this world.

If we look at past researches, they showed mixed results as many researchers such as Angrist and Lavy (2002), Goolsbee and Guryan (2006), and Banerjee, Cole, Duflo, and Linden (2004) found that there were no evidences of key roles of information and communication technology in higher education. They also supported the traditional teacher-oriented pattern of teaching and learning. Whereas, Kulik (1994); Sosin, Blecha, Agawal, Bartlett, and Daniel (2004);

Fuchs and Woessmann (2004); and Coates et al. (2004) believed that there is a real impact of ICT on students learning at this level. They are strongly in favor of using ICT in education at higher level.

Hargreaves (2003) advocated that,

A knowledge society has three dimension; first expanded scientific, technical and educational sphere, second complex ways of processing and circulating knowledge and information in service based economy and third entailing basic changes for continuous innovation in products and services by developing systems, teams and cultures that promote mutual benefit and learning. Education becomes essential to answer the needs of technology and society. (p. 11)

Selwyn (2002) called technology as techno-romance as people are trying hard to avail and adopt technology in educational setting. Bennett and Bennett (2003) concluded that willingness to integrate ICT in education by the faculty members was not found, so it was not due to absence of technological infrastructure but less motivation on the side of the faculty in using ICT. The areas that have to be focused are motivation and satisfaction of the faculty to adopt and use ICT in their daily work.

Medlin (2001) analyzed that personal motivation is mandatory to incorporate ICT in teaching–learning process. The authorities can be successful when they motivate students and teachers to implement technology-based teaching–learning processes. Ma, Andersson, and Streith (2005) concluded that "teachers' perceived usefulness of computer technology had a direct effect on their intention to use with ease of use" (p. 388). If motivation and facilities of training are provided to faculty, they will easily follow technology gadgets.

Research Method

This article was descriptive, and it followed the below-mentioned procedure. The population of the study consisted of all the teachers and students of education departments of public and private universities of Islamabad and Rawalpindi.

Sample

Two groups were randomly selected as samples for the study; one of the groups consists of teachers teaching at university level, and the other group was selected among students of graduate level in education departments of public and private universities offering education programs. The researcher selected six universities from Islamabad and Rawalpindi cities with three universities from the public sector and three from the private sector offering education programs, and 50% teachers and 10% students were selected from each university.

Two questionnaires (one for teachers and the other for students) were developed on the basis of literature review.

Table 1. Sample of the Study (Teacher).

Public universities	No. of teachers	Private universities	No. of teachers
Islamic International University, Islamabad	19	Foundation University, Islamabad	05
NUML, Islamabad	23	WISH, Islamabad	04
ARID Agriculture University, Rawalpindi	O6	Bilquis College of Education	15
Total no.	48	Total no.	24
Selected sample	50% (24) selected	Selected sample	50% (12) selected

Table 2. Sample of the Study (Students).

Public universities	No. of students	Selected sample	Private universities	No. of students	Selected sample
Islamic International University, Islamabad	300	10% (30)	Foundation University, Islamabad	80	10% (08)
NUML, Islamabad	200	10% (20)	WISH, Islamabad	50	10% (05)
ARID Agriculture University, Rawalpindi	160	10% (16)	Bilquis College of Education	600	10% (60)
Total no.	660	66	Total no.	730	73

Questionnaire for Teachers

It consisted of 42 items on a 2-point scale out of which the researcher has selected only six items for this article. The university teachers of education departments were selected as a sample, and they were given the following questionnaire. The first two items were used for analysis in this article.

Teacher's questionnaire had the following breakup:

- Availability of equipments, 11
- Usage of information technology, 16
- Students learning, 08
- Teacher training and skills, 07

Student Questionnaire

The questionnaire for students consisted of 25 items keeping the same above-mentioned heads. The questionnaire was distributed among teachers respondents to assess their opinion about ICT access, use, and availability.

Validation of the Research Instrument

The researcher conducted pilot testing for validity of the instruments. Ten teachers and 20 students were selected for pilot testing; as a result, the changes in the format and language of the questionnaires were made. The suggestions given by teachers were incorporated in the final version of the questionnaires.

In teacher questionnaire, question numbers 9, 10, and 22 were modified according to the suggestions of teachers. Data were collected in quantitative form by questionnaire which

was administered by the researcher personally. It was tabulated, analyzed, and discussed category wise. For statistical treatment, t test was applied at .05 level of significance. If it is below .05 level, it is not significant. The last item was analyzed by using simple percentages of the respondents. The data have been presented item wise. As mentioned in the research questions, t test was used to analyze the data, which were presented in the form of tables.

Table 3 presented the calculated t value as 1.748, which is more than 1.68 at table value at $df(35)$. It indicates a significant difference between availability of equipment between public and private universities.

Table 4 pointed out a calculated t value (0.748), which is less than 1.68 at table value at $df(35)$. It displayed no significant difference between the usage of information technology between public and private universities.

Table 5 founded a calculated t value (0.547), which is less than 1.68 at table value at $df(35)$. It suggested no significant difference between the student learning between public and private universities

Table 6 indicated calculated t value (0.547), which is less than 1.68 at table value at $df(35)$. It means no significant difference between the teachers training and skills between public and private universities.

Table 7 found calculated t value (4.581), which is more than 1.96 at table value at $df(138)$. It indicated a significant difference between the usage of information technology between public and private universities.

Table 8 pointed out calculated t value (1.506), which is less than 1.96 at table value at $df(138)$. It can be interpreted as no significant difference between the teachers learning and skills between public and private universities.

Table 3. Availability of Equipment.

Category	N	M	SD	t	p
Public	24	1.5758	0.04378	1.748*	.008
Private	12	1.8042	0.10396		

Note. df = 35.
*p < .05.

Table 4. t Test: Usage of Information Technology.

Category	N	M	SD	t	p
Public	24	1.6562	0.15639	0.748*	.460
Private	12	1.6971	0.16307		

Note. df = 35.
*p < .05.

Table 5. t Test: Student Learning.

Category	N	M	SD	t	p
Public	24	1.7812	0.16587	0.547*	.588
Private	12	1.7404	0.29075		

Note. df = 35.
*p < .05.

Table 6. t Test: Teacher Training and Skills.

Category	N	M	SD	t	p
Public	24	1.6488	0.17862	0.547*	.008
Private	12	1.7216	0.14913		

Note. df = 35.
*p < .05.

Table 7. t Test: Usage of Information Technology.

Category	N	M	SD	t	p
Public	66	1.5974	0.17270	4.581	.008
Private	73	1.7219	0.14868		

Note. df = 138.

Table 9 shows percentage of frequencies of the problems mentioned in the questionnaire by both teachers and students. Seventy-one percent of respondents stated that their university had no updated computers. The high speed Internet connection is another problem as 67% marked they did not have high speed Internet connection. The next was enough time to use computer lab: Majority of them said they had no time for the lab due to one or the other reason. The item numbers 4 and 5 were responded to negatively as 82% respondents said they did not have video conferencing and interactive white board facilities in the university. The last

Table 8. t Test: Students Learning and Skills.

Category	N	M	SD	t	p
Public	66	1.8019	0.15929	1.506	.134
Private	73	1.7598	0.17073		

Note. df = 138.

item was about multimedia availability in the class as 59% said they did not have it.

Findings

1. Table 3 is about the data on availability of equipment in public and private universities. Calculated t value in the data is 1.748, which is more than table value (1.68) at .05 level of significance, showing a significant difference in the availability of equipment in public and private universities. It is evident that the private universities are better equipped with respect to the availability of equipment.
2. Table 4 is about the data on use of information technology in public and private universities. Calculated t value in the data is 0.748, which is less than table value (1.68) at .05 level of significance. It means that there is no significant difference in the use of information technology in public and private universities. It is evident that both students of public and private universities equally use information technology.
3. Table 5 is about the data on learning of students by using information technology in public and private universities. Calculated t value in the data is 0.547, which is less than table value 1.68 at .05 level of significance. It means that there is no significant difference in the learning of students by using information technology in public and private universities. It is evident that both public and private university teachers need information technology for better students' learning.
4. Table 6 is about the data on teacher training and skills in public and private universities. Calculated t value in the data is 0.547, which is less than table value (1.68) at .05 level of significance. It means that there is no significant difference in the teachers training and skills in public and private universities. It is evident that both public and private universities need teacher training and skills.
5. Table 7 is about the data on use of information technology in public and private universities by teachers. Calculated t value in the data is 4.58, which is more than the table value (1.96) at .05 level of significance, which means that there is significant difference in the use of information technology by teachers in public and private universities. It is evident that private

Table 9. Problems in Use of ICT Facilities.

Items about problems in use of ICT facilities	Yes (%)	No (%)	Do not know (%)
1. Updated computers	27	71	2
2. High speed Internet connection	23	67	10
3. Enough time for lab use	32	62	3
4. Video conferencing	12	82	6
5. Interactive Boards	9	82	9
6. Multimedia in each class	38	59	3

Note. ITC = information and communication technology.

university teachers are better learned with respect to the use of information technology.

6. Table 8 is about the data on teachers learning and skills in public and private universities. Calculated *t* value in the data is 1.506, which is less than table value (1.96) at .05 level of significance. It means that there is no significant difference in teacher learning and skills in public and private universities. It is evident that teacher learning and skills in public and private universities are equal with respect to information technology.

7. Table 9 found that the students had many problems related to availability of information technology as no updated computers are available in the computer labs, no high speed Internet connection was available, they do not have enough time to use labs, no video conferencing facility, no interactive board, and no multimedia available in the class. Although the percentage varies, it is crystal clear that they had problems availing these facilities.

Conclusion

The data analysis, interpretation, and findings of the study can be concluded in the following lines.

- It was concluded that there was a significant difference between the availability of equipments in education departments of public and private universities of Islamabad and Rawalpindi and that the availability of equipment in education departments of private universities are better than public universities

- It was revealed through analysis of the data that there was no significant difference between the usage of information technology among the students of education departments in public and private universities.

- The university teachers are not provided training in the university. As it is observed in Table 6, there is no significant difference in the teacher training and skills in public and private universities. It is evident that both public and private universities need teacher training and skills.

- Teachers of education departments in public and private universities have the ability to use computer, Internet, multimedia, and overhead projectors, and used these technologies for teaching, for office work, and for other academic purposes.

- Table 7 shows that private sector university teachers are well aware of computer technology as compared with teachers of public sector universities.

- Another important factor is observed among teachers learning skills in Table 8. As teachers learning skills in public and private universities are equal with respect to information technology, so both type of teachers use information technology in limited scale.

- In Table 9, it could be concluded that the students face difficulties in availability of updated computers, high speed Internet connection, enough time, video conferencing facility, interactive board, and multimedia in the classrooms

Recommendations

The following recommendations were made in the light of above conclusions:

- According to research, it was proven that the availability of equipment in the public sector is less than that of the private sector; hence, it is recommended that university authorities provide equipment of technology to education departments, similar to their own departmental computer labs, with fully equipped latest computers, and that these are accessible to every student.

- High speed Internet facility may also be provided for the departments of education in public and private universities.

- The students may also be provided first of all, training of information technology; then, they may be given IT-based assignments and projects so that the students pay more attention to it and use it frequently.

- The teachers need to be provided IT training so that they become better users of it and could enhance student learning as well.

- A collaboration of public and private university teachers may also strengthen their knowledge, skills, and computer abilities.

- Updated computers and high speed Internet connection need to be made available to the students and teachers.

- Access to video conferencing facility may also be provided to students, especially research students, so that they could get benefit of national and international expertise.

- The authorities may provide interactive boards and multimedia to the education departments to make the teaching–learning process effective.

Declaration of Conflicting Interests

The author(s) declared no potential conflicts of interest with respect to the research, authorship, and/or publication of this article.

Funding

The author(s) received no financial support for the research and/or authorship of this article.

References

Angrist, J. D., & Lavy, V. (2002). New evidence on classroom computers and pupil learning. *Economic Journal, 112*, 735-765.

Banerjee, A., Cole, S., Duflo, E., & Linden, L. (2004). *Remedying education: Evidence from two randomized experiments in India* (Mimeo). Cambridge, MA: MIT.

Bennett, J., & Bennett, L. (2003). A review of factors that influence the diffusion of innovation when structuring a faculty training program. *Internet and Higher Education, 6*, 53-63.

Coates, D., Humphreys, B. R., Kane, J., Vachris, M., Agarwal, R., & Day, E. (2004). "No significant distance" between face-to-face and online instruction: Evidence from principles of economics. *Economics of Education Review, 23*, 533-546.

Fuchs, T., & Woessmann, L. (2004). What accounts for international differences in student performance? Are-examination using PISA data (No. 1235). CESifo working paper.

Goolsbee, A., & Guryan, J. (2006). The impact of Internet subsidies in public schools. *The Review of Economics and Statistics, 88*, 336-347.

Gorgone, J., Davis, G. B., Valacich, J. S., Topi, H., Feinstein, D. L., & Longenecker, H. E. (2003). IS 2002 model curriculum and guidelines for undergraduate degree programs in information systems. Communications of the Association for Information Systems, 11(1), 1.

Hargreaves, A. (2003). *Teaching in the knowledge society: Education in the age of insecurity*. Maidenhead, UK: Open University Press.

Krotoski, A. (2004). White Paper: Chicks and Joysticks: An Exploration of Women and Gaming. Entertainment & Leisure Software Publishers Association (ELSPA).

Kulik, J. A. (1994). Meta-analysis study of findings on computer-based instruction. In E. L. Baker & H. F. O'Neil (Eds.), *Technology Assessment in education and training* (pp. 9-34). Hillsdale, NJ: Lawrence Erlbaum.

Leu, E., & Price-Rom, A. (2006). Quality of education and teacher learning: A review of the literature. Washington, DC: USAID educational quality improvement project, 1.

Ma, W. W., Andersson, R., & Streith, K.-O. (2005). Examining user acceptance of computer technology: An empirical study of student teachers. *Journal of Computer Assisted Learning, 21*, 387-395.

Medlin, B. D. (2001). *The factors that may influence a faculty member's decision to adopt electronic technologies in instruction* (Doctoral dissertation). Virginia Polytechnic Institute and State University, Blacksburg, VA.

Selwyn, N. (2002). Learning to love the micro: The discursive construction of "educational" computing in the UK, 1979-89. *British Journal of Sociology in Education, 23*, 427-443.

Sivin-Kachala, J., & Bialo, E. R. (1994). Report on the Effectiveness of Technology in Schools, 1990-1994.

Sosin, K., Blecha, B. J., Agawal, R., Bartlett, R. L., & Daniel, J. I. (2004). Efficiency in the use of technology in economic education: Some preliminary results. *American Economic Review, 94*(2), 253-258.

Udvari-Solner, A., & Thousand, J. S. (1996). Creating a responsive curriculum for inclusive schools. *Remedial and Special Education, 17*(3), 182-191.

Author Biographies

Saima Yasmeen, MPhil, is a teacher of education at college.

Muhammad Tayyab Alam is a professor of education at Foundation University Islamabad. He has 20 years of teaching experience.

Muhammad Mushtaq is a lecturer at Foundation University Rawalpindi Campus. He is a PhD scholar as well.

Maqsud Alam Bukhari is a retired professor of education. He earned his PhD in 1970 from Dhaka University. He has 40 years of teaching and supervising experience.

18

Impact of Technical Support on Customer Satisfaction: Case of Automotive Paints

Natasa Gajic[1] and Mehraz Boolaky[2,3]

Abstract

Technical support through co-creation of value in automotive paint processes and activities is essential to the success of paint manufacturing companies. This study aimed to explore the impact of technical support on customer satisfaction through value-in-use in the automotive paint market. A quantitative questionnaire survey involving a convenience sample of 169 respondents was used for data collection. The questionnaire design bore on the SERVPERF instrument with embedded value-in-use attributes. The data were analyzed by using SPSS 21 statistical methods exploratory factor analysis (EFA) and multiple regression analysis (MRA). The findings of this study revealed that the key value-in-use attributes were relationship quality (trust), knowledge required for providing help in getting maximum product benefits, sharing of knowledge, and a range of product and service offerings that satisfy customer needs. Trust had the greatest impact on customer satisfaction. The results also revealed that service quality dimension assurance exerted the greatest positive impact on customer satisfaction.

Keywords

technical support, automotive paint market, service-dominant logic, value-in-use, service quality

Introduction

Services in manufacturing have emerged as a response to slow down of sales growth, fall down of product margins, and typically higher service margins and return on invested capital (Bundschuh & Dezvane, 2003). The automotive paint market is specific in that it involves and requires skilled, knowledgeable, well-informed, and experienced professionals who deal with chemical products and apply them properly and promptly in such a way that damaged car has final finish of new car. The very nature of automotive paint job requires that paint manufacturers permanently inform, educate, train, and support customers. Significant interactions with customers are also needed and are essential for mutual success and growth.

Defining Technical Support (TS)

The importance of product support has long been recognized. Scholars have revealed that product support is essential for customers to obtain maximum value from products, increase sales revenue, and positively affect customer satisfaction (e.g., Lele, 1986; Lele & Karmarkar, 1983; Saccani, Songini, & Gaiardelli, 2006). However, the marketing literature on TS is ambiguous in that different terms such as product support, service support, customer service, customer support, product service, and after-sales service are used. TS has also been identified in the software and chemical industries and has somehow been separated from customer service. Using TS in these two industries is probably due to the fact that the software and (IT) products are sophisticated just like special paints from the chemical industry that are more scientifically product development–oriented (Cooper & Kleinschmidt, 1993) and require more technical assistance to customers than in other industries (Chemical Business, 2013).

In this study, TS has the meaning of such activities that take place in interactions between customers, employees, other individuals, and products with aim to assist and support customers' everyday practices partly taken from the definition of service defined by Grönroos (2008). TS activities include complex interactions and relationships with customers and products through which the customers get maximum product benefits, optimize product usage and business risks related to product utilization. The TS activities can be viewed as informative, educational, and relational, and TS as such can be seen as an element of integrated product-service design in manufacturing. Although customer relationship marketing points out customer service as the link between firms and customers (Christopher, Payne, & Ballantyne, 1991), TS is neither separated from customer service nor

[1]KAPCI Coatings, Port Said, Egypt
[2]University of Liverpool/Laureate, Amsterdam, The Netherlands
[3]Asia Pacific Institute of Management, New Delhi, India

Corresponding Author:
Natasa Gajic, KAPCI Coatings, P.O. Box 118, Port Said, 42511, Egypt.
Email: natasagajic13@gmail.com

strictly divided into pre-sale and after-sale activities. TS has economic, societal, and ecological potentials. It ensures higher profit margins to firms, better productivity to customers, lesser customer's operating costs, and higher dissemination (including internationally) of latest development and information related to ecological product usage (Aurich, Fuchs, & Wagenknecht, 2006). Therefore, TS is critical for both successful global marketing and technology transfer "Automotive Coatings: Technologies and Global Markets," 2014; (Kosenko & Samli, 1985). According to Automotive Paint Market (2015), automotive paint manufacturers are driven to provide the high-quality service to their customers through dedicated teams of professionals, implying that TS is likely to play more strategic roles in the future.

Service-Dominant (S-D) Logic and TS

Widely acknowledged by marketing academics, S-D logic proposed by Vargo and Lusch (2004, 2008b) and further reviewed by Lusch and Nambisan (2015) has shifted the marketing view from goods-dominant (G-D) logic emphasizing products and transactions to a more service-oriented view that considers all businesses to be service businesses, emphasizing interactions and relationships with customers. However, despite an increasing interest in S-D logic and a strong conceptual background of value-in-use in the marketing literature, there is little empirical evidence on how customers experience value-in-use and how value-in-use affects customer satisfaction in particular in the automotive paint market. A gap also exists in that most of the studies have been conducted in Western countries, whereas the paint industry encompasses manufacturers all over the world, in both developed and developing countries. Valuing value-in-use by customers from different walks of lives especially from developing countries might shed some more light on the value co-creation process from the S-D logic stance.

Recent evidence suggests that there are seven key attributes of value-in-use (Raja, Bourne, Goffin, Çakkol, & Martinez, 2013)—knowledge, accessibility, relational dynamics, a range of product and service offerings, delivery, price, and locality, and that these attributes have positive impact on customer satisfaction. The findings of the key value-in-use attributes can help manufacturers to improve their service strategies as they can better understand needs of their customers, increase customer satisfaction, maintain loyal and gain new customers, and increase revenue. The literature on value co-creation activities also suggests that active participation of customers in the value-creation process positively affects customer satisfaction (Vega-Vazquez, Revilla-Camacho, & Cossío-Silva, 2013). Considering the aforementioned TS activities, it is evident that TS may affect customer satisfaction through knowledge and expertise, accessibility, relationships, providing help in product choices, and being involved in broadening of product offerings through the initiation of new product development (NPD).

Aim and Objectives

The aim of this study is to investigate the relationship between TS and customer satisfaction. The concepts discussed above and the findings of Raja et al. (2013) provide a base for the following two research questions:

Research Question 1: What are the key attributes of value-in-use in the automotive paint market?
Research Question 2: How do these attributes affect customer satisfaction?

Therefore, this study aims to contribute to the studies on TS and impact of TS in manufacturing companies on customer satisfaction. The research findings can extend the underlying S-D logic view and the growing area on value-in-use. They may also lead to changes and improvements in service strategies and TS activities in the paint industry for achieving competitive advantages.

Review of the Literature

The Importance and Key Dimensions of TS

Traditionally, product support has been differentiated as pre-sale and after-sale support (e.g., Burger & Cann, 1995; Morris & Davis, 1992). A large volume of published studies has acknowledged that product support can be a source of revenue (Cohen, Agrawal, & Agrawal, 2006; Goffin & New, 2001), is essential for achieving customer satisfaction (Burger & Cann, 1995; Innis & La Londe, 1994; Raja et al., 2013; Saccani et al., 2006) and increasing customer loyalty and enhancing customer relationship (Ahn & Sohn, 2009; Pan & Nguyen, 2015), is essential for obtaining maximum value from products (Das, 2003; Lele & Karmarkar, 1983), and is a key for achieving competitive advantages (Goffin, 1998; Oliva & Kallenberg, 2003) and increasing success rate of new products in market (Cherubini, Iasevoli, & Michelini, 2015; Cooper & Kleinschmidt, 1993). However, Raithel, Sarstedt, Scharf, and Schwaiger (2012) addressed the value relevance of customer satisfaction and found that the dealer service quality is not an important factor that investors would look for but rather the perceived product quality.

Over the past years, manufacturers have been called for adoption of servitization, the process of shifting from the position of "pure manufacturer to that of service provider" (Redding, Tiwari, Roy, Phillips, & Shaw, 2014, p. 2) to offer bundles of combinations of products, services, support, and knowledge (Vandermerwe & Rada, 1988) instead of a single product or a service. Attention is also focused on new business models such as product service systems (PSSs; for example, Baines et al., 2007; Tukker, 2004). Service systems are value co-creators and networks of resources such as people, information, and technology connected with other systems by value propositions (Vargo, Maglio, & Akaka, 2008). PSSs provide integrated and customized offerings that fulfill customer needs, build unique relationships with customers enhancing customer's loyalty, and provide better following of customer needs, thus enabling faster innovations (Tukker, 2004). Similarly, Baines et al. (2007) found that PSSs offer value-in-use. In product-oriented PSSs, customer support enhances functionality and durability of products and optimizes product application through training and consultation.

One study by Cooper and Kleinschmidt (1993) has revealed that NPD projects in the chemical industry utilizing TS and customer service are more successful and that firm's technical competences perceived by the customer increase success of products in markets. As noted by Aurich et al. (2006), the potentials of TS services are economic, societal and ecological—economic potential lies in higher profit margins of TS being service, providing higher productivity to customers who use products as resources for their own processes, and reducing customer's operating costs through training which assists in economical use of products; societal potential lies in building and securing knowledge intensive work and spreading it geographically considering that support is provided at product usage location; ecological potential relates to providing customers with more conscious product usage. Flint, Blocker, and Boutin (2011) also opined that as customer value perceptions are also in a state of constant flux, it is important for suppliers to anticipate the requirements of users and propose to do so as to ensure continued satisfaction and develop loyalty.

Knowledge. There is an unambiguous relationship between TS and knowledge as knowledge is integrated part of TS work. Das (2003) emphasized that TS is an "exemplar of knowledge work" (p. 416) and that heuristic nature of knowledge underlies problem formulation and problem solving. In the chemical industry, hence in the paint industry as well, where the science involvement is high and technical uncertainty is greater (Cooper & Kleinschmidt, 1993), solving problems requires that the TS personnel locate, adapt, and generate the problem (Das, 2003). In TS work, the main problems that customers face are during and after application of automotive paints and related products or during the research process in research and development (R&D). Of particular importance for TS is tacit knowledge. Tacit knowledge should be renewed, and the renewal is possible if there is trust between employees who share knowledge (Ballantyne & Varey, 2006). However, it is sharing of knowledge between TS personnel and customer that is critical because "knowledge sharing and application is a hidden source of competitive advantage" (Vargo & Lusch, 2004, cited in Ballantyne & Varey, 2006, p. 341). The knowledge as a key dimension of TS may be best understood from six perspectives suggested by Alavi and Leindler (2001, cited in Wang & Lu, 2010) as a kind of comprehended or personalized information regarding data and personal information, a status of understanding, stored and utilized personal or group-owned asset, a means of utilizing experience and expertise, a

condition of access to information, and a capability to perform effective actions.

Understanding customer needs. According to Tyagi, Yang, Tyagi, and Verma (2012), the competitiveness of any manufacturing industry is much dependent on its capability to readily cater to the requirements of market niches through the production of a variety of high-utility and low-cost products targeted for them. When TS personnel support the customer, the customer's support expectations and responsiveness help the TS personnel in defining customer needs and assessing customer's expectations (Bowen, Siehl, & Schneider, 1989). Information that a TS team gathers from the automotive paint market is used as inputs or ideas for the NPD and for modifications of existing products in R&D to customize the products where needed and satisfy customer needs in many countries. Empirical studies by Cooper and Kleinschmidt (1993), Goffin (1998), and Goffin and New (2001) have revealed that manufacturers engaging customer support in early stages of the NPD process are able to increase the success rate of new products in markets.

Field visit is another area widely used for understating customer needs and supporting customer requirements. For an automotive paint manufacturer, field visits are integral part of TS service and work performed in the automotive paint market because field service improves customer satisfaction (Ahn & Sohn, 2009) and maintains customer loyalty (Hull & Cox, 1994). Field service is a communication link from the customer back to product design, manufacturing, and marketing (Hull & Cox, 1994), a source of new information for understanding customer needs and supporting customer's actions. Knowing customer needs would be a base for tailored after-sale activities. Such tailored after-sale activities increase customer's loyalty and strengthen customer relationship (Ahn & Sohn, 2009).

Relationships. Several studies draw our attention on the importance of relationship between manufacturers and customers (e.g., Bowen et al., 1989; Dorsch, Swanson, & Kelley, 1998; Flint et al., 2011; Mathieu, 2001). Trust has been identified as an essential element of relationship (e.g., Flint et al., 2011; Jap, 1999) and a key factor for successful and sustainable relationship (Huang & Wilkinson, 2014) together with ethics (Goffin, Lemke, & Szwejczewski, 2006). A study by Goffin et al. (2006) investigating close or partnership-like relationship has revealed trust as commitment. Dorsch et al. (1998) argued that trust and commitment together with satisfaction, opportunism, customer orientation, and ethics are dimensions of relationship quality.

More recent attention has focused on relationship quality (e.g., Macdonald, Wilson, Martinez, & Toossi, 2011; Rauyruen & Miller, 2007; Segoro, 2013). This shift from service quality toward relationship quality likely comes from the changes of the quality perspective in American Society of Quality (ASQ). The ASQ has shifted toward the definition of quality as "the ongoing process of building and sustaining relationships by assessing, anticipating and fulfilling stated and implied needs" (Judd, 1994, cited in Macdonald et al., 2011, p. 672). Therefore, relationship quality underpins building of sustainable relationship. Flint et al. (2011) used the term *collaborative relationship* with customers that enables firms to acquire skills and insights what their customers value, and also enables both firms and customers to achieve competitive advantages (Jap, 1999). Customized and close relationship support customer's actions (Mathieu, 2001). In a study that set out to determine value creation in business relationships from S-D logic, Ulaga (2003) found that product support was one among key relationship value drivers along with product-related additional services including the right information at the right time, manufacturer's availability/presence, quick response, and providing appropriate information.

Value Co-Creation and Value-in-Use From S-D Logic

One study by Woodruff (1997) suggested customer value as *value-in-use*, which is perceived preference to and evaluation of product attributes, its performances, and results of the use that facilitate or block achieving the customer's goals and purposes. Similarly, recent S-D logic proposed by Vargo and Lusch (2004, 2008b) has introduced the concept of value-in-use calling firms for a new view of marketing. According to S-D logic, products and services assist customers in their own value-creation processes, where value is not something produced and sold, but something that is co-created with the customer and all other participants engaged in the co-creation process (Vargo & Lusch, 2008a). The authors assert that customers are resources and part of firm's network and that firm's resources are knowledge and skills, therefore emphasizing interactions as the central concept in value creation and value co-creation.

The stance that TS department is a service provider is supported by Vargo and Lusch's (2008a) premises that service is a collaborative process of value creation among participants and that products need some actions to be valuable. Service provided directly or through a product represents the knowledge and skills of a service provider and is an essential source of value and competitive advantage. According to S-D logic, manufacturers can only propose offerings whereas the value becomes *value-in-use* evaluated by the customer in the consumption process and through use (Vargo & Lusch, 2004). Evaluation of value-in-use is an outcome of co-creation process and "experience and perception are essential to value evaluation" (Vargo et al., 2008, p. 148). According to the latter view, it might be that more experienced customers in the automotive paint market in developed counties would assess value-in-use in a different way than their counterparts in developing countries.

Prahalad and Ramaswamy (2004) hold the view that value is co-created through customers' unique experience. Such experience is based on high-quality interactions between the customer and the firm. They have proposed the DART model (Dialog, Access, Risk-Benefits, and Transparency) of interactions for co-creation of value that is built: from dialog and access to as much information to the customer whereby access and transparency are critical for meaningful dialog; and from risk-benefits assessments of actions and decisions provided through dialog, transparency, and access.One question that needs to be asked, however, is whether and how the quality of information provided by the firm to the customer affects the customer's experience and the quality of (existing) relationship. It might be that such information can lead to losing potential or current customer or having an impact on customer satisfaction. Norman (2001, cited in Payne, Storbacka, & Frow, 2008) points out that customers' ability to create value depends on the amount of information, knowledge, and skills that customers can access and use. This view holds true; however, it is likely linked with prior customer's knowledge and experience as well. Overall, both prior knowledge and experience together with those that the customer can access and use might affect customer's creation of value and evaluation of value-in-use because, as pointed out by Gummesson (2008), customers create the value often on their own and sometimes in contact with suppliers.

However, S-D logic has not been without criticism by some researchers (e.g., Schembri, 2006; Stauss, 2005). Schembri (2006) argued that S-D logic is grounded in rationalistic assumptions proposing marketing as a process of interacting with customers instead of understanding that marketing starts with customers' experience. A large and growing body of S-D logic literature has theoretically investigated value co-creation (e.g., Grönroos, 2008, 2012; Möller, 2006; Payne et al., 2008; Prahalad & Ramaswamy, 2004), and a few studies have examined the concept of value-in-use itself (e.g., Ballantyne & Varey, 2006; Macdonald et al., 2011; Raja et al., 2013). One criticism emerging from the S-D literature on value co-creation and value-in-use is that value co-creation and value-in-use have been largely examined as theoretical concepts; therefore, more empirical research is needed to understand how value-creating activities such as those of TS might affect value-in-use derived by customers.

The SERVPERF Instrument

Parasuraman, Zeithaml, and Berry (1988) have developed widely known SERVQUAL for measuring service quality perceived by customer. SERVQUAL is a multi-item scale instrument having 10 service quality attributes, that is, 10 dimensions. Parasuraman et al. have found that there is a high degree of correlation between some items (i.e., variables), which have been placed under corresponding dimensions. The final result has given the five dimensions whose detailed descriptions are adapted by Lovelock and Wirtz (2011):

(1) *Tangibles*—appearance of physical facilities, equipment, communicating materials, and personnel;

(2) *Reliability*—ability to perform promised service dependably and accurately;

(3) *Responsiveness*—willingness to help customer and provide prompt service;

(4) *Assurance*—the knowledge and courtesy of personnel and their ability to inspire trust and confidence. Assurance contains items including *credibility* (trustworthiness, believability, and honesty), *security* (freedom from risk, danger, or doubt), *competence* (possession of the skills and knowledge required to perform the service), and *courtesy* (politeness, respect, consideration, and friendliness);

(5) *Empathy*—The provision of caring individualized attention to customers. Empathy includes items like *access* (approachability and ease of contact), *communication* (listening to customers and keeping them informed), and *understanding the customer* (making the efforts to know customers and their needs). (p. 385)

SERVQUAL involves customer's perceived service quality by measuring the gaps of the five dimensions between the customer's expectations of the service to be performed and the customer's perceptions about the actual performance of the service. Several authors have used SERVQUAL in various service sectors with modifications of the scale items (i.e., the questions) according to the specific context (e.g., Banu & Gül, 2012; Freeman & Dart, 1993; Haque, 2013; Kettinger & Lee, 1994; Nadiri & Hussain, 2005).

However, despite the acknowledgment of the SERVQUAL instrument and all the benefits of the service quality theory, SERVQUAL has been subjected to both theoretical and operational criticisms (e.g., Buttle, 1996; Cronin & Taylor, 1992, 1994). Cronin and Taylor (1992) have criticized SERVQUAL arguing that customer satisfaction is an antecedent of service quality whereby service quality is strongly affected by existing service performances. Accordingly, they have developed an instrument that measures only service performances called SERVPERF based on the same five dimensions as SERVQUAL and with a reduced number of items from 44 to 22. Furthermore, because customers' expectations may differ within different regions and populations based on, for example, customers' experience and performance in accomplishing tasks, actions, or processes, SERVPERF may be more reliable as a measure of service quality.

Customer Satisfaction

Customer satisfaction is one of business performance metrics linked with service quality, product evaluation, repeat purchase, customer behavior, brand loyalty, and revenue (e.g., Bell, Auh, & Smalley, 2005; Finkelman & Goland, 1990; Griffin, Gleason, Preiss, & Shevenaugh, 1995; Vega-Vazquez

et al., 2013). Woodruff (1997) suggested that customer value and product utilization have impact on customer satisfaction. He argues that product usage plays a critical role in customer's evaluations and desires, and proposes customer value hierarchy model consisting of levels of customer satisfaction. Consistent with Woodruff's (1997) model are findings by Macdonald et al. (2011) that customers can assess the usage process in terms of whether their goals, purposes, and objectives were supported or hindered therefore being satisfied or dissatisfied.

Individual communication improves customer satisfaction not only because of its immediate responses and faster solving of customer problems but also because direct communication of technically trained people assists in gathering the right information directly from customers who have them. Accordingly, this helps to minimize possible misinterpretations and wrong judgments about the importance of relevant aspects (Griffin et al., 1995). Yi and Gong (2013) argued that customer co-creation behavior consists of information seeking and information sharing with employees supplying them with information following employees' guidelines and orientations; personal interactions with the employees based on courtesy, friendliness, and respect, which are fundamental for value co-creation; feedback leading to long-term improvement of the service provision; the recommendations to other individuals (families and friends); willingness to advice and give information to other customers; and tolerance when customers' expectations were not met. Adopting S-D logic conceptual framework in the personal care sector, an empirical study by Vega-Vazquez et al. (2013) has revealed that customer involvement in the value co-creation process increases customer satisfaction with firm's service.

Method

A quantitative research design was conceptualized based on the theoretical framework developed from the theories and concepts discussed in review of the literature. In quantitative research process, the researcher should "try to maintain objectivity and detachment from the research process" (Gray, 2014, p. 175). In this quantitative research, the main author took a posture of an observer with social constructionism as the epistemological paradigm, which means that the author's personal experience from the paint industry and automotive paint market can assist in general understating of the TS work and its impact on customer satisfaction. Namely, social constructionism stems from the view that reality is "socially constructed and given meaning by people" (Esterby-Smith, Thorpe, & Jackson, 2012, p. 23). In social constructivism, the researcher is part of what is being observed and explanations are aimed to increase general understanding of a particular situation (Esterby-Smith et al., 2012). Because the TS team participates in value co-creating processes, and assists and affects customers' own processes, the main author being in charge of managing the TS department was interested to investigate the impact of the TS activities perceived by customers on customer satisfaction through value-in-use from the S-D logic perspective.

Participants in this study were customers of an Egyptian paint manufacturer who were located in Egypt and India. The customers from Egypt and India were chosen because the paint manufacturer was located in Egypt with a subsidiary in India and due to similarities of the automotive paint market in both countries. The customers from Egypt and India were chosen by using non-probability convenience sampling method considering that the participants were customers of the Egyptian paint manufacturer who could easily be reached. The survey was conducted face-to-face during a 6-week period (July/August 2014), and the data were gathered using a paper-based survey questionnaire administered by hand as this method provided the fastest and most reliable way to reach the participants. The initial sample consisted of 169 customers. Of 169 participants, 11 had no variations in responses, and they were excluded from the analysis. The remaining 158 participants provided usable responses used for the analysis.

The questionnaire design bore on the performance-based SERVPERF instrument consisting of the same five dimensions as the SERVQUAL instrument. The original SERVPERF instrument was adapted to the TS activities. The total number of items was 25, 24 items for measuring TS service quality performance perception by customer and a separated item for measuring overall customer satisfaction (see Appendix A). The adapted SERVPERF instrument also contained embedded value-in-use attributes in order to identify them in the TS activities. The 25 items were measured on a 5-point Likert-type scale. The pilot test of the questionnaire in both Arabic and English language was conducted with five professional colleagues and five potential users of data to check whether the questionnaire gathered all needed information relevant to the TS activities and whether the questions were understood well. Based on the feedback from the respondents, a few changes were made and the final questionnaire (see Appendix B) administered to the respondents.

Results and Analysis

Results

Descriptive statistics. Demographic breakdown of the sample showed that from the 158 respondents, 139 were from Egypt and 19 were from India. Table 1 presents frequency distribution of kind of jobs of the participants. Of the 158 participants, 32.9% were retailers, 25.3% were painters, 22.8% were agents, 12.7% were technicians, 5.1% were "others" (i.e., body shop managers as stated by the participants), and 1.3% did not respond to the question about job specification.

Table 2 shows frequency distribution of working experience in the automotive paint market. Of the 158 participants,

Table 1. Frequency Distribution of Kind of Jobs.

	Frequency	%
Valid		
Agent	36	22.8
Retailer	52	32.9
Technician	20	12.7
Painter	40	25.3
Others	8	5.1
Total	156	98.7
Missing		
system	2	1.3
Total	158	100.0

Table 2. Frequency of Working Experience.

	Frequency	%
Valid		
Less than 12 months	2	13
Less than 5 years	12	7.6
Between 5 and 10 years	31	19.6
More than 10 years	109	69.0
Total	154	97.5
Missing system	4	2.5
Total	158	100.0

69% had more than 10 years of experience, 19.6% had between 5 and 10 years of experience, 1.3% had less than a year, 7.6% had less than 5 years of experience, and 2.5% did not respond to the question.

The results of the mean from Table 3 construe that the respondents from Egypt and India had relatively higher perceptions for reliability, assurance, and empathy, and relatively lower perceptions for tangibles and responsiveness. Considering that tangibles do not comprise value-in-use attributes, this service quality dimension was not taken into consideration in the analysis. Relatively higher mean values ($M \geq 4.02$) show that the respondents had relatively higher perceptions for solving customers' problems accurately and showing sincere interest in solving problems. The ability of TS personnel to tackle customers' specific problems (such as color matching, product problems, application problems, etc.) and the trust placed in the TS personnel are also rated on the higher side. Finally, Table 3 shows that building sustainable relationship with customers assists in providing and maintaining satisfactory service quality.

As mentioned earlier, responsiveness plays important role in customer satisfaction and helps in improving it. The results of relatively low perception scores ($M \leq 3.29$) for informing customer about the exact time for performing the TS activities such as field visits, training, and so on (P9), suggest that TS failed somewhat in ensuring responsiveness. There are several possible explanations for this result. It might be that the TS activities were not well-organized, the TS team could

Table 3. TS Service Quality Perception (*N* = 158).

Item	*M*	*SD*
Tangibles		
P1	3.53	1.41
P2	3.45	1.40
P3	4.32	0.91
P4	3.61	1.22
Reliability		
P5	3.74	1.01
P6	4.21	0.86
P7	3.65	1.08
P8	4.06	0.91
Responsiveness		
P9	3.29	1.08
P10	3.60	1.13
P11	3.91	1.05
P12	3.61	1.12
Assurance		
P13	4.05	0.72
P14	3.96	0.85
P15	4.12	0.79
P16	4.09	0.84
P17	3.92	0.82
P18	3.96	0.84
Empathy		
P19	3.68	0.85
P20	4.02	0.83
P21	3.35	1.18
P22	3.94	0.86
P23	3.20	1.18
P24	3.91	1.11

Note. TS = technical support.

not provide the right information, or there was a lack of (adequate) employees earmarked for providing TS.

The assessment of sampling adequacy. The suitability of the respondent data was checked by generation of correlation matrix shown in Table 4. The pattern of relationships of the correlated items is used to investigate whether the items are correlated too low ($R < .1$) or too high ($R > .9$; Institute for Digital Research and Education–University of California, Los Angeles [IDRE-UCLA], 2014). From Table 4, it is noted that none of the items correlated above .9 or below .1 suggesting possibility to determine unique contribution of items onto a factor. However, it was also necessary to check whether the items did not correlate too low by using Kaiser–Meyer–Olkin (KMO) and the Bartlett's test of sphericity. If KMO is higher than 0.50 and the Bartlett's test of sphericity is significant ($p < .001$), then the data for exploratory factor analysis (EFA) are suitable (B. Williams, Brown, & Onsman, 2012). The results of KMO of 0.942 and significance $p < .001$ for the Bartlett's test proved the suitability of the sample size and data set for EFA.

Reliability. Reliability coefficient Cronbach's alpha varies between 0 and 1, where values of .7 and above are considered acceptable (Gray, 2014). The results of reliability value of Cronbach's alpha for the initial SERVPERF scale shown in Table 5 interpret that that Cronbach's alpha was acceptable at .959. Cronbach's alpha for each of the five dimensions was also acceptable being above .7. The results of reliability for the resulting SERVPERF scale presented also in Table 5 show that the overall Cronbach's alpha for the resulting SERVPERF slightly decreased to .942 because the number of the items decreased from 24 to 17, while all resulting dimensions had increased Cronbach's alpha values.

Construct validity. Table 6 shows that P3 item with the lowest correlation value of .546 from *tangibles* would have increased $\alpha = .875$ if P3 was deleted, and P12 with the correlation value of .577 from *responsiveness* would have increased to $\alpha = .853$ if the item was deleted whereby all other items showed higher correlations with their corresponding items and higher values of Cronbach's alpha. Because the correlations ranged from .546 to .803 and were above Lai, Hutchinson, Li, and Bai's (2007) recommendation of .20 for the inclusion of items, all items were included.

Value-in-use attributes. Principal axis factoring (PAF) with promax rotation was used for EFA as it produced the results that could easily be interpreted following Fabrigar, Wegener, MacCallum, and Strahan's (1999) recommendation that the researcher can choose the procedure, a number of factors and methods for rotating to get more readily interpreted data. As suggested by Field (n.d.), the initial EFA included the eigenvalue greater than 1 which extracted three factors. To keep as much data and because the next two factors F4 and F5 had eigenvalues close to 1, that is, 0.882 for F4 and 0.734 for F5 as shown in Table 7, it was decided to choose five factors as the SERVPERF instrument had five dimensions. For easier reading and interpretation of factors, factor loadings with values lees than 0.4 were suppressed.

After 12 iterations, PAF produced the final results that could easily be interpreted, presented in Table 8. Following B. Williams et al.'s (2012) suggestion, items attributable to a factor were examined and the factor labeled by a proper name. Table 8 shows that the items loaded on F1 were P13, P14, P17, and P18 from *assurance* and P19, P20, and P22 from *empathy*, whereas P18 and P17 were the most attributable to F1 having the highest factor loadings at the values of 0.915 and 0.877, respectively. Therefore, Factor F1 was named *assurance*. F1 contained, along with the original item P13 "you can trust TS personnel," new items added in this dimension implying the importance of *assurance* in TS service quality. This finding confirms the strong relationship between TS work, knowledge, and assurance, as mentioned earlier in the literature review. The Items P6 from *reliability*, P11 and P12 from *responsiveness*, and P24 from *empathy* loaded on Factor F2 whereby P24 was the most attributable

Table 4. Correlation Matrix for the Initial SERVPERF.

	P1	P2	P3	P4	P5	P6	P7	P8	P9	P10	P11	P12	P13	P14	P15	P16	P17	P18	P19	P20	P21	P22	P23	P24
P1																								
P2	.870																							
P3	.483	.506																						
P4	.640	.612	.602																					
P5	.389	.482	.572	.573																				
P6	.303	.368	.609	.488	.585																			
P7	.575	.613	.538	.697	.578	.536																		
P8	.524	.491	.629	.684	.530	.556	.696																	
P9	.679	.704	.445	.727	.567	.467	.637	.586																
P10	.295	.266	.243	.281	.228	.152	.323	.219	.387															
P11	.477	.504	.609	.630	.600	.708	.605	.593	.611	.236														
P12	.317	.366	.558	.549	.583	.642	.600	.521	.472	.131	.697													
P13	.417	.431	.603	.613	.620	.622	.574	.632	.571	.203	.642	.607												
P14	.478	.442	.621	.661	.481	.558	.576	.671	.572	.225	.680	.611	.697											
P15	.347	.334	.564	.515	.508	.592	.514	.583	.467	.144	.563	.648	.703	.654										
P16	.324	.309	.494	.505	.440	.508	.477	.496	.472	.142	.518	.534	.704	.548	.756									
P17	.418	.428	.596	.662	.532	.469	.540	.560	.515	.294	.570	.476	.631	.655	.579	.578								
P18	.437	.445	.586	.639	.499	.428	.587	.546	.512	.288	.569	.530	.697	.653	.630	.666	.759							
P19	.421	.453	.469	.668	.481	.451	.531	.569	.627	.223	.616	.539	.663	.634	.606	.633	.646	.702						
P20	.510	.507	.627	.604	.479	.502	.543	.624	.546	.125	.643	.534	.721	.698	.602	.588	.615	.683	.709					
P21	.640	.637	.505	.602	.474	.478	.625	.470	.535	.225	.597	.578	.522	.576	.445	.410	.498	.544	.518	.530				
P22	.506	.503	.607	.637	.556	.595	.589	.615	.517	.210	.665	.574	.717	.737	.634	.568	.670	.660	.674	.723	.656			
P23	.563	.596	.353	.606	.346	.267	.536	.439	.483	.166	.372	.323	.434	.474	.398	.391	.505	.522	.539	.460	.621	.553		
P24	.414	.466	.554	.521	.564	.769	.589	.539	.506	.124	.808	.721	.602	.602	.588	.529	.452	.465	.505	.592	.594	.625	.362	

to this factor having the highest factor loading of 1.01. Considering that *empathy* includes attributes such as accessibility included with P11 and P12, understanding customer and customer needs included with P6, F2 was named *empathy*. Factor F3 was named *tangibles* including Items P1 and P2 with high factor loadings of 0.981 and 0.859, respectively. Factor F4 was named *reliability* including P16 with higher factor loading of 0.726 and P15 with lower factor loading of 0.519, the items that had initially been in *assurance*. Nevertheless, both items may relate to *reliability* because this service quality dimension emphasizes ability to perform promised service dependably and accurately. Because Items P9 with factor loading of 0.877 and P10 with factor loading of 0.513 from the original *responsiveness* dimension loaded on F5, this factor kept the same name *responsiveness*.

Table 9 shows the resulting SERVPERF scale containing 17 items in total with five service quality dimensions including *assurance, empathy, tangibles, reliability*, and *responsiveness* and embedded value-in-use attributes in them. Table 10 interprets that the value-in-use attributes embedded in corresponding items and four dimensions including assurance, empathy, reliability, and responsiveness were *knowledge, relationship, providing a range of product and service offerings according to customer needs*, and *accessibility*, and that TS service quality can be explained by five dimensions with 73.38% of the total variance.

The key value-in-use attributes and customer satisfaction. The key value-in-use attributes and their impact on customer satisfaction were determined by conducting multiple regression analysis (MRA). The R^2 shows the "percentage of variance in the dependent variable explained by the independent variable(s)" and ranges from 0 to 1 (Gray, 2014, p. 591), that is, up to 100%. Table 11 shows that the regression coefficient R^2 was .717, which means that 71.7% of customer satisfaction can be predicted by *tangibles, reliability, responsiveness, assurance*, and *empathy*. Adjusted R^2 was .683 (68.3%) and did not differ significantly from the R^2 value of 71.7% implying that there were not too many predictor variables used in the model.

The *F* test from ANOVA presented in Table 12 interprets that the model was statistically significant and there was statistically significant relationship between all 17 predictor variables and customer satisfaction as the *p* value was lower than .001 at the *F* value of 20.907. From Table 13, it is noted that the largest value of standardized coefficient beta (β) had relationship (β = .403 for P13) followed by providing help to customers in getting maximum product benefits (β = .215 for P18), providing a choice of product and service offerings according to customer needs (β = .202 for P17), and knowledge sharing (β = −.163 for P14). As can be seen from the *t* scores, trust (*t* = 5.200, Sig. = .000), providing help in getting maximum product benefits (*t* = 2.707, Sig. = .008), providing

Table 5. Cronbach's Alpha for Each Dimension and the Initial and Resulting SERVPERF.

The initial SERVPERF Scale	Cronbach's α	The resulting SERVPERF Scale	Cronbach's α
Tangibles		Tangibles	
P1		P1	
P2	.855	P2	.928
P3		P3	
P4		P4	
Reliability		Reliability	
P5		P15	
P6	.818	P16	.834
P7			
P8			
Responsiveness		Responsiveness	
P9		P9	
P10	.848	P10	.878
P11			
P12			
Assurance		Assurance	
P13		P13	
P14		P14	
P15		P17	
P16	.899	P18	.920
P17		P19	
P18		P20	
		P22	
Empathy		Empathy	
P19		P6	
P20		P11	
P21	.861	P12	.902
P22		P24	
P23			
P24			
The overall scale	.959	The overall scale	.942

a choice of product and service offerings according to customer needs ($t = 2.829$, Sig. = .005), and knowledge sharing ($t = -2.171$, Sig. = .032) were all significant at a significance level $p < .05$. The positive β values for trust, providing help to customers in getting maximum product benefits, and providing a choice of product and service offerings according to customer needs had a positive relationship with customer satisfaction, whereas the negative β value for knowledge sharing had a negative relationship with customer satisfaction. A possible explanation for this might be that highly knowledgeable personnel providing TS might have been assuming that customers were aware of developments in the area when such might not have been the case.

Although P22 had the value of standardized coefficient β = .142 and was statistically nonsignificant ($t = 1.793$, Sig. = 0.075), this item should be taken into consideration as predictor of customer satisfaction as its significance p level was close to .05. That P22 may be important can be seen from the correlation matrix in Table 4 where P22, "technical support personnel give you personal attention," was relatively highly

correlated with P13, "you can trust TS personnel," at the value of .717 and with P14, "technical support personnel share their knowledge with customer," at the value of .737. The results for P22 from *empathy* should be interpreted with caution because personal attention in relationships may also be viewed as a value-in-use attribute for prediction of customer satisfaction. The overall results also show that *assurance* was the single most important dimension affecting TS-perceived service quality.

Analysis

The key value-in-use attributes. The first question in the study sought to determine the most important value-in-use attributes from perspective of customer who uses automotive paints and related products. Obtained service quality dimensions assurance, empathy, reliability, and responsiveness determined value-in-use attributes built in each corresponding item from these dimensions. The results show that the key attributes of value-in-use were trust, providing help to

Table 6. Item–Total Statistics for the Initial SERVPERF.

	Corrected item–total correlation	Cronbach's α if item is deleted
Tangibles (overall Cronbach's α = .855)		
P1	0.803	.768
P2	0.798	.771
P3	0.546	.875
P4	0.687	.819
Reliability (overall Cronbach's α = .818)		
P5	0.605	.788
P6	0.604	.788
P7	0.694	.745
P8	0.669	.759
Responsiveness (overall Cronbach's α = .848)		
P9	0.683	.808
P10	0.744	.781
P11	0.747	.782
P12	0.577	.853
Assurance (overall Cronbach's α = .899)		
P13	0.757	.877
P14	0.689	.887
P15	0.730	.880
P16	0.714	.883
P17	0.703	.884
P18	0.769	.874
Empathy (overall Cronbach's α = .861)		
P19	0.669	.837
P20	0.698	.834
P21	0.711	.828
P22	0.777	.820
P23	0.555	.860
P24	0.607	.848

customers in getting maximum product benefits, providing a choice of product and service offerings according to customer needs, and knowledge sharing. These results corroborate the findings of a great deal of Raja et al.'s (2013) findings, which showed that the key attributes of value-in-use were relational dynamics, access, knowledge, and providing a range of product and service offerings. In contrasts to their findings, access was not detected as a key attribute. Previous study has also stressed the importance of fast responses and instant attention. However, the result suggests that access to information on time or when the information was needed was not of significance to customers. A possible explanation for this result may be that relationship quality and knowledge are considered more important in customer usage processes than accessibility by customers who are well experienced. The results of demographic breakdown show that most of the respondents were retailers and end users/ painters with more than 10 years of experience in the automotive paint market being able to solve common product problems satisfactory. Another possible explanation may dwell in customer participation behavior in value co-creation who can express tolerance when service provision does not meet their expectations (Yi & Gong, 2013).

The study found that trust was the most important value-in-use attribute. It is the quality of the relationship that is important for creation and application of knowledge. Previous study contributing to S-D logic has found that product support was one among the key relationship value-creation drivers (Ulaga, 2003). Some authors have noted that value is co-created through customers' unique experience of interactions in which risk-benefits assessments can be obtained through communication, shared learning, shared problem solving, and transparency of and access to information. It can thus be suggested that TS through product support and its key dimensions assists in customers' everyday practice, particularly when such assistance is ensured by experienced personnel providing TS.

The study also found that helping customers in getting maximum product benefits and providing a choice of product and service offerings according to customer needs are key value-in-use attributes. These findings together show the importance of engagement of TS in NPD, field visits, and

Table 7. Eigenvalues for the Initial SERVPERF.

Factor	Initial eigenvalues			Rotation sums of squared loadings
	Total	% of variance	Cumulative %	Total
1	12.845	53.521	53.521	10.713
2	1.991	8.295	61.816	9.614
3	1.415	5.894	67.710	9.771
4	0.882	3.676	71.386	
5	0.734	3.058	74.444	
6	0.664	2.766	77.210	
7	0.649	2.703	79.914	
8	0.593	2.472	82.385	
9	0.506	2.108	84.493	
10	0.437	1.821	86.315	
11	0.422	1.757	88.072	
12	0.370	1.540	89.612	
13	0.314	1.308	90.920	
14	0.303	1.264	92.184	
15	0.255	1.061	93.245	
16	0.231	0.963	94.208	
17	0.227	0.946	95.154	
18	0.213	0.889	96.044	
19	0.202	0.844	96.887	
20	0.192	0.801	97.689	
21	0.178	0.743	98.432	
22	0.153	0.637	99.069	
23	0.136	0.566	99.635	
24	0.088	0.365	100.000	

Table 8. Factor Loadings for the Resulting SERVPERF.

Item	Factor				
	1	2	3	4	5
P18	0.915				
P17	0.877				
P22	0.770				
P20	0.712				
P14	0.686				
P19	0.661				
P13	0.505				
P24		1.011			
P6		0.874			
P11		0.729			
P12		0.687			
P1			0.981		
P2			0.859		
P16				0.726	
P15				0.519	
P9					0.877
P10					0.513

training to customers. Without field visits, engagement in NPD, and training, it would not have been possible to help customers in getting maximum product benefits and offer products and services that satisfy customer needs. In reinforcement to this, P. Williams and Nauman (2011) observed that there are strong links between customer satisfaction and

Table 9. The Resulting SERVPERF: The Service Quality Dimensions and Value-in-Use Attributes.

Dimension	Factor	Item	Factor loadings	Value-in-use attribute
Assurance	F1	P18	0.915	Knowledge
		P17	0.877	A range of product and service offerings
		P22	0.770	Relationship
		P20	0.712	Relationship
		P14	0.686	Knowledge
		P19	0.661	Knowledge
		P13	0.505	Relationship
Empathy	F2	P24	1.011	Accessibility
		P6	0.874	Relationship
		P11	0.729	Accessibility
		P12	0.687	Accessibility
Tangibles	F3	P1	0.981	NA
		P2	0.859	NA
Reliability	F4	P16	0.726	Knowledge
		P15	0.519	Knowledge
Responsiveness	F5	P9	0.877	Accessibility
		P10	0.513	Accessibility

Note. NA = not applicable.

Table 10. Total Variance of Factors for the Resulting SERVPERF.

	Initial eigenvalues			Extraction sums of squared loadings			Rotation sums of squared loadings
Factor	Total	% of variance	Cumulative %	Total	% of variance	Cumulative %	Total
1	9.380	55.176	55.176	9.108	53.578	53.578	8.233
2	1.746	10.268	65.444	1.571	9.243	62.820	7.059
3	1.264	7.436	72.879	0.990	5.824	68.644	4.832
4	0.669	3.936	76.815	0.415	2.443	71.087	3.736
5	0.558	3.281	80.096	0.390	2.292	73.379	5.369
6	0.497	2.922	83.018				
7	0.460	2.707	85.726				
8	0.425	2.497	88.223				
9	0.349	2.052	90.275				
10	0.309	1.817	92.092				
11	0.259	1.521	93.613				
12	0.234	1.377	94.989				
13	0.219	1.286	96.276				
14	0.203	1.194	97.470				
15	0.169	0.996	98.466				
16	0.157	0.921	99.387				
17	0.104	0.613	100.000				

Table 11. Regression Coefficients for the Model With 17 Predictors.

Model	R	R^2	Adjusted R^2	SE of the estimate
1	.847[a]	.717	.683	.43810

[a]Predictors: (Constant), P10, P16, P6, P1, P17, P12, P19, P14, P20, P13, P15, P22, P18, P11, P9, P24, P2.

Table 12. ANOVA: F Test and Statistical Significance for the Model With 17 Predictors.

Model 1	Sum of squares	df	M^2	F	Significance
Regression	68.218	17	4.013	20.907	.000[b]
Residual	26.871	140	0.192		
Total	95.089	157			

Dependent variable: P25 (Customer satisfaction).
[b]Predictors: (Constant), P10, P16, P6, P1, P17, P12, P19, P14, P20, P13, P15, P22, P18, P11, P9, P24, P2.

Table 13. Standardized Coefficients Beta (β), t Value, and Statistical Significance for the Model With 17 Predictors.

		Unstandardized coefficients		Standardized coefficients			Correlations
Model 1	Value-in-use attributes	B	SE	β	t	Significance	Zero-order
Constant		.124	.244		0.507	.613	
P1	NA	.025	.054	.045	0.464	.643	.414
P2	NA	−.017	.056	−.030	−0.297	.767	.428
P15	Knowledge	.076	.075	.077	1.008	.315	.593
P16	Knowledge	−.062	.069	−.067	−0.903	.368	.574
P9	Access	.036	.062	.050	0.591	.555	.512
P10	Access	.003	.061	.005	0.056	.956	.550
P18	Knowledge	.200	.074	.215	2.707	.008	.706
P17	A range of product and service offerings	.191	.068	.202	2.829	.005	.668
P22	Relationship	.129	.072	.142	1.793	.075	.680
P20	Relationship	.023	.071	.025	0.328	.743	.641
P14	Knowledge	−.151	.069	−.163	−2.171	.032	.568
P19	Knowledge	.023	.068	.026	0.343	.732	.616
P13	Relationship	.435	.084	.403	5.200	.000	.756
P24	Access	.110	.065	.157	1.693	.093	.546
P6	Relationship	−.038	.068	−.042	−0.561	.576	.481
P12	Access	−.024	.051	−.034	−0.464	.643	.515
P11	Access	.001	.068	.002	0.021	.983	.583

Note. NA = not applicable.
Dependent Variable: P25 (Customer Satisfaction).

retention with respect to revenue and earnings per share, and hence, better financial performance is more likely to occur.

Previous studies on S-D logic have emphasized knowledge sharing as a source of competitive advantages and noted interconnectedness between relationship and knowledge. Of particular note in the study is the importance of knowledge sharing that led customers to determine it as value-in-use attribute. This finding further supports the idea of Yi and Gong (2013) that customer participation behavior in value co-creation is to seek information and share them with employees following their guidance and orientations. This finding also confirms relationship between knowledge sharing and trust in value co-creation activities. It is encouraging to compare this result with that found by Cai, Goh, de Souza, and Li (2013) that trust was a key determinant of knowledge sharing in collaborative supply–chain relationship. This result may be explained by the fact that TS transfers knowledge to customers but also acquires new knowledge from them. Sharing of knowledge between customers and the TS personnel correlates with sharing of knowledge between the TS team members. The finding also suggests correlation with other three key attributes—trust, help in providing maximum value from a product, and providing products and services that satisfy customer needs. Taken together, these findings support Prahalad and Ramaswamy's (2004) DART model of interactions in the value co-creation process where shared learning, communication and shared problem solving with customers, and transparency of and access to information together assist customers to perceive possible business risks and accordingly make decisions and take actions.

The key value-in-use attributes and customer satisfaction. The second question sought to determine whether the customer was satisfied or dissatisfied with the key value-in-use attributes. The results show that trust was the strongest predictor of customer satisfaction having the greatest positive impact.

The importance of interactions between TS personnel and customers that led to trust to be valued as value-in-use attribute corroborated the idea of Huang and Wilkinson (2014) who suggested that trust was not fostering trusting environment in relationships, but it was fostering trusting actions that reflected decisions "to rely on trust to deal with perceived risks" (p. 272) and the way trusted actions were used. Accordingly, the result suggests that the customer is satisfied when trusted actions were previously experienced and proved to be true in assisting or improving of the customer's processes or operations.

Next strong predictors with positive relationship with customer satisfaction were providing help to customers in getting maximum product benefits and providing a choice of product and service offerings according to the customer needs. As mentioned earlier, a customer is satisfied or dissatisfied depending on whether goals, objectives, and purposes of the usage process were met. It can therefore be assumed that relationship, knowledge, and a range of products and services offerings that satisfy customer needs are the most significant attributes of the usage process for a prediction of customer satisfaction. The results also show that knowledge sharing was the least significant predictor with negative relationship with customer satisfaction. This result suggests that without improvements in quality of the relationship, providing help in getting maximum product benefits, and better understanding customer needs (probably due to ever-increased customer needs), a customer would be dissatisfied if there is only a bundle of new information, which is not properly used and/or does not meet customer goals. Previous studies have noted that value co-creating activities are relating, communicating, and knowing, and that effective knowledge sharing in collaborations between suppliers requires understanding, valuing, and absorbing the partner's knowledge. Therefore, sharing of knowledge requires evidence that information and learning received by customers were supportive in meeting customer needs and goals leading to satisfied customers.

Conclusions and Recommendations

Summary of conclusions

The conclusion about the first research question is that the key attributes of value-in-use in the automotive paint market are trust, providing help to customers in getting maximum product benefits, providing a choice of product and service offerings according to customer needs, and knowledge sharing. Trust, as an essential element of relationship, is the most important key value-in-use attribute in the automotive paint market; therefore, TS affects customer satisfaction through relationship quality. TS as an after-sales strategy builds relationships with customers and assists in product support and customers' own processes reflecting trust as the most important attribute. It comes as no surprise considering that TS

personnel train customers and spend most of their working (and non-working) time on-site and in personal interactions supporting customer's activities and business performances. Customers value those product and service attributes *in use* affected by TS, which are perceived to generate and fulfill desired outcomes of customer's own processes. However, customers did not perceive of higher importance time availability of TS personnel and prompt TS service. Although the availability is perhaps because such activities require that customers organize themselves thus affecting their everyday's activities, the latter can be explained by the customers' tolerance and experience as most of the respondents had more than 10 years of experience and were able to solve product problems without affecting their own processes.

Regarding the second research question, the factors such as trust, knowledge that ensures help to customers in getting maximum product benefits, providing a choice of product and service offerings according to customer needs, and sharing of knowledge with them are good predictors of customers' satisfaction. Among these attributes, trust has the greatest impact on customer satisfaction. Therefore, the conclusion emerging from this study is that TS is perceived by customers as TS personnel possessing capabilities to assess well customer usage processes, understanding customer needs, objectives, and goals, and improving customers' performances.

Another conclusion emerging from the study is that *assurance* has the strongest impact on customer satisfaction because it comprises the key TS dimensions. TS should maintain training to customers and field visits as they are sources of sustainable relationship, knowledge, and understanding customer needs. Both field visits and training ensure active participation of customers on what is important for value co-creation enhancing relationships. They also build customer trust assisting in understanding customer needs because in such interactions, TS personnel can comprehend customer expectations and responses to product and service offerings. Therefore, TS should focus not only on hard skills but on soft skills as well.

Benefits of the Study

The findings from this study make several noteworthy contributions to the existing literature. The current empirical findings add to a growing body of literature on S-D logic and value-in-use, which have strong conceptual background, yet still little empirical evidence. The study provides additional evidence of the importance of relationship quality and the recent shift from service quality to quality of relationships with customers. The present study also enhances our knowledge on TS in manufacturing. It contributes to our knowledge on TS in the paint industry and automotive paint market. In addition, the study contributes to existing knowledge on customer satisfaction by providing empirical evidence on the impact of TS on customer satisfaction.

Implications of the Study

This study has several implications for the automotive paint industry in general and the provision of TS in particular. First, the industry is an evolving one with new products developing at an increasing pace. Second, assumptions regarding customers of automotive paints are fully aware of developments, and new products might be wrong, and hence, it requires a dynamic team to visit customers as often as required to share with them knowledge and expertise to accommodate new techniques. Last, relationship building and trust arising from such relationship are key drivers for ensuring competitive advantage and growth.

Limitations of the study

The findings in this study are subjected to a number of limitations. First limitation is that small sample size and the use of the non-probability convenience sampling method limited the research results to generalizability. Second, the SERVPERF instrument was adapted to the research context and may not be applicable in other industries. Third, the value-in-use attributes were included in the SERVPERF instrument based on the main author's assumption from the literature review that the four key value-in-use attributes may be included in each corresponding item from the questionnaire. The fourth limitation is that most of customers were from Egypt and far less

from India producing results more related to the customers from Egypt than from India. Last, customers' prior knowledge and experience were not taken into consideration although they might likely be positively associated with how customers performed and determined value-in-use and customer satisfaction as well.

Recommendations for Further Research

The findings of this study may be tested in other industries to explore impact of TS on customer satisfaction through attributes found in a specific industry. The use of the SERVPERF instrument with embedded value-in-use attributes can also be investigated in a further research. A further study might use in-depth interviews to investigate the identified attributes and their relationship. Another possible area of further research may be longitudinal research to understand how value-in-use attributes may change over time with changes of customers' goals and with TS service quality as well. Finally, with respect to the finding "knowledge sharing" is significantly negatively related to customer satisfaction, it might be useful to look into this further by conducting research in other sectors to shed light on the same. The assumption that some respondents did not like to share their knowledge, expertise, and experience or be told what to do might not hold true.

Appendix A

Table A1. The Adapted SERVPERF Instrument and Value-in-Use Attributes.

Item		Value-in-use attributes
Tangibles		
P1	The paint manufacturing company has up-to-date equipment (such as spray booth, etc.) used for technical support of customers.	NA
P2	Physical facilities used for training of customers (such as training centers, etc.) in the paint manufacturing company are visually appealing.	NA
P3	Technical support personnel appear neat.	NA
P4	Materials associated with technical support (such as Technical Data Sheets/TDSs, etc.) are in keeping with the type of services provided.	NA
Reliability		
P5	When technical support personnel promise to do something by a certain time, they do so.	Relationship
P6	When you have problems, technical support shows sincere interest in solving them.	Relationship
P7	The paint manufacturing company provides technical support to customer at the time it promises to do so.	Access
P8	Technical support personnel provide accurate technical data.	Knowledge
Responsiveness		
P9	Technical support personnel tell customer exactly when services (such as field visits, training, presentations, seminars, etc.) will be performed.	Access
P10	You receive prompt technical support from the paint manufacturing company.	Access
P11	Technical support personnel are always willing to help customer.	Access
P12	Technical support personnel are not too busy to respond to customer request promptly.	Access
Assurance		
P13	You can trust technical support personnel.	Relationship
P14	Technical support personnel share their knowledge with customer.	Knowledge

(continued)

Appendix A (continued)

Item		Value-in-use attributes
P15	Technical support personnel resolve customer complaints accurately.	Knowledge
P16	Technical support personnel are able to tackle your specific problems (products, colors, application, etc.).	Knowledge
P17	Technical support personnel provide you with choice of product and service offerings according to your needs.	A range of product and service offerings
P18	Technical support personnel help you in getting maximum product benefits.	Knowledge
Empathy		
P19	Technical support personnel know what your needs are.	Knowledge
P20	Technical support personnel build sustaining relationship with customer.	Relationship
P21	The paint manufacturing company gives you individual attention.	Relationship
P22	Technical support personnel give you personal attention.	Relationship
P23	The paint manufacturing company has your best interest at heart.	Knowledge
P24	Technical support personnel are available at any time when you need them.	Access

Note. NA = not applicable.

Appendix B

Survey Questionnaire

This survey deals with your feelings about technical support offered by a paint manufacturing company producing car refinish products. For each statement, please show the extent to which you believe the paint manufacturing company has the feature described by the statement. If you strongly agree with the feature described by the statement, choose 5 and if you strongly disagree, choose 1. You may use any of the numbers in the middle as well to show how strong your feelings are. There are no right or wrong answers—all we are interested in is a number that best shows your perceptions about the paint manufacturing company whether you use its technical support or not.

		Strongly disagree	Disagree	Neutral	Agree	Strongly agree
1	The paint manufacturing company has up-to-date equipment (such as spray booth, etc.) used for technical support of customers.	1	2	3	4	5
2	Physical facilities used for training of customers (such as training centers, etc.) in the paint manufacturing company are visually appealing.	1	2	3	4	5
3	Technical support personnel appear neat.	1	2	3	4	5
4	Materials associated with technical support (such as Technical Data Sheets/TDSs, etc.) are in keeping with the type of services provided.	1	2	3	4	5
5	When technical support personnel promise to do something by a certain time, they do so.	1	2	3	4	5
6	When you have problems, technical support shows sincere interest in solving them.	1	2	3	4	5
7	The paint manufacturing company provides technical support to customer at the time it promises to do so.	1	2	3	4	5
8	Technical support personnel provide accurate technical data.	1	2	3	4	5
9	Technical support personnel tell customer exactly when services (such as field visits, training, presentations, seminars, etc.) will be performed.	1	2	3	4	5
10	You receive prompt technical support from the paint manufacturing company.	1	2	3	4	5
11	Technical support personnel are always willing to help customer.	1	2	3	4	5
12	Technical support personnel are not too busy to respond to customer request promptly.	1	2	3	4	5

(continued)

Appendix B (continued)

	Strongly disagree	Disagree	Neutral	Agree	Strongly agree
13 You can trust technical support personnel.	1 ☐	2 ☐	3 ☐	4 ☐	5 ☐
14 Technical support personnel share their knowledge with customer.	1 ☐	2 ☐	3 ☐	4 ☐	5 ☐
15 Technical support personnel resolve customer complaints accurately.	1 ☐	2 ☐	3 ☐	4 ☐	5 ☐
16 Technical support personnel are able to tackle your specific problems (products, colors, application, etc.).	1 ☐	2 ☐	3 ☐	4 ☐	5 ☐
17 Technical support personnel provide you with choice of product and service offerings according to your needs.	1 ☐	2 ☐	3 ☐	4 ☐	5 ☐
18 Technical support personnel help you in getting maximum product benefits.	1 ☐	2 ☐	3 ☐	4 ☐	5 ☐
19 Technical support personnel know what your needs are.	1 ☐	2 ☐	3 ☐	4 ☐	5 ☐
20 Technical support personnel build sustaining relationship with customer.	1 ☐	2 ☐	3 ☐	4 ☐	5 ☐
21 The paint manufacturing company gives you individual attention.	1 ☐	2 ☐	3 ☐	4 ☐	5 ☐
22 Technical support personnel give you personal attention.	1 ☐	2 ☐	3 ☐	4 ☐	5 ☐
23 The paint manufacturing company has your best interest at heart.	1 ☐	2 ☐	3 ☐	4 ☐	5 ☐
24 Technical support personnel are available at any time when you need them.	1 ☐	2 ☐	3 ☐	4 ☐	5 ☐

Customer Satisfaction

To what extent do you agree or disagree with the following?

		Strongly disagree	Disagree	Neutral	Agree	Strongly agree
25	Overall, I am very satisfied with technical support.	1 ☐	2 ☐	3 ☐	4 ☐	5 ☐

Authors' Note

The study was carried out as part of Master in Business Administration of the lead author at the University of Liverpool/ Laureate.

Declaration of Conflicting Interests

The author(s) declared no potential conflicts of interest with respect to the research, authorship, and/or publication of this article.

Funding

The author(s) received no financial support for the research, authorship, and/or publication of this article.

References

Ahn, J. S., & Sohn, S. Y. (2009). Customer pattern search for after-sales service in manufacturing. *Expert Systems With Applications, 36*, 5371-5375.

Aurich, J. C., Fuchs, C., & Wagenknecht, C. (2006). Life cycle oriented design of technical product-service systems. *Journal of Cleaner Production, 14*, 1480-1494.

Automotive coatings: Technologies and global markets. (2014, September 17). *PR Newswire US*, 1-197.

Automotive Paint Market. (2015). Global industry analysis and forecast to 2020. *Paintindia, 65*, 154-155.

Baines, T. S., Lightfoot, H. W., Evans, S., Neely, A., Greenough, R., Peppard, J., . . . Wilson, H. (2007). State-of-the-art in product-service systems. *Journal of Engineering Manufacture, 221*, 1543-1552.

Ballantyne, D., & Varey, R. J. (2006). Creating value-in-use through marketing interaction: The exchange logic of relating, communicating and knowing. *Marketing Theory, 6*, 335-348.

Banu, A., & Gül, B. (2012). Is there a need to develop a separate service quality scale for every service sector? Verification of SERVQUAL in higher education services. *Journal of Faculty of Economics and Administrative Sciences, 17*, 423-440.

Bell, S. J., Auh, S., & Smalley, K. (2005). Customer relationship dynamics: Service quality and customer loyalty in the context of varying levels of customer expertise and switching costs. *Journal of the Academy of Marketing Science, 33*, 169-183.

Bowen, D. E., Siehl, C., & Schneider, B. (1989). A framework for analyzing customer service orientations in manufacturing. *Academy of Management Review, 14*, 75-95.

Bundschuh, R. G., & Dezvane, T. M. (2003). How to make after-sales services pay off. *McKinsey Quarterly, 4*, 116-127.

Burger, P. C., & Cann, C. W. (1995). Post-purchase strategy: A key to successful industrial marketing and customer satisfaction. *Industrial Marketing Management, 24*, 91-98.

Buttle, F. (1996). SERVQUAL: Review, critique, research agenda. *European Journal of Marketing, 30*(1), 8-32.

Cai, S., Goh, M., de Souza, R., & Li, G. (2013). Knowledge sharing in collaborative supply chains: Twin effects of trust and power. *International Journal of Production Research, 51*, 2060-2076.

Chemical Business. (2013, December). WACKER expands range of services offered at its Singapore Technical Center. *Chemical Business, 27*(12), 83.

Cherubini, S., Iasevoli, G., & Michelini, L. (2015). Product-service systems in the electric car industry: Critical success factors in marketing. *Journal of Cleaner Production, 97*, 40-49.

Christopher, M., Payne, A., & Ballantyne, D. (1991). *Relationship marketing: Bringing quality customer service and marketing together.* Retrieved from https://dspace.lib.cranfield.ac.uk/handle/1826/621

Cohen, M. A., Agrawal, N., & Agrawal, V. (2006). Winning in the aftermarket. *Harvard Business Review, 84*, 129-138.

Cooper, R. G., & Kleinschmidt, E. J. (1993). Major new products: What distinguishes the winners in the chemical industry? *Journal of Product Innovation Management, 10*, 90-111.

Cronin, J. J., & Taylor, S. A. (1992). Measuring service quality: A reexamination and extension. *Journal of Marketing, 56*(3), 55-68.

Cronin, J. J., & Taylor, S. A. (1994). SERVPERF versus SERVQUAL: Reconciling performance-based and perceptions-minus-expectations measurement of service quality. *Journal of Marketing, 58*(1), 125-131.

Das, A. (2003). Knowledge and productivity in technical support work. *Management Science, 49*, 416-431.

Dorsch, M. J., Swanson, S. R., & Kelley, S. W. (1998). The role of relationship quality in the stratification of vendors as perceived by customers. *Journal of the Academy of Marketing Science, 26*, 128-142.

Esterby-Smith, M., Thorpe, R., & Jackson, P. (2012). *Management research* (4th ed.). London, England: SAGE.

Fabrigar, L. R., Wegener, D. T., MacCallum, R. C., & Strahan, E. J. (1999). Evaluating the use of exploratory factor analysis in psychological research. *Psychological Methods, 4*, 272-299.

Field, A. P. (n.d.). *Discovering statistics using SPSS.* Retrieved from www.sagepub.com/field4e/study/smartalex/chapter17.pdf

Finkelman, D. P., & Goland, A. R. (1990). How not to satisfy your customers. *McKinsey Quarterly, 1*, 2-12.

Flint, D. J., Blocker, C. P., & Boutin, P. J. (2011). Customer value anticipation, customer satisfaction and loyalty: An empirical examination. *Industrial Marketing Management, 40*, 219-230.

Freeman, K. D., & Dart, J. (1993). Measuring the perceived quality of professional business services. *Journal of Professional Service Marketing, 9*, 27-47.

Goffin, K. (1998). Evaluating customer support during new product development—An exploratory study. *Journal of Product Innovation Management, 15*, 42-56.

Goffin, K., Lemke, F., & Szwejczewski, M. (2006). An exploratory study of "close" supplier–manufacturer relationships. *Journal of Operations Management, 24*, 189-209.

Goffin, K., & New, C. (2001). Customer support and new product development—An exploratory study. *International Journal of Operations & Production Management, 21*, 275-301.

Gray, D. E. (2014). *Doing research in the real world* (3rd ed.). London, England: SAGE.

Griffin, A., Gleason, G., Preiss, R., & Shevenaugh, D. (1995). Best practice for customer satisfaction in manufacturing firms. *Sloan Management Review, 36*(2), 87-98.

Grönroos, C. (2008). Service logic revisited: Who creates value? And who co-creates? *European Business Review, 20*, 298-314.

Grönroos, C. (2012). Conceptualising value co-creation: A journey to the 1970s and back to the future. *Journal of Marketing Management, 28*, 1520-1534.

Gummesson, E. (2008). Quality, service-dominant logic and many-to-many marketing. *The TQM Journal, 20*, 143-153.

Haque, M. I. (2013). Assessing the adequacy of SERVQUAL dimensions in retail banking. *International Journal of Academic Research, 5*, 99-104.

Huang, Y., & Wilkinson, F. (2014). A case study of the development of trust in a business relation: Implications for a dynamic theory of trust. *Journal of Business Market Management, 7*, 254-279.

Hull, D. J., & Cox, J. F. (1994). The field service function in the electronics industry: Providing a link between customers and production/marketing. *International Journal of Production Economics, 37*, 115-126.

Innis, D. E., & La Londe, B. J. (1994). Customer service: The key to customer satisfaction, customer loyalty, and market share. *Journal of Business Logistics, 15*(1), 1-27.

Institute for Digital Research and Education-University of California, Los Angeles. (2014). *SAS annotated SPSS output-factor analysis.* Retrieved from http://www.ats.ucla.edu/stat/sas/output/factor.htm

Jap, S. D. (1999). Pie-expansion efforts: Collaboration processes in buyer-supplier relationships. *Journal of Marketing Research, 36*, 461-475.

Kettinger, W. J., & Lee, C. C. (1994). Perceived service quality and user satisfaction with the information services function. *Decision Sciences, 25*, 737-766.

Kosenko, R., & Samli, A. C. (1985). China's four modernization programs and technology transfer. In A. C. Samli (Ed.), *Technology transfer* (pp. 107-131). New York, NY: Quorum.

Lai, F., Hutchinson, J., Li, D., & Bai, C. (2007). An empirical assessment and application of SERVQUAL in mainland China's mobile communications industry. *International Journal of Quality & Reliability Management, 24*, 244-262.

Lele, M. M. (1986). How service needs influence product strategy. *Sloan Management Review, 28*, 63-70.

Lele, M. M., & Karmarkar, U. S. (1983). Good product support is smart marketing. *Harvard Business Review, 61*, 124-132.

Lovelock, C. H., & Wirtz, J. (2011). *Services marketing: People, technology, strategy* (7th ed.). Upper Saddle River, NJ: Pearson Prentice Hall.

Lusch, R. F., & Nambisan, S. (2015). Service innovation: A service-dominant logic perspective. *MIS Quarterly, 39*, 155-176.

Macdonald, E. K., Wilson, H., Martinez, V., & Toossi, A. (2011). Assessing value-in-use: A conceptual framework and exploratory study. *Industrial Marketing Management, 40*, 671-682.

Mathieu, V. (2001). Product services: From a service supporting the product to a service supporting the client. *Journal of Business and Industrial Marketing, 16*, 39-61.

Möller, K. (2006). Role of competences in creating customer value: A value-creation logic approach. *Industrial Marketing Management, 35*, 913-924.

Morris, M. H., & Davis, D. L. (1992). Measuring and managing customer service in industrial firms. *Industrial Marketing Management, 21*, 343-353.

Nadiri, H., & Hussain, K. (2005). Perceptions of service quality in North Cyprus hotels. *International Journal of Contemporary Hospitality Management, 17*, 469-480.

Oliva, R., & Kallenberg, R. (2003). Managing the transition from products to services. *International Journal of Service Industry Management, 14*, 160-172.

Pan, J., & Nguyen, H. (2015). Decision support: Achieving customer satisfaction through product–service systems. *European Journal of Operational Research, 247*, 179-190.

Parasuraman, A., Zeithaml, V. A., & Berry, L. L. (1988). SERVQUAL: A multiple-item scale for measuring consumer perceptions of service quality. *Journal of Retailing, 64*, 12-40.

Payne, A. F., Storbacka, K., & Frow, P. (2008). Managing the co-creation of value. *Journal of the Academy of Marketing Science, 36*, 83-96.

Prahalad, C. K., & Ramaswamy, V. (2004). Co-creation experiences: The next practice in value creation. *Journal of Interactive Marketing, 18*(3), 5-14.

Raithel, S., Sarstedt, M., Scharf, S., & Schwaiger, M. (2012). On the value relevance of customer satisfaction. Multiple drivers and multiple markets. *Journal of the Academy of Marketing Science, 40*, 509-525.

Raja, J. Z., Bourne, D., Goffin, K., Çakkol, M., & Martinez, V. (2013). Achieving customer satisfaction through integrated products and services: An exploratory study. *Journal of Product Innovation Management, 30*, 1128-1144.

Rauyruen, P., & Miller, K. E. (2007). Relationship quality as a predictor of B2B customer loyalty. *Journal of Business Research, 60*, 21-31.

Redding, L. E., Tiwari, A., Roy, R., Phillips, P., & Shaw, A. (2014). The adoption and use of through-life engineering services within UK manufacturing organisations. *Proceedings of the Institution of Mechanical Engineers, Part B: Journal of Engineering Manufacture*. Advance online publication. doi:10.1177/0954405414539931

Saccani, N., Songini, L., & Gaiardelli, P. (2006). The role and performance measurement of after-sales in the durable consumer goods industries: An empirical study. *International Journal of Productivity and Performance Management, 55*, 259-283.

Schembri, S. (2006). Rationalizing service logic, or understanding services as experience? *Marketing Theory, 6*, 381-392.

Segoro, W. (2013). The influence of perceived service quality, mooring factor, and relationship quality on customer satisfaction and loyalty. *Procedia—Social and Behavioral Sciences, 81*, 306-310.

Stauss, B. (2005). A Pyrrhic victory: The implications of an unlimited broadening of the concept of services. *Managing Service Quality, 15*, 219-229.

Tukker, A. (2004). Eight types of product–service system: Eight ways to sustainability? Experiences from SusProNet. *Business Strategy and the Environment, 13*, 246-260.

Tyagi, S., Yang, K., Tyagi, A., & Verma, A. (2012). A fuzzy goal programming approach for optimal product family design of mobile phones and multiple-platform architecture. *IEEE Transactions on Systems, Man, and Cybernetics–Part C: Applications and Reviews, 42*, 1519-1530.

Ulaga, W. (2003). Capturing value creation in business relationships: A customer perspective. *Industrial Marketing Management, 32*, 677-693.

Vandermerwe, S., & Rada, J. (1988). Servitization of business: Adding value by adding services. *European Management Journal, 6*, 314-324.

Vargo, S. L., & Lusch, R. F. (2004). Evolving to a new dominant logic for marketing. *Journal of Marketing, 68*(1), 1-17.

Vargo, S. L., & Lusch, R. F. (2008a). From goods to service(s): Divergences and convergences of logics. *Industrial Marketing Management, 37*, 254-259.

Vargo, S. L., & Lusch, R. F. (2008b). Service-dominant logic: Continuing the evolution. *Journal of the Academy of Marketing Science, 36*, 1-10.

Vargo, S. L., Maglio, P. P., & Akaka, M. A. (2008). On value and value co-creation: A service systems and service logic perspective. *European Management Journal, 26*, 145-152.

Vega-Vazquez, M., Revilla-Camacho, M. A., & Cossío-Silva, F. J. (2013). The value co-creation process as a determinant of customer satisfaction. *Management Decision, 51*, 1945-1953.

Wang, W.-T., & Lu, Y.-C. (2010). Knowledge transfer in response to organizational crises: An exploratory study. *Expert Systems With Applications, 37*, 3934-3942.

Williams, B., Brown, T., & Onsman, A. (2012). Exploratory factor analysis: A five-step guide for novices. *Australasian Journal of Paramedicine, 8*(3), 1-13.

Williams, P., & Naumann, E. (2011). Customer satisfaction and business performance: A firm-level analysis. *Journal of Services Marketing, 25*, 20-32.

Woodruff, R. B. (1997). Customer value: The next source for competitive advantage. *Journal of the Academy of Marketing Science, 25*, 139-153.

Yi, Y., & Gong, T. (2013). Customer value co-creation behavior: Scale development and validation. *Journal of Business Research, 66*, 1279-1284.

Author Biographies

Natasa Gajic earned her MBA at the University of Liverpool/Laureate and Bachelor of Chemical Engineering from the University of Belgrade. She spent more than ten years in paint formulation in R&D. She has been engaged in the automotive paint market for more than eleven years. She is a Technical Support Manager in KAPCI Coatings, a manufacturer of car refinish products, wood coatings, and decorative paints in Egypt. She is accountable for managing Technical Support Department providing technical support to users of car refinish and wood coatings products in more than 50 counties all over the world.

Mehraz Boolaky earned his MBA and PhD from the University of Mauritius (UOM) and Bachelor of Chemical Engineering from UICT, Mumbai. He was a Dean of the FLM at the University of Mauritius. He held visiting/full time academic positions in Kazakhstan, France, Madagascar, Malaysia, Morocco, etc. He has widely published and presented papers in conferences and workshops. He spent more than twenty five years in senior managerial positions after graduation. He is a Professor of Marketing at the Asia Pacific Institute of Management in New Delhi and a Honorary Lecturer and Dissertation Adviser at the University of Liverpool/Laureate.

19

A Novel Approach for Allocating Mathematical Expressions to Visual Speech Signals

Mohammad Hossein Sadaghiani[1],
Niusha Shafiabady[1], and Dino Isa[1]

Abstract

In this article, visual speech information modeling analysis by explicit mathematical expressions coupled with words' phonemic structure is presented. The visual information is obtained from deformation of lips' dimensions during articulation of a set of words that is called visual speech sample set. The continuous interpretation of the lips' movement has been provided using Barycentric Lagrange Interpolation producing a unique mathematical expression named visual speech signal. Hierarchical analysis of the phoneme sequences has been applied for words' categorization to organize the database properly. The visual samples were extracted from three visual feature points chosen on the lips via an experiment in which two individuals pronounced the aforementioned words. The simulation results show that each individual word can be represented by a mathematical expression or visual speech signal whereas the sample sets can also be derived from the same mathematical expression, and this is a significant improvement over the popular statistical methods.

Keywords

viseme, visual speech signal, Barycentric Lagrange Interpolation

Introduction

The audiovisual speech recognition and visual speech synthesizers are two interfaces for human and machine interaction (Chin, Seng, & Ang, 2012). Such interaction relies on the analysis and synthesis of both audio and visual information, which humans use for face-to-face communication. It has been shown that the visual cues of speech also enhance the transparency of speech when they are degraded by environmental noise (Sumby & Pollack, 1954).

The visual speech information is the bimodal component of audio speech signal. These modalities are strongly correlated that could affect the perception of lip movement and audio speech signal (McGurk & MacDonald, 1976). In audiovisual speech recognition systems, the visual information is used as complementary tool for enhancing the perception of speech signal modality or independently for lip reading. The visual recognition module is called interchangeably as lip-reading, visual speech recognition, speech reading, or visual-only automatic speech recognition (Petajan, 1984; Potamianos, Neti, Luettin, & Matthews, 2004).

In lip-reading systems, the dynamics of visual speech in image sequences are extracted by focusing on the appearance of the articulator organs like lips' geometry. Further processing could be conducted to be compatible for fusing to the audio-only automatic speech recognition modality.

A basic component related to phoneme in the visual domain was determined and called viseme (visual phoneme; Fisher, 1968). Since then, no globally accepted model has been suggested for formulating the visual speech components. The concept of visual phoneme does not suggest an explicit definition of lips' structure during phoneme utterance. The visemes are formed based on human perceptions which are categorized using confusion matrix where the most accurately detected visemes form a phoneme–viseme table (Williams, Rutledge, Garstecki, & Katsaggelos, 1997). The deficiency of this method can be observed by the fact that there are various phoneme–viseme tables used (Goldschen, Garcia, & Petajan, 1994; Hazen, Saenko, La, & Glass, 2004; Jiang, Alwan, Auer, & Bernstein, 2001). The next drawback of the viseme concept was the lack of accuracy in visual phonemes perception caused by human interference. In other words, depending on the observers, whether they are hearing impaired or not, the results were changing. In addition, the concept of viseme represents the visual information

---

[1]University of Nottingham, Malaysia Campus, Semenyih, Malaysia

Corresponding Author:
Niusha Shafiabady, School of Electrical and Electronic Engineering, University of Nottingham, Malaysia Campus, Semenyih 43500, Malaysia.
Email: niusha.shafiabady@nottingham.edu.my

as discrete components. Therefore, the continuity of visual speech components must be conducted by complementary methods (Cohen & Massaro, 1993).

To eliminate the additional methods for preserving the continuity of visemes, the most dominant approaches are statistical and probabilistic models (e.g., Hidden Markov Model [HMM]; Heracleous, Tran, Nagai, & Shikano, 2010; Kosmopoulos & Chatzis, 2010; Potamianos & Graf, 1998; Saenko, Livescu, Glass, & Darrell, 2009). The advantage of this approach is its ability to apply modifications and adjustments to the model using optimization methods and training algorithms (Huang & Povey, 2005; Lee & Park, 2006). However, those approaches do not have the ability to formulate the relations between viseme sequences. Therefore, currently there is no standard method which is globally accepted for representing the lips' movement during articulation.

In Revéret and Benoît (1998), a 3D lip model was fitted to lip images, and the amount of similarities between the feature points on the model and actual lips were measured. Their results did not conclude the dynamic of articulating lips with a mathematical expression. In the suggested lip-reading system by Potamianos and Graf (1998), the visual features related to the width, height, and area of lip were tracked over frame or time domain during articulating a sequence of digits. Such trajectories of visual features were used for training a Hidden Markov Model that was not mathematically formulated. The acquired data (from movement of the lips) in Lucero (2002) was formulated by a sixth-order B-spline curve (Mortenson, 1997). The main disadvantage of using a B-spline curve for formulating the lips' movement is the lack of definition of a single expression for the sample sets as B-spline curve is defined on segmented intervals (Mortenson, 1997). The visual information was extracted from tongue tip, tongue blade, tongue dorsum, velum, upper and lower lip, and lower jaw by Electromagnetic Articulography (Ananthakrishnan & Engwall, 2008).

The method suggested in Birkholz, Kröger, and Neuschaefer-Rube (2011) is a time-variant dynamic system used for modeling the articulatory organ to relate facial parts mentioned above by using Electromagnetic Articulography approach. A combination of phoneme was chosen with normal and low speaking rates. The input–output relations of the systems were obtained by solving differential equation in time domain. Although the proposed method tried to allocate a transfer function to describe the dynamics of lips and the related organs, finding the solution of the system in time via differential equations caused computational overhead and do not represent a single expression for the lip movement that is considered as a disadvantage for this method.

The extraction of feature points using the appearance-based approach in Zhao, Barnard, and Pietikainen (2009) was suggested using a statistical visual feature modeling by concatenating the image sequence portions in block volume. Applying an image processing approach called Local Binary Pattern (LBP), the lips' textures between successive images in gray scale are used and transformed to binary codes. Afterward, the LBP is applied on the other two orthogonal planes to the image plane, which codes the spatial and temporal axis. The histogram of these binary orthogonal planes was calculated and represented as the visual feature. By concatenating of the histograms, two important characteristics as lips' appearance and also the motion of lips are extracted. The sequences of histograms were joined to represent the overall statistical property of lip movement. In other words, their method not only extracts the visual feature but also expresses the deformation of visual features dynamically. Interpreting the dynamic of lips is determined based on analyzing the extracted data as appearance-based methods including statistical models (Adjoudani & Benoît, 1996; Bregler & Konig, 1994; Erber, Sachs, & DeFilippo, 1979) and shape-based methods or geometric approaches (Chiou & Hwang, 1997; Rogozan & Deléglise, 1998; Teissier, Robert-Ribes, & Schwartz, 1999) and a combination of appearance-based and shape-based methods (Cootes, Edwards, & Taylor, 1998). The necessity for suggesting a globally defined visual phoneme concept supports the mathematical modeling of the visual information. Therefore, the focus of this article is on extracting the geometry of lips during articulation to suggest a mathematical model of lips' dynamics as there is no specific attempt for expressing the visual speech data during articulation by an explicit mathematical formula. In this article, a straightforward method for formulating the visual speech data is represented, and the above-mentioned data is extracted from transcribed words which are designed based on phonemic structures.

There are two varieties of human–machine interactions, where in the first one, the machine plays a role as interface in human-to-human interaction (audiovisual speech recognition) or animation (visual speech synthesis). In the audiovisual automatic speech recognition (AVASR) systems, the visual information is the input of identification for enhancing the perception of speech signal modality. This visual recognition module is called interchangeably as lip-reading, visual speech recognition (Petajan, 1984), speech reading, or visual-only automatic speech recognition (Potamianos et al., 2004). In the lip-reading systems, by focusing on the appearance of the lip's geometry or pixels color in image sequences, the dynamic of articulating lip is extracted. This information has to be processed to be compatible for fusing to the audio-only automatic speech recognition (ASR) modality. The usage of viseme appears in this stage. However, in animation or visual speech synthesis, the speech signal (Massaro, Beskow, Cohen, Fry, & Rodriguez, 1999) or transcribed information of words (Ezzat & Poggio, 1998) is used for lip's movement identification. In both cases, the lip movement is interpreted from phoneme–viseme table. Furthermore, in audiovisual speech processing, the visual expressions can also be used as audio speech anticipators. The anticipation effect has been represented in Kim and Davis (2003), where in Conrey and Pisoni (2003), the asynchrony of visual speech and audio speech information perception are examined. It has been shown that the tolerance of perceivers to visual speech cues preceding the audio speech cues is higher.

Referring to viseme, its framework has not been explained systematically. The main issue in viseme concept was lack of globally accepted definition for relation between phoneme and its visual appearance, for example, there are varieties of phoneme–viseme tables. The next drawback of the viseme concept is the lack of accuracy in perception of visual phonemes due to dependency on human perception of visual phonemes. In other words, depending on the observers whether they are hearing impaired or not, the results will vary. The next important issue is lack of explicit definition of the lip's deformation to make a standard model of articulating lips. The concept of visual phoneme does not suggest an explicit definition of lip's structure during phoneme utterance. In addition, the main factor of human perception in evaluating and selecting viseme could vary from one observer to another. Therefore, there is not any globally agreed and standardized phoneme–viseme table reference.

To interpret the dynamic of lips, many approaches have been employed. These approaches belong to three main groups depending on analyzing the extracted data as appearance-based methods including statistical methods (Adjoudani & Benoît, 1996; Bregler & Konig, 1994; Erber et al., 1979) and shape-based methods or geometric approaches (Chiou & Hwang, 1997; Rogozan & Deléglise, 1998; Teissier et al., 1999) and combinations of appearance-based and shape-based methods (Cootes et al., 1998). The necessity for suggesting a globally defined visual phoneme concept suggests expressing of the visual information by mathematical models. Therefore, the focus of this article is on extracting the geometry of lip during articulation to suggest a mathematical model for lips' dynamics.

A 3D lip model was fitted to lips' image (Revéret & Benoît, 1998), and the amount of similarities between the feature point on the model and actual lip has been measured. Their results did not conclude the dynamic of articulating lip with a mathematical expression. In the suggested lip-reading system by Potamianos and Graf (1998), the visual features related to the width, height, and area of lip were tracked over frame/time domain during articulating a sequence of digits "81926." Such trajectories of visual features were used for training a HMM model but were not mathematically formulated. The acquired data (from movement of lip) in Lucero (2002) was formulated by B-spline cure (Mortenson, 1997). The main disadvantage of using B-spline for formulating the lip's movement was lack of defining a single expressions for the sample sets as B-spline curve functions on segmented intervals. The extraction of feature points using the appearance-based approach in Zhao et al. (2009) was suggested by a statistical visual feature modeling by concatenating the image sequence portions in block depicted volumes. Using an image processing approach called Local Binary Pattern (LBP), the lip texture between successive images in gray scale are used to transform to binary codes. Afterward, the LBP is applied to the other two orthogonal planes to the image plane, which codes the spatial and temporal axis. The histogram of these binary orthogonal planes was calculated and represented as visual feature. By concatenating of the histograms, the two important characteristics (lips' appearance and also the motion of lip) are extracted. The sequences of histograms are joined to represent the overall statistical property of lip's movements. In other words, their method not only extracts the visual feature but also expresses the deformation of visual features dynamically. The method suggested in Birkholz et al. (2011) is a time-variant dynamic system used for modeling the articulatory relation of facial parts as tongue tip, tongue blade tongue, jaw, dorsum, upper lips, and lower lips by using electromagnetic articulography approach. The combination of phoneme was chosen as [CVCVCVCV] sequence with normal and low speaking rates. The input–output relation of the systems was obtained by solving differential equations in time domain. Although the proposed method tried to allocate a transfer function to describe the dynamics of lips and the related organs, the solution of the system in time via differential equations was computationally expensive and does not represent a single expression for the lip movement. According to the literatures, there is not any specific attempt for expressing the visual speech data during articulation by an explicit mathematical formula. In this article, a straightforward method for formulating the visual speech data is represented. The visual speech data is extracted from a transcribed words which are designed based on phonemic structure.

Having seen the need for a standard approach to model the visual phoneme, the emphasis of this research is on introducing a mathematical model for human visual speech. In this article, a novel method has been introduced for the analysis of the visual speech that has been developed based on both lips' movements and linguistic clues in English language. The visual speech information was related to a set of words chosen based on phoneme sequences from a transcribed database.

This article is organized as follows. "Visual Feature Determination" section explains visual feature determination procedure. "Corpus Design" section discusses corpus design. Visual speech sample extraction is described in "Visual Speech Sample Extraction" section. The modeling of visual speech signals is shown in "Visual Speech Signal" section, and the results are given in "Results and Discussion" section. Finally the conclusion is discussed in "Conclusion" section.

Visual Feature Determination

In this section, a visual sample set which describes the lip geometry is defined. The process is also referred as parameterizing the lip. The visual samples extracted from these visual feature points are the upper lip position, the lower lip position, and the corner lip position. These positions are

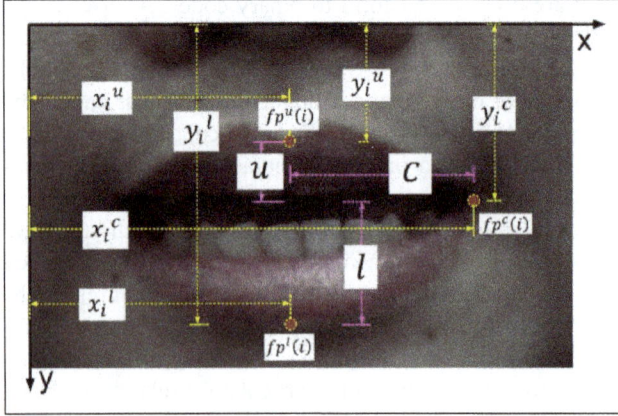

Figure 1. The geometrical representation of lip in a frame's region of interest (ROI) and visual features pixels relations in Cartesian coordinates.

denoted by superscripts: u, l, and c for upper, lower, and corner lips' features sample sets, respectively. The desirable measurements of lips are depicted in Figure 1 with pink dashed lines, and red points are representing the set markers on speaker's lips.

The samples from upper feature point $fp^u(i)$ are determined by subtracting y_i^c from y_i^u and referred to by variable u. The distance of lower feature point $fp^l(i)$ from the horizontal line connecting lips' corners is determined by subtracting y_i^c from y_i^l and addressed by variable l. These parameters show the amount of lips opening in vertical directions. Similarly, the distance of corner feature point from the vertical line which connects the upper and lower feature points is determined by subtracting x_i^u from x_i^c and shown by variable c. If the upper and lower feature points pixels are collinear, then x_i^u will be interchangeable with x_i^l. Due to the jaw movement, unwanted fluctuations may appear more in lower feature point during articulation in comparison with the upper feature point; as unlike the upper lip which is fixed to the mandible, the lower lip is controlled by the jaw. Therefore, x_i^u is chosen to extract the corner visual vector. The visual features are represented in Equations 1 to 3.

$$fp^u(i) = P\left(x_i^u, y_i^c - y_i^u\right), i = \{0, 1, 2, \ldots, (F-1)\}, \tag{1}$$

$$fp^l(i) = P\left(x_i^l, y_i^l - y_i^c\right), i = \{0, 1, 2, \ldots, (F-1)\}, \tag{2}$$

$$fp^c(i) = P\left(x_i^c - x_i^u, y_i^c\right), i = \{0, 1, 2, \ldots, (F-1)\}. \tag{3}$$

Here, F indicates the constant number of frames during utterance of each word. The pixel differences are shown by $u = \{u_i\}$, $l = \{l_i\}$, and $c = \{c_i\}$, forming the visual speech sample sets for the upper, lower, and corner feature points sample sets $FP^u(i)$, $FP^l(i)$, and $FP^c(i)$. These vectors

measure the visual features according to the Cartesian coordinate system that is defined within the lips' geometry. The displacement of the visual features in Equations 1 to 3 leads to defining the visual speech sample sets as follows.

$$FP_{W_m}^u(i) = \left\{ u_0^{W_m}, u_1^{W_m}, u_2^{W_m}, \ldots, u_{F_m}^{W_m} \right\}, \tag{4}$$

$$FP_{W_m}^l(i) = \left\{ l_0^{W_m}, l_1^{W_m}, l_2^{W_m}, \ldots, l_{F_m}^{W_m} \right\}, \tag{5}$$

$$FP_{W_m}^c(i) = \left\{ l_0^{W_m}, l_1^{W_m}, l_2^{W_m}, \ldots, l_{F_m}^{W_m} \right\}, \tag{6}$$

where w is representing a word with specific number of frames m. These sample sets are used for describing lips' movements during articulating a set of words. Now that the lips are parameterized, we move on to the next section where a method for designing the transcribed data is represented.

Corpus Design

The arrangements of phonemes in transcribed form (corpus) are also considered as one of the important issues in visual speech analyzers and synthesizers. It addresses the structures of phonemes used in lip-reading systems and visual animation of speech. There are several different corpora that differ in the structure of phonemes for generating lip movements. The structures of the corpora are shown in Kim and Davis (2003) where C and V refer to consonant and vowel phonemes, respectively. Each set corresponds to specific combinations of isolated phonemes.

In the first category of transcribed data, the isolated phoneme was used for lip reading (Petajan, 1984) or animation (Bregler, Covell, & Slaney, 1997). In the second scenario, a specific arrangement of vowel or consonant was fixed for visual analysis of speech. A corpora developed by Montgomery (1980) used the /CVCVC/ synthesizer while Adjoudani and Benoît (1996) used /V_1CV$_2$CV$_1$/ corpus. The random combination of phonemes as /aCa/ was addressed in the corpora structure by Su and Silsbee (1996). The phoneme arrangements as a part of a French sentence "C'est pas / VCVCVz/?" was also examined by Revéret and Benoît (1998). Czup (2000) used triphone sequences /V_1CV$_1$/ and /C_1VC$_1$/ corpora. For a Brazilian Portuguese facial animation system proposed by De Martino, Magalhaes, and Violaro (2006), the corpora consisted of two types of phonemes combinations as /CV$_1$CV$_2$/ and /V_1V$_2$/ while the /VCV/ structure was used by Cox, Harvey, and Newman (2008) and Alothmany, Boston, Li, Shaiman, and Durrant (2010). This model was used for training a HMM in ASR system with different four conditions under seven levels of noise.

The suggested structure of corpus of Castilian Spanish used by Melenchón, Martínez, De La Torre, and Montero (2009) consisted of /CVCV/. Their purpose for using such structure was supported by a strong statement, which claims

Table 1. The Structure of Phonemes Used as Corpora.

Corpora	Authors	Language
/CVCVC/	Montgomery (1980)	English
$/V_1CV_2CV_1/$	Adjoudani and Benoît (1996)	English
/aCa/	Su and Silsbee (1996)	English
"C'est pas /VCVCVz/?"	Revéret and Benoît (1998)	French
$/V_1CV_1/$ and $/C_1VC_1/$	Czup (2000)	Hungarian
$/CV_1CV_2/$ and $/V_1V_2/$	De Martino, Magalhaes, and Violaro (2006)	Brazilian Portuguese
/VCV/	Cox, Harvey, and Newman (2008)	English
/CVCV/	Melenchón, Martínez, De La Torre, and Montero (2009)	Castilian Spanish
/VCV/	Alothmany, Boston, Li, Shaiman, and Durrant (2010)	English
/CVCVCVCV/	Birkholz, Kröger, and Neuschaefer-Rube (2011)	Not stated

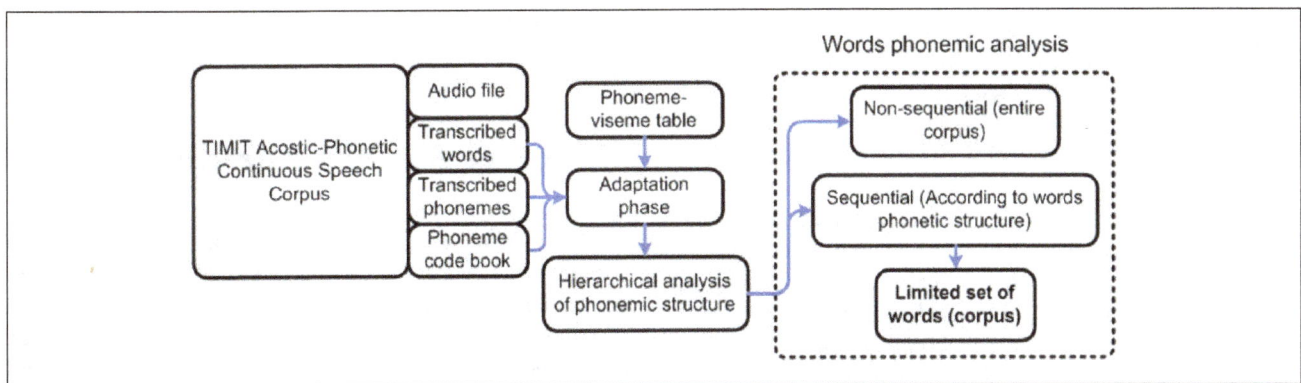

Figure 2. The designing steps toward extracting a limited set of words.
Note. TIMIT = Texas Instrument and Massachusetts Institute of Technology.

more than 80% of Castilian Spanish words flow /CV/ structure (de Vega, Álvarez, & Carreiras, 1992). A suggested phoneme arrangements by Birkholz et al. (2011) produced /CVCVCVCV/ sequences. These methods are also noted in Table 1 where accordingly no evidence for choosing such arrangements of phoneme sequences was presented. The main disadvantage of such corpora arrangement is ignoring the actual sequences of consonants and vowels in existing words, in contrary with the above considered sequences.

Hierarchical analysis of the phoneme sequences has been applied for words' categorization to organize the database properly that covers the above-mentioned disadvantage. The sequences of phonemic symbols are the subject of such analysis. Driving visual speech signals for visual speech samples will be more comprehensive if the signals are produced from a specific family of words. Accordingly, the most frequent phonemic alphabets that lead to the most frequent sets of words which have similar phoneme sequences can be extracted. The phonemic structures make the words families. This procedure is called phoneme-based analysis of words in the current study.

In reality, words have specific uniquely defined sequences of phonemes. Generating the combination of phoneme sequence starts from the beginning of a word. This eliminates

the need to study the random combination of phonemes. Obviously, many words share the same phoneme sequence in their initial parts. One of the significant advantages of phoneme-based analysis of words is the ability of controlling the mathematical expressions in the visual domain. In this work, the common phonemes in the initial parts of the words are found based on the search in three sequences of phonemes (triphones), and this is the basis of designing the corpus.

In Figure 2, detail about the hierarchical approach for obtaining a limited set of words is illustrated. One of the most well-known sources of ASR, which provides varieties of the audio-only databases bodies, is the Linguistic Data Consortium (LDC). Among the databases, the Texas Instrument and Massachusetts Institute of Technology (TIMIT) Acoustic-Phonetic Continues Speech Corpus is chosen for this study. This database was widely used in the ASR systems like Hidden Markov Model Toolkit (HTK).

The hierarchical analysis is applied to entire corpus to derive more information about the behavior of phonemes in the corpus.

This concept provides the relation between the mathematically expressed visual speech signals called visual words and the meaningful combinations of sounds that are marked by words. Consequently, the mathematical expressions, which would be driven for the visual words, are also categorized as a group. Therefore, the words are grouped mathematically. In mathematical form, the difference between each member of group will be revealed after the third phoneme.

Therefore, the visual speech signals demonstrate a systematic relation between a member of the word's family and its family members. This method could be used in lip-reading and also animation systems. Toward establishing the signals expression for structural words, a set of words were chosen by the criteria of maximum phonemic sequence occurring among all the transcribed words. This method can be applied to any word in dictionary, but for this study, a collection of words is chosen in English language for synthesizing the visual word database.

Adaptation Phase

In the field of speech processing, it is possible to select the appropriate viseme table that has been chosen as the visual domain representation reference. Selecting appropriate phoneme–viseme mapping table for analysis of phonemic structure would be beneficial in deriving the practical visual speech signal expressions that are adaptable to audiovisual speech recognition systems. In other words, the method of deriving visual speech signals could be applied to the speech recognizer system as its collaborating visual domain. According to Potamianos et al. (2004), a phoneme set used in HTK was selected as the phoneme–viseme mapping table. In this mapping scheme, the many-to-one strategy was applied based on the manner of articulation. The consonants consist of eight subgroups of phonemes that have visual similarities during articulation, while vowels are categorized into four subgroups.

Phoneme-Based Hierarchical Analysis

In this part, the hierarchical analysis is used for studying the combination of phonemic components in the words. In the hierarchical analysis of words phonemic structure, the goal is grouping all possible sequences of consonants and vowels. This process schematically is depicted in Figure 3.

Words Phonemic Pattern Analysis

The hierarchical analysis provides the information for systematic analysis of organized data. Using this approach to the phonemic representation of words, the hierarchical probability information of phonemes sequences is extracted as they appear in words. A decision tree is defined in sequentially arranged levels. The nodes relations were defined by connecting lines from their current level to the other nodes in the successive level. The nodes represent a particular state, group, event, or position in each level that exists in the data set. The connecting lines represent conditional rules that define the transition among the nodes between two

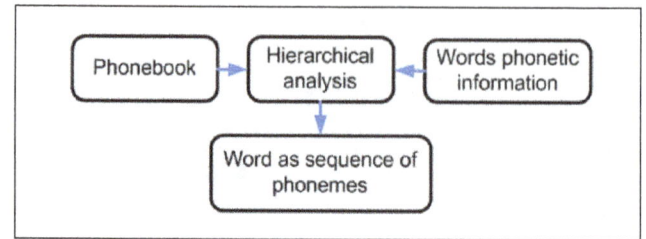

Figure 3. Searching for word's phonemic rules.

consecutive levels. These rules are stated by the maximum numbers of repetitions mapped to phonemes in each level. In this work, the nodes of the decision tree are denoted by the consonants and vowels. The tree representation of phonemes sequences is shown in Table 2 where 13 words are selected and categorized having the same phonemic sequence for three levels.

Visual Speech Sample Extraction

The practical phase of this work was represented in Figure 4. After designing a corpus, the test subjects were asked to articulate the words. A camera captured their lips' movements. The lips' geometry was parameterized on the region of interest by three static feature points located on upper, lower, and corner outer contour of lip. Therefore, each word has three sets of samples in visual domain. The movement of these feature points can describe the lips' movement. The visual data was extracted automatically and manually.

The speakers were two non-native women aged between 25 and 31 with no speech disorders or hearing disabilities. The visual speech sample sets $FP_{W_m}^u(i)$, $FP_{W_m}^l(i)$, and $FP_{W_m}^c(i)$ were extracted manually from the upper $fp^u(i)$, lower $fp^l(i)$, and the corner $fp^c(i)$ visual feature points on two speakers' lips demonstrated in Figure 5a to 5c, respectively, for $i = \{0, 1, 2, \ldots, F_{W_m} - 1\}$ where F is substituted by F_{W_m}.

Visual Speech Signal

The Lagrange (Lagrange-Newton) interpolation was a method of polynomial formulation, which constructs a continuous-time polynomial $L(x)$ from N number of samples taken from a function $y = f(x_i)$ and $i \in \mathbb{N}$ over an equidistance interval x_i, where $x_i \in [0 : N-1]$.

The corresponding amplitudes of samples are defined as $a_i = f_i$. The variable x was defined via changing the polynomial's sampling frequency β. More specifically, by recalling the definition of continuous-time interval (where $\beta \to \infty$), the interval x is defined as

$$x = \{x_v\}, x_v = \frac{i}{\beta}, i = 0, 1, 2, \ldots, \beta(N-1) \in \mathbb{N} \qquad (7)$$

The Lagrange interpolation is expressed as

Table 2. The Tree Representation of Phonemes Sequences Categorizing All 13 Words as Branches of One Root.

1	2	3	4	5	6	7	8	9	10	11	12	13	Word groups
/s/	/ih/	/m/	P	/l/	/ax/	/t/	—	—	—	—	—	—	Simple
						—	—	—	—	—	—	—	Simpler
					/iy/	—	—	—	—	—	—	—	Simplest
				/el/	—	—	—	—	—	—	—	—	Simply
				/ax/	/th/	/eh/	/t/	/ih/	/k/	/ax/	/l/	/iy/	Sympathetically
				/ow/	/z/	/z/	/iy/	/ax/	/m/	—	—	—	Symposium
				/t/	/ax/	/m/	—	—	—	—	—	—	Symptom
			B	/ax/	/l/	/ih/	—	—	—	—	—	—	Symbolic
						/ay/	—	—	—	—	—	—	Symbolism
				/aa/	/l/	/ih/	—	—	—	—	—	—	Symbolize
				/el/	/z/		—	—	—	—	—	—	Symbols
			Ax	—	—		—	—	—	—	—	—	Simmer
			Eh	/n/	/t/	—	—	—	—	—	—	—	Cement

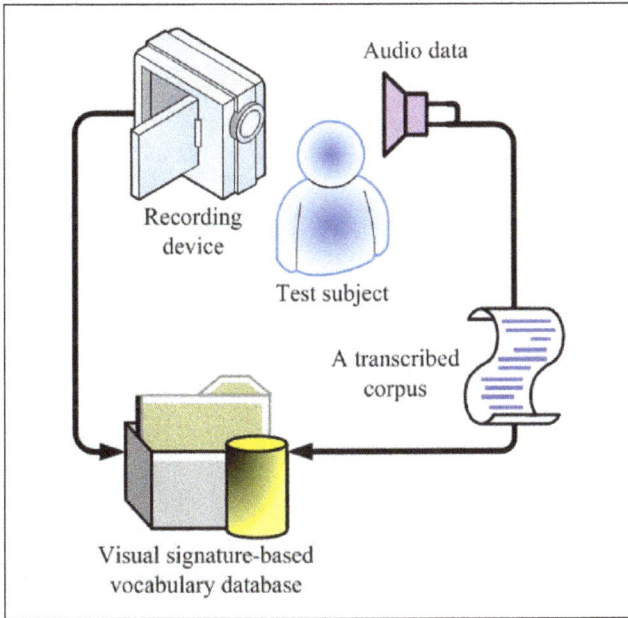

Figure 4. The practical phase of the work.

$$L(x) = \sum_{i=0}^{N-1} f_i l_i(x) \qquad (8)$$

where $l_i(x)$ is the basic function corresponding to node $\{x_i\}$ that is defined as

$$l_i(x) = \frac{\prod_{\substack{k=0 \\ k \neq i}}^{N-1}(x-x_k)}{\prod_{\substack{k=0 \\ k \neq i}}^{N-1}(x_i-x_k)} \qquad (9)$$

The oscillation of Lagrange polynomial is addressed as Runge (1901) phenomenon. This degrading effect can be well eliminated by repositioning sample's nodes $\{x_i\}$ and interval modification.

To apply the proposed procedure, starting with the basic function modification in Equation 9 and rewriting it we have

$$l_i(x) = \frac{l(x)w_i}{(x-x_i)}, \qquad (10)$$

where $l(x)$ is

$$l(x) = \prod_{i=0}^{N-1}(x-x_i). \qquad (11)$$

In parallel, the Barycentric weight function $w_i(x)$ is

$$w_i(x) = \frac{1}{\prod_{\substack{k=0 \\ k \neq i}}^{N-1}(x_i-x_k)}, i = \{0,1,2,\ldots,(N-1)\}. \qquad (12)$$

Substituting Equation 10 in Equation 8 gives

$$L(x) = l(x)\sum_{i=0}^{N-1} f_i \frac{w_i}{(x-x_i)}. \qquad (13)$$

If Equation 13 is used for interpolating constant amplitudes equal to 1, then the resulted expression is

$$l(x)\sum_{i=0}^{N-1}\frac{w_i}{(x-x_i)} = 1. \qquad (14)$$

Finally, the Barycentric Lagrange interpolation (BLI) can be formulated by substituting Equation 14 into Equation 13 which leads to

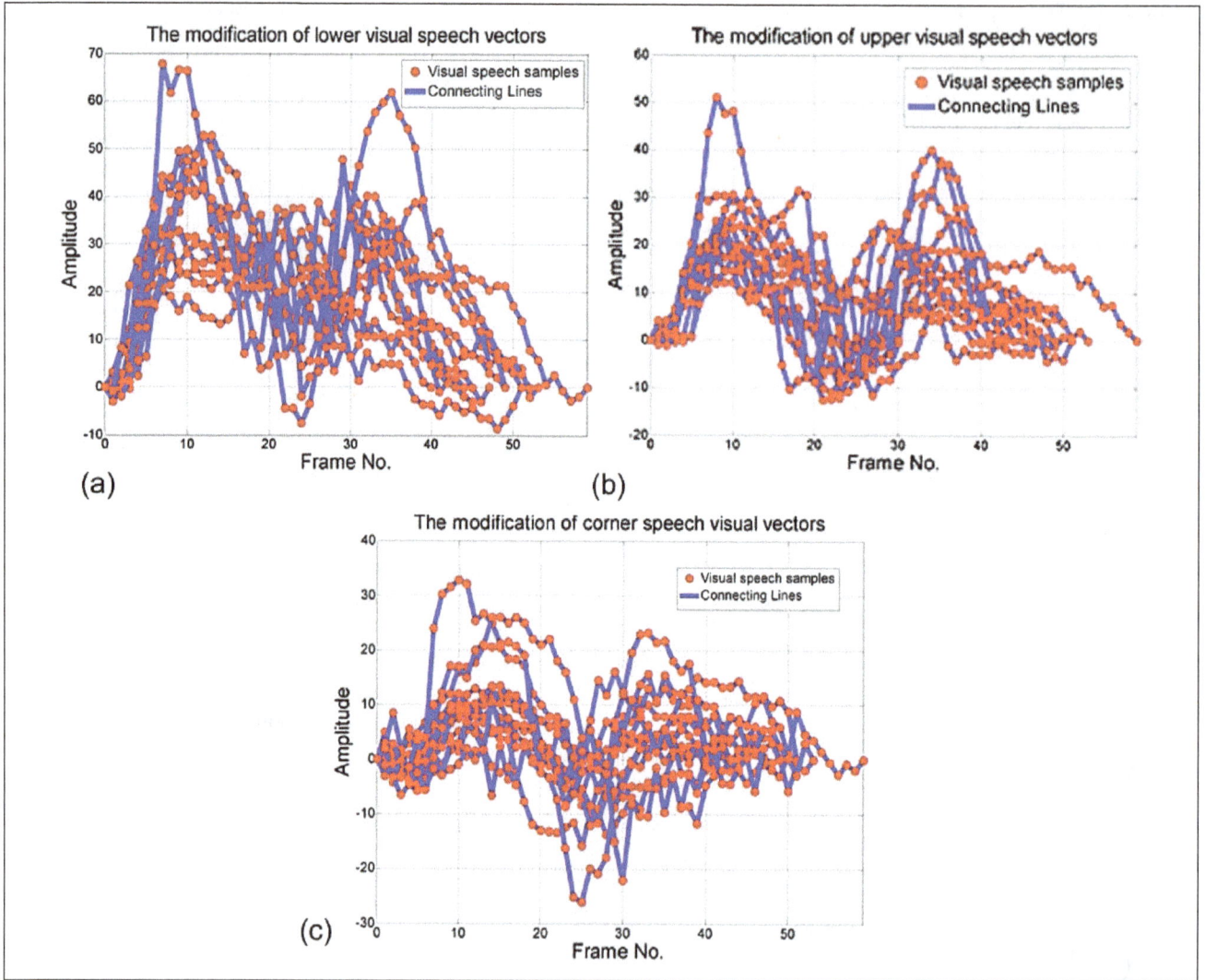

Figure 5. The (a) upper, (b) lower, and (c) corner visual speech sample sets after adjusting the end points.

$$L_B(x) = \frac{\sum_{i=0}^{N-1} f_i \dfrac{w_i}{(x-x_i)}}{\sum_{i=0}^{N-1} \dfrac{w_i}{(x-x_i)}}. \tag{15}$$

This modification tackles the problem of destructive oscillations on the interval boundaries by a transformation to another domain. The transformation is possible by Chebyshev points of the second kind as

$$x_i^{ch} = \cos\frac{i\pi}{(N-1)}, i = \{0,1,2,\ldots,(N-1)\}. \tag{16}$$

Equation 16 can be thought as a mapping scheme that transfers a uniformly spanned interval x_i to another interval called Chebyshev interval x_i^{ch}.

The overall representations of the visual words, which are separated according to the three visual features, are shown in

Figure 6: the upper (a), lower (b), and corner (c) visual speech signals.

Regarding to the speech processing field, it is possible to select the appropriate viseme table that has been selected as the visual domain representation reference. Selecting appropriate phoneme–viseme mapping table for analysis of phonemic structure would be beneficial in deriving the practical visual speech signal expressions that are adaptable to audio-visual speech recognition systems. In other words, the method of driving visual speech signal signature could be applied to the speech recognizer system as its collaborating visual domain. The mapping scheme, which is represented in Table 3, is a sample presented that can be applied based on the manner of articulation.

The consonants consist of eight subgroups of phonemes that have visual similarities during articulation, whereas vowels categorized into four subgroups. Although the English alphabet consists of 26 letters, there are 43 phonemes. This fact shows the difference between possible sound in words and the

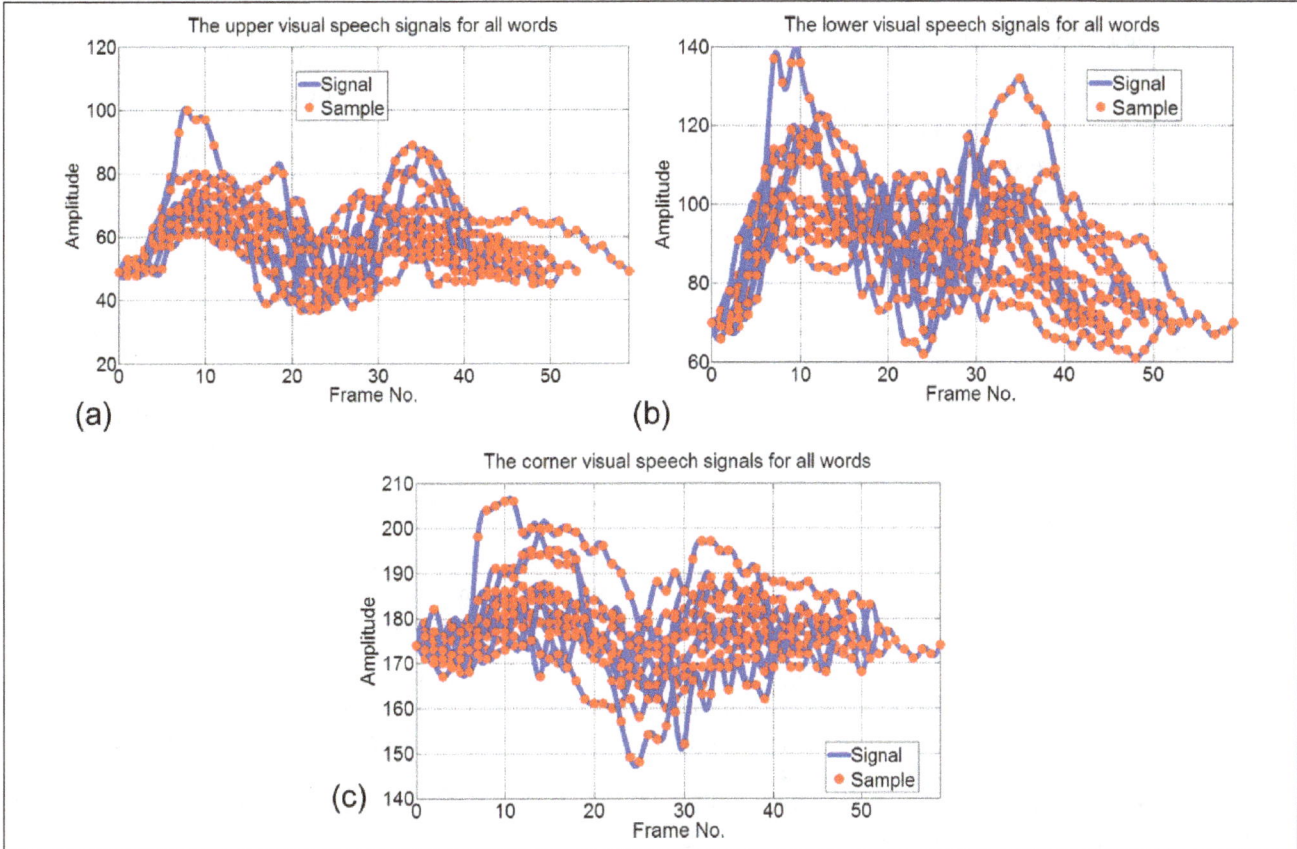

Figure 6. The visual speech signals of the upper (a), lower (b), and corner (c) visual speech sample sets.

Table 3. The Phoneme–Viseme Table (Potamianos, Neti, Luettin, & Matthews, 2004), Phonemes are Grouped Into Viseme Classes.

	Consonants	Vowels
I	/p, b, m/	/ih, iy, ix, ax/
2	/f, v/	/ae, eh, ey, ay/
3	/l, el, r, y/	/uw, uh, ow/
4	/s, z/	/ao, ah, aa, er, oy, aw, hh/
5	/sh, zh, ch, jh/	
6	/t, d, n, en/	
7	/th, dh/	
8	/ng, k, g, w/	

letters. Now it was time to find the transcribed words text that was marked according to this phonemic table (Table 3.).

Results and Discussion

In Figure 6a to 6c, the visual speech signal $VS_{BLI}^{u}{}^{W_m}(f_x)$, $VS_{BLI}^{l}{}^{W_m}(f_x)$, and $VS_{BLI}^{C}{}^{W_m}(f_x)$, $m = 1,2,3,\ldots,13$, are represented. The visual speech signals are processed for allocation of mathematical expressions to words. The words are categorized based on the hierarchical relations in phoneme

sequences. A word is represented in visual domain with specific mathematical expressions. The mathematical expression of a visual word consists of three visual speech signals. These visual speech signals will be interpolated from three visual features located on the upper, lower, and corner of outer lip contour. The mathematical expressions were derived by the BLI method. The number of visual speech signals is set to three as the three visual features include the necessary information for lip dimension during articulation. The reason for using BLI is the ability of formulating the visual speech signals which involve the sample sets extracted from the visual feature points without Rounge effect.

The mathematical expressions of the visual speech data obtained from the upper, lower, and corner visual speech sample sets were called the visual speech signals. As it was mentioned earlier, the extracted samples from a visual feature were called visual speech sample set. Collection of three visual speech samples sets corresponding to the upper, lower, and corner visual feature points are representing a word. The first observation from Figure 6 suggests a coherence incremental pattern. It is shared with upper and lower visual speech sample sets for $F_W \in [0:10]$ or from 0 to 379 milliseconds. However, the corner visual speech sample sets show more steady incremental changes. The second segment starts in 12th frame until the 25th frame where almost all visual sample sets are obtained by upper, lower, and corner visual features reach to their minimum. From the 25th to the 34th frame, the trends of visual speech sample sets in all three cases are again increasing. Afterward, from 35th frame until the ending frame in each sample set, the trends are decreasing. The visual speech samples are sharing similarities in a portion of beginning frames. This is because the words' phonemic structure has the same sequence of three phonemes (/s/+/ih/+/m/+ . . .).

Conclusion

In this article, a new method for mathematical formulation of visual speech information has been suggested. Since the concept of viseme does not provide a unique representation of visual phoneme, in addition to the unavailability of an explicit representation of the dynamics of lips during articulation of the words. The lips' movements during articulating a specifically designed set of words were mathematically formulated in this work. The visual information is corresponding to a set of words which have the same sequence of triphones in their initial parts. The words are extracted from TIMIT database with hierarchical analysis. The mathematical expressions of the visual speech obtained from the visual features were called the visual speech signals. Normalized rational relations of visual speech sample sets were chosen for providing more compact versions of visual speech signals as well as preserving their scaling ability. Assigning a mathematical unique expression to visual speech information is a novel method to create a more precise database for words in comparison with the existing methods.

Declaration of Conflicting Interests

The author(s) declared no potential conflicts of interest with respect to the research, authorship, and/or publication of this article.

Funding

The author(s) received no financial support for the research and/or authorship of this article.

References

Adjoudani, A., & Benoît, C. (1996). *On the integration of auditory and visual parameters in an HMM-based ASR.* Berlin, Germany: Springer.

Alothmany, N., Boston, R., Li, C., Shaiman, S., & Durrant, J. (2010, September 15-17). *Classification of visemes using visual cues.* Proceedings of the International Symposium ELMAR, Zadar, Croatia.

Ananthakrishnan, G., & Engwall, O. (2008, December 8-12). *Important regions in the articulator trajectory.* Proceedings of the 8th International Seminar on Speech Production, Strasbourg, France.

Birkholz, P., Kröger, B. J., & Neuschaefer-Rube, C. (2011). Model-based reproduction of articulatory trajectories for consonant-vowel sequences. *IEEE Transactions on Audio, Speech & Language Processing, 19,* 1422-1433.

Bregler, C., Covell, M., & Slaney, M. (1997). Video rewrite: Driving visual speech with audio. In *Proceedings of the 24th annual conference on SIGGRAPH 97* (pp. 353-360). New York, NY: Association for Computing Machinery.

Bregler, C., & Konig, Y. (1994, April 19-22). *Eigenlips for robust speech recognition.* IEEE International Conference on Acoustic Speech and Signal Processing, Adelaide, Australia.

Chin, S. W., Seng, K. P., & Ang, L.-M. (2012). Audio-visual speech processing for human computer interaction. In T. Gulrez & A. E. Hassanien (Eds.), *Advances in robotics and virtual reality* (pp. 135-165). Berlin, Germany: Springer.

Chiou, G., & Hwang, J. N. (1997). Lipreading from color video. *IEEE Transactions on Image Processing, 6,* 1192-1195.

Cohen, M., & Massaro, D. (1993). Modelling coarticulation in synthetic visual speech. In N. M. Thalman & D. Thalman (Eds.), *Models and techniques in computer animation* (pp. 139-156). Tokyo, Japan: Springer-Verlag.

Conrey, B. L., & Pisoni, D. B. (2003, September 4-7). *Audiovisual asynchrony detection for speech and nonspeech signals.* Proceedings of the ISCA Tutorial and Research Workshop, St. Jorioz, France.

Cootes, T. F., Edwards, G. J., & Taylor, C. J. (1998, June 2-6). *Active appearance models.* European Conference on Computer, Freiburg, Germany.

Cox, S., Harvey, R., & Newman, J. (2008). *The challenge of multispeaker lip-reading.* Queensland, Australia: International Conference on Auditory-Visual Speech Processing.

Czup, L. (2000, October 16-20). *Lip representation by image ellipse.* International Conference on Spoken Language Processing, Beijing, China.

De Martino, J. M., Magalhaes, L. P., & Violaro, F. (2006). Facial animation based on context-dependent visemes. *Journal of Computers and Graphics, 30,* 971-980.

de Vega, M., Álvarez, C., & Carreiras, M. (1992). Estudio estadístico de la ortografía castellana: La frecuencia silábica. *Cognitiva, 4*(1), 75-114.

Erber, N. P., Sachs, R. M., & DeFilippo, C. L. (1979). *Optical synthesis of articulatory images for lipreading evaluation and instruction.* Advances in Prosthetic Devices for the Deaf: A Technical Workshop, Rochester, NY.

Ezzat, T., & Poggio, T. (1998, June 8-10). *MikeTalk: A talking facial display based on morphing visemes.* Proceeding of Computer Animation Conference, Philadelphia, PA.

Fisher, C. G. (1968). Confusions among visually perceived consonants. *Journal of Speech and Hearing Research, 11,* 796-804.

Goldschen, A. J. (1993). *Continuous automatic speech recognition by lipreading* (Doctoral dissertation). George Washington University, Washington, DC.

Goldschen, A. J., Garcia, O. N., & Petajan, E. (1994). *Continuous optical automatic speech recognition by lipreading.* Pacific Grove, CA: IEEE Computer Society Press.

Hazen, T. J., Saenko, K., La, C. H., & Glass, J. R. (2004). *A segmentbased audio-visual speech recognizer: Data collection, development, and initial experiments.* New York, NY: Association for Computing Machinery.

Heracleous, P., Tran, V. A., Nagai, T., & Shikano, K. (2010). Analysis and recognition of NAM speech using HMM distances and visual information. *IEEE Transactions on Audio, Speech, and Language Processing, 18,* 1528-1538.

Huang, J., & Povey, D. (2005). *Discriminatively trained features using fMPE for multi-stream audio-visual speech recognition.*

In INTERSPEECH-2005 (pp. 777-780). Retrieved from http://www.isca-speech.org/archive/interspeech_2005/i05_0777.html

Jiang, J., Alwan, A., Auer, E. T., & Bernstein, L. E. (2001). *Predicting visual consonant perception from physical measures.* Retrieved from http://www.seas.ucla.edu/spapl/paper/jiang_eurospeech01.pdf

Kim, J., & Davis, C. (2003, September 4-7). *Testing the cuing hypothesis for the AV speech detection.* International Conference on Audio-Visual Speech Processing, St Jorioz, France.

Kosmopoulos, D., & Chatzis, S. P. (2010). Robust visual behavior recognition. In *IEEE Signal Processing Magazine* (pp. 34-45). New York, NY: Institute of Electrical and Electronics Engineers.

Lee, J. S., & Park, C. H. (2006). *Training hidden Markov models by hybrid simulated annealing for visual speech recognition.* Taipei, Taiwan: IEEE.

Lucero, J. C. (2002). Identifying a differential equation for lip motion. *Medical Engineering & Physics, 24,* 521-528.

Massaro, D. W., Beskow, J., Cohen, M. M., Fry, C. L., & Rodriguez, T. (1999, August 7-9). *Picture my voice: Audio to visual speech synthesis using artificial neural networks.* International Conference on Auditory-Visual Speech Processing, Santa Cruz, CA.

McGurk, H., & MacDonald, J. (1976). Hearing lips and seeing voices. *Nature, 264,* 746-748.

Melenchón, J., Martínez, E., De La Torre, F., & Montero, J. A. (2009). Emphatic visual speech synthesis. *IEEE Transactions on Audio, Speech, and Language Processing, 17,* 459-468.

Montgomery, A. A. (1980). Development of a model for generating synthetic animated lip shapes. *Journal of the Acoustical Society of America, 68*(S1), S58.

Mortenson, M. E. (1997). B-spline curves. In M. Spencer (Ed.), *Geometric modeling* (pp. 113-142). New York, NY: Wiley.

Petajan, E. D. (1984). *Automatic lipreading to enhance speech recognition.* Atlanta, GA: IEEE Global Telecommunications.

Potamianos, G., & Graf, H. P. (1998, May 12-15). *Discriminative training of HMM stream exponents for audio-visual speech recognition.* International Conference on Acoustics, Speech and Signal Processing, Seattle, WA.

Potamianos, G., Neti, C., Luettin, J., & Matthews, I. (2004). *Audio-visual automatic speech recognition. An overview.* Cambridge, MA: MIT Press.

Revéret, L., & Benoît, C. (1998). *A new 3D lip model for analysis and synthesis of lip motion in speech production.* Terrigal, Australia: Auditory-Visual Speech Processing.

Rogozan, A., & Deléglise, P. P. (1998). Adaptive fusion of acoustic and visual sources for automatic speech recognition. *Speech Communication, 26,* 149-161.

Runge, C. (1901). Uber empirische Funktionen und die Interpolation zwischen aquidistanten Ordinaten. *Zeitschrift fur Mathematik und Physik,* 224-243.

Saenko, K., Livescu, K., Glass, J., & Darrell, T. (2009). Multistream articulatory feature-based models for visual speech recognition. *IEEE Transactions on Pattern Analysis and Machine Intelligence, 31,* 1700-1707.

Su, Q., & Silsbee, P. L. (1996). *Robust audiovisual integration using semicontinuous hidden Markov models.* Philadelphia, PA: International Science Congress Association.

Sumby, W. H., & Pollack, I. (1954). Visual contribution to speech intelligibility in noise. *Journal of the Acoustical Society of America, 26,* 212-215.

Teissier, P., Robert-Ribes, J., & Schwartz, J. L. (1999). Comparing models for audiovisual fusion in a noisy-vowel recognition task. *IEEE Transactions on Speech and Audio Processing, 7,* 629-642.

Williams, J., Rutledge, J., Garstecki, D., & Katsaggelos, A. (1997, June 23-25). *Frame rate and viseme analysis for multimedia applications.* Proceeding of IEEE Multimedia and Signal Processing Conference, Princeton, NJ.

Zhao, G., Barnard, M., & Pietikainen, M. (2009). Lipreading with local spatiotemporal descriptors. *IEEE Transactions on Multimedia, 11,* 1254-1265.

Author Biographies

Mohammad Hossein Sadaghiani has received his PhD in the University of Nottingham, Malaysia Campus, and still continuing his research on voice recognition.

Niusha Shafiabady is currently a faculty member at the University of Nottingham, Malaysia Campus. Her research interests are applications of artificial intelligence in different areas of science.

Dino Isa is the professor of intelligent systems at the University of Nottingham, Malaysia Campus. His research interest is application of artificial intelligence in different areas of science and technology.

20

Retention in Online Courses: Exploring Issues and Solutions—A Literature Review

Papia Bawa[1]

Abstract

Despite increasing enrollment percentages from earlier years, online courses continue to show receding student retention rates. To reduce attrition and ensure continual growth in online courses, it is important to continue to review current and updated literature to understand the changing behaviors of online learners and faculty in the 21st century and examine how they fit together as a cohesive educational unit. This article reviews literature to ascertain critical reasons for high attrition rates in online classes, as well as explore solutions to boost retention rates. This will help create a starting point and foundation for a more, in-depth research and analysis of retention issues in online courses. Examining these issues is critical to contemporary learning environments.

Keywords

online courses, student retention models, social and motivational issues, technology in online courses, online learners and faculty, computer-mediated communications, online course design

Online courses are a revolutionary trend of educational technology today. With the rapid rise in online course enrollment comes a growing concern for low retention rates in many online courses and programs. Heyman (2010) points out that one of the biggest concerns in online education emanates from the excessively high attrition rates in fully online programs compared with traditional classes. Online courses have a 10% to 20% higher failed retention rate than traditional classroom environments (Herbert, 2006). Totally, 40% to 80% online students drop out of online classes (B. Smith, 2010). Review of existing literature indicates that online courses have several social, technological, and motivational issues existing from both the learners' and the faculty's perspectives.

The Importance of Studying Retention Issues in Online Courses

The online delivery system has revolutionized educational technology and has provided easy access to learning for multitudes of students, including many who were unable to go to school prior to this revolution. Today, online education is one of the top industries in the world, providing support, knowledge, and jobs to a large segment of the world's population. Allen and Seaman (2011) report that more than 6 million students were taking at least one online course in the year 2010 and that there is a steady 10% growth in online course enrollments. Online learning is also becoming an integral part of corporate training. Organizations that utilize this platform

have better chances at business and financial gains, as it provides a positive impact on workplace motivation. Access to electronic data and a self-paced learning environment may increase the interest and value of on-the-job training (Overton, 2007).

Despite all these benefits, online classes continue to display serious retention issues, which need to be addressed. A good place to start this is by examining why online learners leave, when in their academic careers are they most prone to leave, and what can be done to eliminate or mitigate these causes. Literature reviews indicate that the online attrition pattern is not limited to any specific period or level of graduation. Students may withdraw from online classes anytime in the semester and at any level of their learning process. Several studies have been conducted specifically to observe when and why students withdraw from graduate programs. The study conducted by Perry, Boman, Care, Edwards, and Park (2008) indicated that out of a group of 113 students who had withdrawn from the graduate program, 17 had been accepted, but they did not begin any class work prior to withdrawing. The balance consisted of students who had been registered and had started attending their classes, and yet decided to drop the program. These students stayed in the

[1]Purdue University, West Lafayette, IN, USA

Corresponding Author:
Papia Bawa, Purdue University, 1623 Lionheart Lane, West Lafayette, IN 47906, USA.
Email: pbawa@ivytech.edu

program anywhere between 2 months and 2 years. Willging and Johnson's (2009) study indicated that although students were less likely to leave after investing in several semesters, there was no dominant reason for dropping out. Most students dropped out of a program due to personal, job-related, and program-related reasons. Perry et al. (2008) mention the Canadian Association of Graduate Schools Report of 2004, which indicates that withdrawals from programs may occur even after several semesters.

Jaggars (2011) refers to several research reviews indicating that the mid-semester withdrawal rates for online courses may be higher than face-to-face courses. Levy's (2007) study indicates that students at a lower learning level at college are at a higher risk of dropping out than upper level students. Students who are less experienced and at an earlier semester of their program are more likely to drop the program. Levy indicated that students who are in the early stages of their program feel less prepared to deal with the academic rigors. On the contrary, students who have spent longer time in the program may be more motivated to complete the course, because they have already invested considerable time and efforts on it. The input in time and effort is a critical determinant as to when a student is more likely to withdraw.

The fact that students are liable to withdraw at any given stage makes it even more crucial to explore ways and means to mitigate the underlining causes of this phenomenon. Stanford-Bowers (2008) points out that a fall in online attrition rates will benefit students, faculty, and institutions. They believe that this can be accomplished if all those who have a vested interest in online learning recognize the significance of this new trend in the educational industry and examine every aspect of this revolutionary learning medium (Stanford-Bowers, 2008).

Theoretical Backgrounds for Examining Online Learners

A synthesis of literature information pertaining to retention issues and solutions for online environments must begin with a discussion of the theoretical concepts that determine the contexts within which online learning environments and learners are placed. There are several sociological theories, which explain learner behaviors in an online context. These, in turn, can become predictors and precursors of issues and solutions pertaining to online environments. Theories of marginalization or social exclusion have been used in literature to explain decisions of learners to select or reject the online platform. Ball, Davies, David, and Reay (2002) discuss how "the perceptions and choices of prospective HE (higher education) students are constructed within a complex interplay of social factors that are underpinned by basic social class and ethnic differences" (p. 53). Based on their study, Ball et al. determined that learners used cognitive and social criteria to determine their choices.

For the social criteria, the determining factors are the learners' perceptions of social classification of self and institutions. Many times, learners gravitate toward online environments as it provides them with the perceived benefits of "virtual" anonymity and protection from being at the receiving end of discriminatory behavior. However, in the context of marginalization issues, although online environments can provide some protection, under certain circumstances, this environment can become the issue. D. Smith and Ayers (2006) discuss such implications through the lens of community colleges. Their examination points to the critical issue of the prevalence of Westernized curriculum within the United States that places marginalized learners in a situation where "the pro-Western bias inherent in the technological foundations of distance learning presents an obstacle both to access and to understanding" (D. Smith & Ayers, 2006, p. 402). Nuances of cross-cultural communication, coupled with technological impediments, can create untenable learning environments, leading to attrition. " . . . when discourses are intricately nuanced with specific cultural meanings, such meanings may be 'lost in translation' as they are converted to Western-dominated electronic media" (D. Smith & Ayers, 2006, p. 406). Thus, although technology can be considered neutral, there is always the danger of its hegemonic contamination.

Motivational theories of self-determination and self-efficacy are also pertinent to examining learners within online environments. Self-determination is defined as action generated by one's own mind or free will, with no influence from outside situations or entities (Wehmeyer, Abery, Mitaug, & Stancliff, 2003). Chen and Jang (2010) discuss how in the context of online learners, self-determination theory prescribes three needs, namely, a sense of control, feelings of competency for tasks, and sense of inclusion or affiliation with others. Just as the satisfaction of these needs fosters better performance, the absence may actually produce highly negative results. While examining a model of online learner motivation based on Deci and Ryan's (1985) self-determination theory, Chen and Jang demonstrated the direct correlation between contextual support by teachers, need satisfaction of students, motivation, and performance. They concluded that online learners have different reasons to participate in class, including their perceptions of how the three needs of self-determination are met or unmet. Learners belonging to marginalized groups will need special consideration if these needs are to be satisfied, which means teachers need to be cognizant of the student backgrounds and design their contextual support strategies accordingly. In fact, the study suggested, "haphazard and aimless supports without addressing students' needs are likely to lead to adverse—even worse than 'no effects'—outcomes" (Chen & Jang, 2010, p. 750).

Unfortunately, the online environments may include culturally unaware faculty who are rapidly being thrown into a situation that they were not historically prepared to face. Although the globalization of education is a relatively new trend, it has escalated exponentially within this decade. This has given birth to a situation wherein many faculty and instructors have had little to no time to increase their own cultural awareness, at least not to the extent required by the cultural rigors of this evolving situation. Due to education globalization, this issue is becoming increasingly pronounced, as more and more foreign students seek to enroll themselves in courses offered by the United States and Europe, attracted by the perceived value that credits and degrees from such courses/institutions may provide for them. In the majority of the cases, such learners enroll in courses led by faculty who may have little to no exposure to the international community (Stewart, 2012). Despite having the best intentions, the lack of cross-cultural interaction also creates a lack of empathy for one another on part of both students and faculty alike (Gelb, 2012; Ruggs & Hebl, 2012). Thus, positive or negative self-determination situations nestled within online environments will affect the retention of online learners.

Proponents of socio-cognitive views and models (Bandura, 1986; Zimmerman & Schunk, 1989) describe how self-efficacy beliefs of learners determine their abilities to persist and self-regulate. Shea and Bidjerano (2010) studied the Community of Inquiry model as proposed by Garrison, Anderson, and Archer (2000), and concluded that human, face to face, interaction may have a more positive effect on learners' self-efficacy beliefs. They indicated that "This result provides support for the assumption that the absence of traditional and familiar classroom conventions may result in additional uncertainty for fully online students" (p. 1727) and sought to argue in favor of the need "to pay more attention to supporting the relationship between teaching presence and self-efficacy in fully online environments" (p. 1727). Thus, they supported blended environments as opposed to fully online ones. The fact remains that not all online programs can afford to provide "live" student–teacher interaction, which means that the underpinning issues of learner demotivation remain at large in fully online environments.

Constructivism and andragogy are closely related concepts and a huge factor in determining the content, structure, and climate of online learning environments. Chu and Tsai (2009) studied the factors that influence adult learners to select online programs/courses. They concluded that even though adult learners have concerns about their Internet efficacies, they find the constructivist approach of self-directed learning prevalent in online environments very attractive. However, this preference may not be enough to sustain such learners within the online environments, given their lower educational technology and Internet usage skills. Most of these learners belong to the "digital immigrant" group (Prensky, 2001), and although they could be technology users for personal things, they may not be equally well informed when it comes to using educational technology. Thus, online educators and course designers have a greater responsibility to give enough time to adult learners to practice online

activities to increase confidence, design content that the learners can connect to their everyday lives, and provide resources to allow the learners to construct their own knowledge of pedagogy and technology (Chu & Tasi, 2009). In the absence of such comprehensive teaching and designing approaches, it is very likely that adult learners may not prevail within an online environment.

Cognitivism and related theories are critical to understanding online education, particularly when viewed through the lens of globalized content creation and management, as is required in many online learning programs that have international students. Silva, Costa, Rogerson, and Prior (2009) conducted a meta-analysis of different learning theories to define knowledge and content with respect to different pedagogical approaches. When defining the boundaries of knowledge with relationship to online learning environments, they state that cognitive appropriation is key to justifiable knowledge. Budin (2008) discussed the significance of content as being culture-bound, meaning that the users of content may come from widely diverse populations. Hence, the process of content management must include a justifiable consideration for cultural factors with respect to content design.

Siemens's (2014) theory of "Connectivism" provides a new spin on traditional learning theories, by addressing the technological and digital aspects of learning. Traditional theories rely on the belief that learning takes place within people and that it is a social process. Siemens argues that learning also occurs outside of people, within the realms of technology and organizations as individual entities. His belief is that given the importance of technology in the learning environs, the focus of discussion and analysis must shift from the actual process of learning to understanding the value that any learning can bring. Siemens defines learning as "actionable knowledge" (p. 5) that can exist outside the realm of human cognition, within organizations and databases. Being able to make connections within and between specialized information that enables us to learn more "are more important than our current state of knowing" (p. 5). In this context, Connectivism is "The ability to draw distinctions between important and unimportant information . . . The ability to recognize when new information alters the landscape based on decisions made yesterday" (p. 5). This line of thought highlights new paradigms and possibly new challenges, which must be taken into cognizance when analyzing online environments.

Examining High Attrition Rates in Online Environments

Misconceptions Relating to Cognitive Load

Online learning may sometimes be a completely new platform for learners, but learners still choose it using several different criteria and assumptions. Common assumptions related to online learning are that because face-to-face presence is not required, an online platform will be less demanding on time, will require less effort to manage workload, and will not disrupt the learners' lifestyle. Schaarsmith (2012) indicates that some of the reasons students have for joining online courses is related to financial factors, such as saving money on transportation, and the ability to continue working while pursuing a degree. Shay and Rees's (2004) research data indicate something similar. When asked about the most important reasons for choosing online classes, students indicated that they chose online courses based on considerations such as convenience, flexibility, opportunity to fuse their current lifestyles to their desire to study, availability of programs, and affordability (Shay & Rees, 2004).

Although most of these reasons could be valid and viable, they are also indicative that learners do not consider the magnitude of workload and the required depth of their involvement in the online courses as reasonable criteria to make the decision to go online. As a result, when they attend the online classes, many of them are unpleasantly surprised to find that the conveniences of flexible hours and lower cost outweigh the inconveniences of excessive demands on lifestyles, technical issues, and concerns related to the attitude and aptitude of learners toward a new platform. Online learning environment is very largely self-driven and dependent on the learners' ability to manage academic responsibilities, with fewer props than those available in face-to-face classes. If learners have not experienced this kind of self-imposed academic discipline before, they are very likely to experience demotivation, forcing them to quit. Another factor in this equation is that many online classes follow constructivist models of teaching, wherein learners are given props and aids to learn, but are left to solve complex problems on their own. If learners are not comfortable with self-learning and constructing knowledge out of their own initiatives, the online environment can become intimidating for them.

Spiro, Coulson, Feltovich, and Anderson (1988) researched the issues and impediments related to "advanced knowledge acquisition in ill-structured domains" (p. 2). While discussing their cognitive flexibility theory, they refer to the need for the learner to "attain a deeper understanding of content material, reason with it, and apply it flexibly in diverse contexts" (p. 2). Driscoll (2005) takes cue from this concept when she discusses how "errors of oversimplification, overgeneralization, and overreliance on context-independent representations" can occur when learners attempt to understand "ill-structured domains," by applying the same information they had used to understand "well-structured domains" (p. 398). When learners use their experiences from the well-structured domains of face-to-face courses as benchmarks for online classes, they may perceive the online environment as ill structured. A review of literature indicates that these misconceptions related to the cognitive loads or overloads may significantly contribute to higher attrition rates. Paas, Renkl, and Sweller (2004) refer to the concept of cognitive load as a situation where learners are intimidated by a large amount of information that

needs to be processed all at once before real learning can begin. Their study discusses the importance of "managing working memory load in order to facilitate the changes in long-term memory associated with schema construction and automation" (Paas et al., 2004, p. 2).

When learners are not familiar with the online educational delivery system, they are more apt to be frustrated with the disparities existing between the long-term memories of their face-to-face course associations and the new realities of online learning that they are forced to face. McQuaid (2009) analyzed the effects of cognitive load on online learners and discussed how important it is for learners to adapt to the online learning environment for meaningful learning to take place, as well as the critical need for instructional designers to adapt to the learners' assumptions about their ability to complete a course.

Another thing to consider is the fact that online courses allow for less student–teacher interaction, as opposed to face-to-face. Even though multiple communication options are available in online setups, they may not be used as extensively as they should be, simply because the usage is largely dependent on the learners' own initiatives. Consequently, online learners tend to communicate with their instructors more to get help with a problem and less to take actual guidance to facilitate their learning. As a result, the online environment can become less guidance-oriented, which in turn may be non-conducive to retention. Kirschner, Sweller, and Clark (2006) indicate that

> the free exploration of a highly complex environment may generate a heavy working memory load that is detrimental to learning. This suggestion is particularly important in the case of novice learners, who lack proper schemas to integrate the new information with their prior knowledge. (p. 80)

Ongoing research also lends support to the fact that cognitive load may be closely related to student satisfaction with online courses. Bradford's (2011) research using his Factor Correlation Matrix and the Principal Components Analysis indicates that there are significant connections between cognitive load and satisfaction and that "approximately 25% of the variance in student satisfaction with learning online can be explained by cognitive load" (p. 217).

Social and Family Factors

The reasons for high attrition rates in online classes could be a combination of social factors, as well as the attitude, aptitude, and motivational threshold of the students. Family commitment and social obligations of the student could be contributing factors in low retention. Evans (2009) discusses how students indicate obligations to their families as a primary and recurring reason for why they drop an online course. Other key studies in this field, for example, the works of Tinto and Summers, indicate the involvement of social factors in retention, and although these authors discussed retention in traditional classrooms, some of the things they propounded may hold true for e-learning as well. For example, Summers (2003) discussed the retention issues in relation to community college students and observed that students who had value orientations that were different from the norm were not able to interact socially with their peers. As a result, such students felt incompatible with the institution's social system and were more likely to drop out. Tinto (2006-2007) emphasized the need to understand the role family and society plays so that it helps institutions to create better and more effective support programs for students with diverse situations and backgrounds.

A turning point in the retention research came with Alfred Rovai's (2003) discussion of the Composite Persistence model, which was designed to gauge factors affecting retention for online students. This model discusses several factors affecting retention, both prior to admission and after, and includes social integration and family responsibilities as applicable factors in the retention equation.

Motivational Factors

Motivational aspects can also cause high attrition rates in online classes. Because online courses are heavily self-directed and self-learned, motivation or lack thereof can be a deciding factor in attrition. Erin Heyman (2010) indicates that motivation and accountability are closely related to student retention in online programs. Motivation in online courses can be directly linked to the overall course design, as well as the students' own aptitude and attitude toward learning and technology. Studies reveal that several factors such as the time needed to complete modules, lack of real world issues and contexts in course materials, and problems with accessibility and availability of resources and support systems create motivational constraints (Smart & Cappell, 2006).

Technological Constraints and the Digital Natives

Prensky (2001) refers to the term "digital natives" to describe learners who may be familiar with popular technology but are not conformable with educational technology. Several studies support this idea and indicate that student satisfaction related to the overall course design is a key concern and determinant in student retention. Weber and Farmer (2012) indicate that students consider satisfaction regarding course delivery as a major cause of continuing or withdrawing in online classes. Another issue relates to the technical expertise of the students in relation to the course design. Although this generation of students may have technical knowledge relating to social media and digital entertainment options such as video games, these skills may not be enough to be successful in an online course. A key flaw when assessing student compatibilities with technology is crediting them with more capabilities than they actually possess in relation

to the online course materials. Overestimating the technology readiness of online students is a mistake (Clark-Ibanez & Scott, 2008). Prensky's research leading to the coining of the phrase "digital natives" to describe learners who live a highly digitized life, surrounded by technology, is critical to understanding the important role this factor plays in e-learning attrition. A key reason for high attrition rates in online courses is related to ineffective course designs that are created based on assumptions about the online learner, which may or may not be true. One such assumption is that if a student is "tech savvy" and is familiar with mobile and/or social media technology, he or she is a perfect fit for online learning. Ng's (2012) study reveals key aspects of this issue and discusses it from a solution point of view, rather than just articulating the problem. Ng contends that the digital natives, who can also be the online learners, prefer to be online for everything including accessing information, getting entertainment, and socializing. They prefer quick delivery and exchange of information, like to multi-task, and respond better to graphics instead of text. The article examines the argument that although such learners can use technology, they do not possess skills required to use them for learning. In essence, a large segment of today's online learners know how to use technology and are familiar with the digital environment; however, it does not necessarily mean that they are equally conversant with educational technology and e-learning environments as envisioned by institutions that offer online courses and programs. The article describes educational technology as the use of materials and processes to facilitate teaching and learning. Such educational technologies could be related to formal and/or informal learning, for example, an online course or self-learning by surfing the Internet. Although almost all participants in the study were familiar and comfortable with sites such as Facebook and YouTube, they were far less conversant with the usage of teaching/learning technologies such as wikis, blogs, Google Docs, Movie Maker, and Photoshop, to name a few. Almost none of them was familiar with widely used educational technologies such as Prezi or VoiceThread. They were also unfamiliar with concepts of ePortfolio or cloud computing (Ng, 2012). Therefore, it is quite possible that students of online classes often experience computer-related issues, especially at the beginning of the semester, and probably during the course of the semester, if they choose to continue in the class. This causes many of them to drop the course well before they get the opportunity to become comfortable in the courses' cyber zones and also after they have made it several weeks into the semester.

Lack of Instructor Understanding of Online Learners

It is not only the learners but also the instructors and course designers who face similar challenges relating to interaction in online classes. Many times, the individual perceptions of the students and the teachers are dramatically different, resulting in overall poorly designed courses that are confusing and dissatisfying for the learners. To do a good job of designing online courses, instructional designers need to understand how an online learner perceives things. Available literature suggests that online instructors find it increasingly challenging to maintain a cohesive learning atmosphere in the class compared with face-to-face classes. Muirhead (2004) points out that online instructors feel challenged to create collaborative learning atmospheres that generate true and meaningful learning. Many times, this difference in perception results in a certain amount of apathy on the instructors' part to recognize student emotions and feelings. Tallent-Runnels et al. (2006) indicate that research results point to the need to create more student-compliant courses. For example, instructors should be more cognizant of the psychological aspects of student reactions as revealed in the student responses to discussions. Knowing why students react the way they do can provide an insight into modulating discussions and other collaborative avenues to make courses more flexible and learner friendly.

Faculty Limitations of Using Technology: The Digital Immigrant Issues

Prensky (2001) refers to the term "digital immigrants" to describe instructors who are unable to keep up or understand the language of the digital native community, stating that "our Digital Immigrant instructors, who speak an outdated language (that of the pre-digital age), are struggling to teach a population that speaks an entirely new language" (p. 2). Ng (2012) makes similar assertions when she points out that educators are responsible for raising awareness of educational technologies in digital natives, so that they can be used to facilitate the digital natives' formal learning. This is exemplified by comparing how children need to be introduced and taught to speak languages or use appliances to facilitate their informal learning. She further contends that digital natives are less likely to self-explore or look to use educational technology, unless they are formally introduced to them.

Based on the results of her study, Ng (2012) inferred the need for educators to be aware of the benefits and possibilities that various technological tools provide for teachers' training and students' learning. Digital natives, although familiar with technology and Internet, may have severe limitations in understanding how technology could support their learning. Therefore, they need constant guidance from their teachers until they become familiar with the educational technologies (Ng, 2012). However, this need for the technical "savviness" of educators is not being met, because the instructors who are teaching online courses are not technically literate to the extent required. Prensky (2001) also points out those digital immigrant instructors have incorrect assumptions that today's online learners are no different from what learners have been in the past, and that the teaching methods that were

successful yesterday will be effective today. As a result, there is a marked dissatisfaction among online learners regarding the lack of technical knowledge of their instructors. The new generation of e-learners is "networked most or all of their lives" and possesses "little patience for lectures, step-by-step logic, and 'tell-test' instruction. Unfortunately, for our Digital Immigrant teachers, the people sitting in their classes grew up on the 'twitch speed' of video games and MTV" (Prensky, 2001, p. 3).

The key contributing factor for this is the paucity of technical resources and expertise available to online faculty and course designers. Liu, Gibby, Quiros, and Demps (2002) highlight the challenges faculty face when trying to keep pace with the ever emerging and rapidly evolving technologies that are necessary to create effective online course designs. They point out that although instructional design courses make students well conversant with the theoretical aspects of the subject, they do not provide the expertise and knowledge required for practical applications of technologies. Another key factor leading to ineffective online course designs is the level of confidence and comfort that the faculty have with respect to online classes and using technology in the classroom. The results of a case study by Osika, Johnson, and Buteau (2009) show that this could be due to a combination of factors such as the faculty's belief that online courses are not equal to face to face with respect to learning quality. Many faculty do not subscribe to the concept of online course delivery as a full-time medium of instruction. A large number of faculty from the study group expressed concern over the lack of support they receive from the institution, indicating that this was a major factor that made online courses unattractive to faculty. Young (2004) refers to a national survey released in 2004 by the Educause Center for Applied Research that reveals that students were very dissatisfied with the way instructors used, or did not use, technology. Young reports that students complained that sometimes professors perform poorly, because of technology, indicating that such professors are better off when they use the chalkboard.

Institution Limitations to Training Faculty

One of the prime reasons for lack of good faculty input in online courses is the lack of drive of educational institutions to create good training programs for their faculty. The emphasis is more on developing and deploying online courses rapidly to increase enrollment, rather than create a body of well-trained faculty to boost retention. Young (2004) points out that although colleges spend top dollars for adding technological components in classrooms, far fewer resources are devoted to train professors to use these technologies. Another key reason that institutions and organizations should spend more time, money, and effort on training faculty is the changing expectations for online courses and course designs that involve the use of many different media and technologies to deliver course content. In their report, Liu et al. (2002) compare the process of instructional designing today with a movie production or conducting a symphony. The authors discuss how instructional designers like to use different media to create a harmonious blend of technology and learning to incite the attention of the students, just as movie directors or symphony conductors do when they try to attract their patrons and viewers (Liu et al., 2002). The modern trends in the changing attitudes and aptitudes in education technology create a need for better trained faculty. As the existing literature review indicates, several factors relating to online course design can cause high attrition rates in online classes. Keeping these factors in mind while designing online courses may help regain and retain students.

Some Solutions to Improve the Online Course Experience

Make Orientation Programs Mandatory

One of the biggest deterrents to online retention is the overestimation of student capabilities with respect to the demands of time, commitment, and technological skills required in online learning. One way to deal with this is through orientation programs that introduce students to the rigors and unique demands of the online classes. However, that in itself can be a challenge. Studies conducted by Bozarth, Chapman, and LaMonica (2004) reveal the need for designers and facilitators to understand that students' own perceptions or misconceptions of their technological skills becomes the biggest challenge as it makes students feel that an online orientation program is not required. As a result, many students show resistance to what they perceive as unnecessary intervention to their course penetration. Instead of feeling frustrated with this attitude, instructors and institutions must think about strategies that will enforce orientation, rather than make it obligatory. Instructors should also evaluate their own technological, communication, and facilitation skills and attempt to update them if necessary.

Using "Live" Interaction and Transparency in Computer Mediated Communication (CMC)

As indicated by the literature reviews above, social factors play a significant part in determining student retention. There is already a preexisting prejudice against online courses relating to the level of interaction between students and teachers. According to Roblyer and Ekhaml (2000), several studies indicate that students and faculty alike have huge doubts regarding the depth of interaction possible in an online environment. This creates a serious discomfort in the minds of learners and educators when it comes to embracing online delivery systems. Literature reviews also support that enhancing the social culture of an online class goes a long way in allowing students to continue with their e-learning and complete their education.

Dow (2008) provides insight as to how designing online courses to foster effective dialogue, ease of the use of media tools, well-structured interactions, and transparency of CMC helped create a better learning environment. Dow's study reveals that not having a "live" component in the interactions was very detrimental to the online learning atmosphere. He lists several areas of concern in this regard such as the absence of live conversations, not having any visible identifiers such as photos of teachers and peers, and a general frustration about the time gaps between communications. Students feel uncomfortable when they are unable to see the people they are conversing with, which in turn hinders how they may gauge the feelings of their peers online. Consequently, online courses should be designed to foster more social interaction between peers and students-teachers.

Creating Classes Structured for Collaborative Learning

In the study conducted by Dow (2008), participants indicated how difficult it was to gauge social presence of their peers and instructors, in the absence of any cohesive working structure and continued interactions. Another research conducted by Moallem (2003) studied the impact of applying an interactive design model for creating an online course that was more structured for collaborative activities, and consequently more amenable to online learning. When applying this model, emphasis was placed on collaborative problem-solving tasks, individual accountability, encouraging commitment to group and its goals, facilitating communication between group members, and providing stability so that group members could work productively together for longer period. The results of Moallem's study indicated that having cohesive and structured tasks, as well as an intuitive design model, might influence positive interactivity and interaction among students in an online course. Muirhead (2004) recommends that instructors develop strategies that will enhance their guidance for the students, such as creating a timeline for feedback and having a specific feedback rubric. This may mitigate the struggle instructors face when trying to establish a meaningful presence in their online classes. This may also facilitate the instructors' own discovery and experimentations to develop strategies for a seamless collaboration with and between the students.

Enhancing Faculty Training and Support

Literature supports the importance faculty training has in online course design and overall retention. Kate (2009) discusses the need to focus more on re-training professors who are taking a huge leap when shifting from face to face to an online environment. Simply having 'good teaching' as part of an institution's mission is not enough, unless it is complemented by having support infrastructures for the faculty. Only then can an institution be able to provide effectuve online course delivery. Even when an institution claims excellence in teaching as its core value, it does not necessarily mean that the institution has appropriate support structures for the teachers. Kate also highlights the importance of including discussions about training faculty whenever institutions discuss educational excellence and quality. Levine and Sun (2002) highlight the connection between effective course design and faculty training when they discuss how faculty training and course deigning are connected, and how these in turn are critical to students. They show concern over the fact that many times faculty do not receive formal training, which they believe is required to create good learning environments in their online classes.

The literature reveals instances of unique and successful solutions to faculty training and support. One such success story is that of University of Illinois where administrators gave faculty a full semester off before teaching online classes, so that faculty could use that time to train and prepare for a smoother transition from face to face to an online environment (Kate, 2009). Another example is that of San Jose State University, where a grant received was used to create a 2-week program that successfully trained instructors using one on one training by professional instructional designers (Kate, 2009). As supported by research, it is possible that faculty will perform better as far as online course design modification and teaching are concerned if some form of training takes place before a faculty teaches an online course for the first time. In a quantitative study, Julie Ray (2009) concluded that training instructors prior to their starting to teach online courses resulted in better preparation for the classes. Similarly, Ray refers to a study of pharmacy instructors that showed how only 3 hr of training resulted in a significant increase in instructors' perceived ability to instruct online. Ray concluded that no matter what and how the training is imparted, it always has a positive influence on the instructor's ability to teach online.

Synthesis of Literature Used

The article provided a synthesis of literature to analyze online learning environments and learners with the intention of highlighting retention issues and recommending solutions. The high attrition rate in online courses is a cause for concern (Herbert, 2006; Heyman, 2010; B. Smith, 2010). This phenomenon needs to be studied in light of the growing demand for online programs in academic and corporate settings, and the fact that a fall in attrition rates will benefit students, institutions, and businesses (Allen & Seaman, 2011; Overton, 2007; Stanford-Bowers, 2008). Several studies have been conducted specifically to observe when and why students withdraw from graduate programs (Jaggars, 2011; Levy, 2007; Perry et al., 2008; Willging & Johnson, 2009). The results of these indicate that students were more apt to drop out during earlier stages of the semester, and there are multiple reasons for doing this including personal preferences, profession-related, and program-related issues.

Learning theories such as social exclusion, self-determination, self-efficacy, cognitivism, constructivism, and connectivism can provide a deeper insight into the workings of online learning environments, including leaners, instructors, and course contents (Bandura, 1986; Budin, 2008; Chen & Jang, 2010; Chu & Tasi, 2009; Deci & Ryan, 1985; Gelb, 2012; Ruggs & Hebl, 2012; Shea & Bidjerano, 2010; Siemens, 2014; Silva et al., 2009; D. Smith & Ayers, 2006; Stewart, 2012; Wehmeyer et al., 2003; Zimmerman & Schunk, 1989).

Several critical factors lead to high attrition rates in online environments. One of them is the misconceptions learners have about the workload, cognitive challenges, and general expectations. Learners may select online classes for personal reasons, without recognizing that they may have issues with their entry-level skills pertaining to the subject or technology being used in online classes. This could place novice-level learners, who are used to the structured forms of face-to-face courses, in the fluid, ill-structured domains of online environments, leading to demotivation and attrition (Bradford, 2011; Driscoll, 2005; Kirschner et al., 2006; McQuaid, 2009; Paas et al., 2004; Schaarsmith, 2012; Shay & Rees, 2004; Spiro et al., 1988).

Family commitment and social obligations of students could be contributing factors in low retention. In addition, students without the norm value orientations may be unable to interact socially with their peers. As a result, such students felt incompatible with the institution's social system and were more likely to drop out (Evans, 2009; Rovai, 2003; Summers, 2003; Tinto, 2006-2007). The constructivist and self-oriented nature of online learning can create issues of motivation, particularly for learners with technological skill limitations. The assumptions of online course designers and educators about the technological compatibility of the digital native learners can lead to issues with course designs. These factors have been known to accelerate attrition rates (Clark-Ibanez & Scott, 2008; Heyman, 2010; Ng, 2012; Prensky, 2001; Smart & Cappell, 2006; Weber & Farmer, 2010).

In addition, the profiles, attitudes, and aptitudes of online faculty could become issues for online learning environments. Research indicates that a large number of online faculty have a low level of understanding of the way online learners learn. Many times, face-to-face faculty are invited to teach or design online courses, with minimal or zero exposure to the pedagogical aspects of online environments. Consequently, they may work under the erroneous assumption that what works for on-ground will work equally well for online. Some of the challenges such faculty face are developing and sustaining interactive and dynamic collaborative climates in their online classes, adjusting their own gaps in technology skills, and falling prey to their inherent prejudice against the perceived lack of value of online classes versus face to face ones. The paucity of training and professional development opportunities compound the problems (Liu et al., 2002; Muirhead, 2004; Ng, 2012; Osika et al.,

2009; Prensky, 2001; Tallent-Runnels et al., 2006; Young, 2004).

Given the magnitude and depth of the issues, researchers have looked for viable solutions. Rigorous orientation programs can help online learners become better prepared for their academic journeys online. Faculty should also evaluate their own technological, communication and facilitation skills and attempt to update them as and when needed so that they can create a more transparent and collaborative online learning environment within their classes and be effective guides of technology for their students. Institutions must find ways to enhance faculty training for online teaching. Using such simple measures can greatly help contain attrition and increase retention rates in online classes/programs (Bozarth et al., 2004; Dow, 2008; Kate, 2009; Levine & Sun, 2002; Moallem, 2003; Ray, 2009; Roblyer & Ekhaml, 2000).

Implications for Future Research

As a survey and review of literature reveals, the causes of poor retention in online courses are many, and although there has been some headway in the area of providing viable solutions to this issue, much deeper and wider studies are required to develop a better understanding of ways and means to solve online course issues and improve online classes and course designs to facilitate and benefit both learners and educators. The most desired outcome of such research should be to help boost retention. At present, there are many emerging trends in the world of e-learning that presents different avenues for future research such as Rovai's (2003) Composite Persistence model or Bradford's (2011) concept of Factor Correlation Matrix and the Principal Components Analysis. However, models and concepts such as these need to be examined in the light of more real world context using larger participant groups. Faculty and institutional involvement, as well as the importance of creating more interactive and better-designed online course content in the retention equation, must also be studied. This article is a small beginning toward a larger and broader scaled research in this field.

Declaration of Conflicting Interests

The author(s) declared no potential conflicts of interest with respect to the research, authorship, and/or publication of this article.

Funding

The author(s) received no financial support for the research and/or authorship of this article.

References

Allen, E. I., & Seaman, J. (2011). *Going the distance: Online education in the United States*. Retrieved from http://www.online-learningsurvey.com/reports/goingthedistance.pdf

Ball, S., Davies, J., David, M., & Reay, D. (2002). "Classification" and "judgement": Social class and the "cognitive structures" of choice of higher education. *British Journal of Sociology of Education, 23*, 51-72. doi:10.1080/01425690120102854

Bozarth, J., Chapman, D. D., & LaMonica, L. (2004). Preparing for distance learning: Designing an online student orientation course. *Educational Technology & Society, 7*(1), 87-106. Retrieved from http://www.ifets.info/journals/7_1/10.pdf

Bradford, G. R. (2011). A relationship study of student satisfaction with learning online and cognitive load: Initial results. *The Internet and Higher Education, 14*, 217-226. Retrieved from http://my-ecoach.com/online/resources/12228/All_docs12.pdf

Budin, G. (2008). Global content management: Challenges and opportunities for creating and using digital translation resources. In E. Y. Rodrigo (Ed.), *Topics in language resources for translation and localisation* (pp. 121-134). Amsterdam, The Netherlands: John Benjamins.

Chen, K., & Jang, S. (2010). Motivation in online learning: Testing a model of self-determination theory. *Computers in Human Behavior, 26*, 741-752. doi:10.1016/j.chb.2010.01.011

Chu, R., & Tsai, C. (2009). Self-directed learning readiness, Internet self-efficacy and preferences towards constructivist Internet-based learning environments among higher-aged adults. *Journal of Computer Assisted Learning, 25*, 489-501. doi:10.1111/j.1365-2729.2009.00324.x

Clark-Ibanez, M., & Scott, L. (2008). Learning to teach online. *Teaching Sociology, 36*, 34-41. Retrieved from http://www.jstor.org.ezproxy.lib.purdue.edu/stable/20058625?seq=3

Deci, E., & Ryan, R. (1985). *Intrinsic motivation and self-determination in human behavior*. New York, NY: Plenum.

Dow, M. (2008). Implications of social presence for online learning: A case study of MLS students. *Journal of Education for Library and Information Science, 49*(4), 238-239. Retrieved from http://www.jstor.org.ezproxy.lib.purdue.edu/stable/40323753?seq=8

Driscoll, M. P. (2005). *Psychology of learning for instruction* (3rd ed.). Boston, MA: Allyn & Bacon.

Evans, T. N. (2009). *An investigative study of factors that influence the retention rates in online programs at selected state, state-affiliated, and private universities* (Doctoral dissertation). Retrieved from ProQuest Dissertations and Theses database. (ProQuest document ID: 1937608371)

Garrison, D., Anderson, T., & Archer, W. (2000). Critical inquiry in a text-based environment: Computer conferencing in higher education. *The Internet and Higher Education, 2*(2), 87-105. Retrieved from http://cde.athabascau.ca/coi_site/documents/Garrison_Anderson_Archer_Critical_Inquiry_model.pdf (accessed 21 September 2014).

Gelb, C. (2012). Cultural issues in the higher education classroom. *Student Pulse, 4*(07). Retrieved from http://www.studentpulse.com/articles/661/3/cultural-issues-in-the-higher-education-classroom

Herbert, M. (2006). Staying the course: A study in online student satisfaction and retention. *Online Journal of Distance Learning Administration, 9*(4). Retrieved from http://www.westga.edu/~distance/ojdla/winter94/herbert94.htm

Heyman, E. (2010). *Overcoming student retention issues in higher education online programs: A Delphi study* (Doctoral dissertation). Retrieved from ProQuest Dissertations and Theses database. (ProQuest document ID: 748309429). Retrieved from http://search.proquest.com/docview/748309429?accountid=13360

Jaggars, S. S. (2011, January). *Online learning: Does it help low-income and underprepared students?* (CCRC Working Paper No. 26) Retrieved from http://ccrc.tc.columbia.edu/media/k2/attachments/online-learning-help-students.pdf

Kate, M. (2009). Learning to teach online: Creating a culture of support for faculty. *Journal of Education for Library and Information Science, 50*(4), 275-292. Retrieved from http://search.proquest.com.ezproxy.lib.purdue.edu/docview/203230521?accountid=13360

Kirschner, P. A., Sweller, J., & Clark, R. E. (2006). Why minimal guidance during instruction does not work: An analysis of the failure of constructivist, discovery, problem-based, experiential, and inquiry-based teaching. *Educational Psychologist, 41*, 75-86. Retrieved from http://igitur-archive.library.uu.nl/fss/2006-1214-211848/kirschner_06_minimal_guidance.pdf

Levine, A., & Sun, J. C. (2002). *Barriers to distance education* (6th ed.). Washington, DC: American Council on Education.

Levy, Y. (2007). Comparing dropouts and persistence in e-learning courses. *Computers & Education, 48*, 185-204. Retrieved from http://www.qou.edu/arabic/researchProgram/eLearningResearchs/eLDropout.pdf

Liu, M., Gibby, S., Quiros, O., & Demps, E. (2002). Challenges of being an instructional designer for new media development: A view from the practitioners. *Journal of Educational Multimedia & Hypermedia, 11*(3), 195-219.

McQuaid, J. W. (2009). *An analysis of the effects of cognitive load on the participation of asynchronous e-learners* (Order No. 3339297). Retrieved from ProQuest Dissertations & Theses A&I; ProQuest Dissertations & Theses Global.

Moallem, M. (2003). An interactive online course: A collaborative design model. *Educational Technology Research & Development, 51*(4), 85-103. Retrieved from http://www.jstor.org.ezproxy.lib.purdue.edu/stable/30221186?seq=1

Muirhead, B. (2004). Encouraging interaction in online classes. *International Journal of Instructional Technology and Distance Learning, 1*(6). Retrieved from http://www.itdl.org/journal/jun_04/editor.htm

Ng, W. (2012). Can we teach digital natives digital literacy? *Computers & Education, 59*, 1065-1078. doi:10.1016/j.compedu.2012.04.016

Osika, E. R., Johnson, R. Y., & Buteau, R. (2009). Factors influencing faculty use of technology in online instruction: A case study. *Online Journal of Distance Learning Administration, 12*(1). Retrieved from http://www.westga.edu/~distance/ojdla/spring121/osika121.html

Overton, L. (2007). Making the case for e-learning. *Adults Learning, 18*(7). Retrieved from http://web.ebscohost.com.ezproxy.lib.purdue.edu/ehost/detail?sid=3ca22271-b517-4263-815b30769f137969%40sessionmgr113&;vid=2&hid=110&bdata=JnNpdGU9ZWhvc3QtbGl2ZQ%3d%3d#db=tfh&AN=24359224

Paas, F., Renkl, A., & Sweller, J. (2004). Cognitive load theory: Instructional implications of the interaction between information structures and cognitive architecture. *Instructional Science, 32*, 1-8. Retrieved from http://www.ucs.mun.ca/~bmann/0_ARTICLES/CogLoad_Paas04.pdf

Perry, B., Boman, J., Care, W. D., Edwards, M., & Park, C. (2008). Why do students withdraw from online graduate nursing and health studies education? *The Journal of Educators Online, 5*(1). Retrieved from http://www.thejeo.com/Archives/Volume5Number1/PerryetalPaper.pdf

Prensky, M. (2001). Digital natives digital immigrants. *On the Horizon, 9*(5). Retrieved from http://www.marcprensky.com/writing/Prensky-DigitalNatives,DigitalImmigrants-Part1.pdf

Ray, J. (2009). Faculty perspective: Training and course development for the online classroom. *Journal of Online Learning and Teaching, 5*. Retrieved from http://jolt.merlot.org/vol5no2/ray_0609.htm

Roblyer, M. D., & Ekhaml, L. (2000). How interactive are your distance courses? A rubric for assessing interaction in distance learning. *Online Journal of Distance Learning Administration, 3*(2). Retrieved from http://www.westga.edu/~distance/roblyer32.html

Rovai, A. P. (2003). In search of higher persistence rates in distance education online programs. *The Internet and Higher Education, 6*, 1-16. doi:10.1016/S1096-7516(02)00158-6

Ruggs, E., & Hebl, M. (2012). Literature overview: Diversity, inclusion, and cultural awareness for classroom and outreach education. In B. Bogue & E. Cady (Eds.), *Apply research to practice (ARP) resources.* Retrieved from https://www.engr.psu.edu/awe/ARPAbstracts/DiversityInclusion/ARP_DiversityInclusionCulturalAwareness_Overview.pdf

Schaarsmith, A. M. (2012, February 16). Growing number of college students choose online courses. *Pittsburgh Post-Gazette.* Retrieved from http://www.post-gazette.com/stories/news/education/growing-number-of-college-students-choose-online-courses-85483/

Shay, J., & Rees, M. (2004). Understanding why students select online courses and criteria they use in making that selection. *International Journal of Instructional Technology and Distance Learning, 1*(5). Retrieved from http://itdl.org/Journal/May_04/article03.htm

Shea, P., & Bidjerano, T. (2010). Learning presence: Towards a theory of self-efficacy, self-regulation, and the development of a communities of inquiry in online and blended learning environments. *Computers & Education, 55*, 1721-1731. doi:10.1016/j.compedu.2010.07.017

Siemens, G. (2014). Connectivism: A learning theory for the digital age. *International Journal of Instructional Technology and Distance Learning*, 1-8. Retrieved from http://er.dut.ac.za/handle/123456789/69

Silva, N., Costa, C., Rogerson, S., & Prior, M. (2009). Knowledge or content? The philosophical boundaries in e-learning pedagogical theories. *Research, Reflections and Innovations in Integrating ICT in Education, 1*, 221-225.

Smart, K. L., & Cappell, J. J. (2006). Students' perceptions of online learning: A comparative study. *Journal of Information Technology Education, 5*. Retrieved from http://jite.org/documents/Vol5/v5p201-219Smart54.pdf

Smith, B. (2010). *E-learning technologies: A comparative study of adult learners enrolled on blended and online campuses engaging in a virtual classroom* (Doctoral dissertation). Retrieved from ProQuest Dissertations and Theses database.

Smith, D., & Ayers, D. (2006). Culturally responsive pedagogy and online learning: Implications for the globalized community college. *Community College Journal of Research and Practice, 30*, 401-415. doi:10.1080/10668920500442125

Spiro, R. J., Coulson, R., Feltovich, P. J., & Anderson, D. K. (1988). *Cognitive flexibility theory: Advanced knowledge acquisi-*

tion in ill-structured domains (Technical Report No. 441, University of Illinois at Urbana-Champaign). Retrieved from https://www.ideals.illinois.edu/bitstream/handle/2142/18011/ctrstreadtechrepv01988i00441_opt.pdf?sequence=1

Stanford-Bowers, D. E. (2008). Persistence in online classes: A study of perceptions among community college stakeholders. *MERLOT Journal of Online Learning and Teaching, 4*, 37-50. Retrieved from http://jolt.merlot.org/vol4no1/stanford-bowers0308.pdf

Stewart, V. (2012). *A world-class education: Learning from international models of excellence and innovation.* Alexandria, VA: Association for Supervision and Curriculum Development (ASCD).

Summers, M. (2003). Attrition research at community colleges. *Community College Review, 30*(4), 64-78. Retrieved from http://search.proquest.com.ezproxy.lib.purdue.edu/docview/213202718/fulltextPDF?source=fedsrch&;accountid=13360

Tallent-Runnels, M. K., Thomas, J. A., Lan, W. Y., Cooper, S., Ahern, T. C., Shaw, S. M., & Liu, X. (2006). Teaching courses online: A review of the research. *Review of Educational Research, 76,* 93-135. Retrieved from http://www.jstor.org.ezproxy.lib.purdue.edu/stable/pdfplus/3700584.pdf?acceptTC=true

Tinto, V. (2006-2007). Research and practice of student retention: What next? *Journal of College Student Retention, 8*(1), 1-19. Retrieved from http://edit.uaa.alaska.edu/governance/faculty-senate/upload/JCSR_Tinto_2006-07_Retention.pdf

Weber, M. J., & Farmer, T. A. (2010). Online Course offerings: Issues of retention and professional relationship skill development. In J. Tareilo & B. Bizzell (Eds.), *NCPEA handbook of online instruction and programs in education leadership.* Retrieved from http://www.ncpeapublications.org/index.php/volume-5-number-1/50-online-course-offerings-issues-of-retention-and-professional-relationship-skill-development

Weber, M. J., & Farmer, T. A. (2012). Online course offerings: Issues of retention and professional relationship skill development. In J. Tareilo & B. Bizzell (Eds.), *NCPEA handbook of online instruction and programs in education leadership.* Retrieved from http://cnx.org/content/col11375/latest/

Wehmeyer, M., Abery, B., Mitaug, D., & Stancliff, R. (2003). *Theory in self-determination: Foundations for educational practice.* Springfield, IL: Charles C. Thomas.

Willging, P. A., & Johnson, S. D. (2009). Factors that influence students' decision to dropout of online courses. *Journal of Asynchronous Learning Networks, 13*(3), 115-127. Retrieved from http://files.eric.ed.gov/fulltext/EJ862360.pdf

Young, J. (2004, November 12). When good technology means bad teaching. *The Chronicle of Higher Education, 51*(12), A31.

Zimmerman, B., & Schunk, D. (1989). *Self-regulated learning and academic achievement theory, research, and practice.* New York, NY: Springer.

Author Biography

Papia Bawa is online educator for more than 10 years. The author has dedicated her research efforts to identifying critical issues within online environments and finding viable solutions. Her current projects involve studying the effects of game based learning on learning outcomes within online environments.

Permissions

All chapters in this book were first published in Media and Communication, by SAGE Publications; hereby published with permission under the Creative Commons Attribution License or equivalent. Every chapter published in this book has been scrutinized by our experts. Their significance has been extensively debated. The topics covered herein carry significant findings which will fuel the growth of the discipline. They may even be implemented as practical applications or may be referred to as a beginning point for another development.

The contributors of this book come from diverse backgrounds, making this book a truly international effort. This book will bring forth new frontiers with its revolutionizing research information and detailed analysis of the nascent developments around the world.

We would like to thank all the contributing authors for lending their expertise to make the book truly unique. They have played a crucial role in the development of this book. Without their invaluable contributions this book wouldn't have been possible. They have made vital efforts to compile up to date information on the varied aspects of this subject to make this book a valuable addition to the collection of many professionals and students.

This book was conceptualized with the vision of imparting up-to-date information and advanced data in this field. To ensure the same, a matchless editorial board was set up. Every individual on the board went through rigorous rounds of assessment to prove their worth. After which they invested a large part of their time researching and compiling the most relevant data for our readers.

The editorial board has been involved in producing this book since its inception. They have spent rigorous hours researching and exploring the diverse topics which have resulted in the successful publishing of this book. They have passed on their knowledge of decades through this book. To expedite this challenging task, the publisher supported the team at every step. A small team of assistant editors was also appointed to further simplify the editing procedure and attain best results for the readers.

Apart from the editorial board, the designing team has also invested a significant amount of their time in understanding the subject and creating the most relevant covers. They scrutinized every image to scout for the most suitable representation of the subject and create an appropriate cover for the book.

The publishing team has been an ardent support to the editorial, designing and production team. Their endless efforts to recruit the best for this project, has resulted in the accomplishment of this book. They are a veteran in the field of academics and their pool of knowledge is as vast as their experience in printing. Their expertise and guidance has proved useful at every step. Their uncompromising quality standards have made this book an exceptional effort. Their encouragement from time to time has been an inspiration for everyone.

The publisher and the editorial board hope that this book will prove to be a valuable piece of knowledge for researchers, students, practitioners and scholars across the globe.

List of Contributors

Evan Ortlieb
Texas A&M University–Corpus Christi, TX, USA

Gillian Warner-Søderholm
Norwegian Business School BI, Oslo, Norway

Caroline Victoria Wamala
Karlstad University, Sweden

Mona Hajin
Stockholm University, Sweden

Simon Pardoe
PublicSpace Ltd., Research Dissemination, Lancaster, UK

Muhammad Shaban Rafi
University of Management and Technology, Lahore, Pakistan

Elnaz Farbod
Kerman University of Applied Sciences and Technology, Iran

Mohammad Ghamari and Mojtaba Amiri Majd
Department of Counseling, Abhar Branch, Islamic Azad University,Abhar, Iran

Shane Tilton
Ohio Northern University, Ada, USA

Brenda L. Hughes
Florida A&M University, Tallahassee, USA

Taher Bahrani
Department of English, Mahshahr Branch, Islamic Azad University, Mahshahr, Iran

Sim Shu Tam and Mohm Don Zuraidah
University Malaya, Kuala Lumpur, Malaysia

Helen Chan
Jean Gilmour and Annette Huntington
Massey University, Wellington, New Zealand

Sue Hanna
Middlesex University, London, UK

Alison Strong
Hawke's Bay District Health Board, Hastings, New Zealand

Donna M. Hughes
University of Rhode Island, Kingston, USA

Amir Manzoor
Bahria University, Karachi

Henry St. Maurice
University of Wisconsin–Stevens Point, USA

Louise Breuer and Chris Barker
University College London, London, UK

Lori Leach
Government of New Brunswick, Fredericton, Canada

Steven Turner
University of New Brunswick, Fredericton, Canada

Saima Yasmeen, Muhammad Tayyab Alam, Muhammad Mushtaq and Maqsud Alam Bukhari
Foundation University, Islamabad, Rawalpindi, Pakistan

Natasa Gajic
KAPCI Coatings, Port Said, Egypt

Mehraz Boolaky
University of Liverpool/Laureate, Amsterdam, The Netherlands
Asia Pacific Institute of Management, New Delhi, India

Mohammad Hossein Sadaghiani, Niusha Shafiabady and Dino Isa
University of Nottingham, Malaysia Campus, Semenyih, Malaysia

Papia Bawa
Purdue University, West Lafayette, IN, USA

Index